Revision of Hawaiian *Nesiomiris*

WAYNE CHARLES GAGNÉ (1942–1988) was an entomologist with the Department of Entomology, Bishop Museum, Honolulu, Hawai‘i

Insular Evolution, Speciation, and Revision of the Hawaiian Genus *Nesiomiris* (Hemiptera: Miridae)

Wayne C. Gagné

Bishop Museum Bulletin in Entomology 7

Bishop Museum Press
Honolulu, 1997

Cover illustration by Alan Hart:
Nesiomiris williamsi Gagné, n. sp.

Frontispiece photograph by William P. Mull:
Nesiomiris curvipes Gagné, n. sp.
from the Big Island of Hawai'i.

Published by
Bishop Museum Press
1525 Bernice Street
Honolulu, Hawai'i 96817, USA

Printed in the United States of America

Library of Congress Catalog Card No. 97-74309
ISBN 0-930897-95-1
ISSN 0893-3146

printed on recycled paper

PREFACE

Wayne Charles Gagné (1942–1988) devoted his life to conservation of Hawai'i's unique biota (e.g., Gagné 1988) and was a conservation biologist long before the term was popular. His interest in, and knowledge of, the Hawaiian biota was rooted in the field work for this study while he was a graduate student at the University of California at Berkeley. He became an authority not only on many different groups of Hawaiian insects, but also on the plants upon which they fed, discovering new species, rediscovering species thought to be extinct, and co-authoring a description of Hawaiian plant communities (Gagné & Cuddihy 1990). His expertise on Hawaiian organisms was often gleaned by visiting specialists, and many of their studies would not have been as successful as they were if Wayne had not assisted, especially with the field work. He also was a gifted story-teller and enthralled audiences of all ages with his accounts of Hawaiian natural history. A strong proponent of environmental education programs, he was one of the principal developers of the 'Ōhi'a Project, an environmental education program compiled by Bishop Museum and the Moanalua Gardens Foundation for grades K-12 currently in use in Hawaiian schools.

This monograph on the systematics and biogeography of the species of the Hawaiian endemic true bug genus *Nesiomiris* was originally prepared as Wayne Gagné's doctoral dissertation, submitted to the University of California at Berkeley in 1975. For various reasons, its publication was delayed several times over the years, and multiple revisions were required before publication. At the time of his unexpected death in May 1988, Wayne was in the process of final revisions of the manuscript for publication, but he had not yet completed some portions, particularly the cladistic and biogeographic discussions.

Following Wayne's death, the manuscript and Wayne's revisions were consolidated onto a word processor. That version was reviewed by Bishop Museum entomologists from the standpoint of the Hawaiian biota, and then went to mirid specialists for review from the standpoint of mirid systematics. All changes were then integrated into the manuscript, and the illustrations and photographs reorganized to match. Some of the older illustrations were recreated because of deterioration of the originals over time. We have done the best we could in this process, but the final product is less than what Wayne would have done had he had opportunity to do so. The reviewers agreed that although the biogeographic and cladistic discussions were outdated, the taxonomic and biological information remained valid and should be published.

Because Wayne was not able to update the biogeographic and cladistic discussions with his own ideas, we have chosen to leave the original text intact as a historical document. The reader should bear in mind that concepts of Hawaiian biogeography, as well as biogeographic theory, have changed in the over 20 years since the original text was written. [For a modern review of Hawaiian biogeography, see Wagner and Funk (1995).] We have also chosen to leave the original plant identifications intact rather than risk mistakes in translating them into modern nomenclature (e.g., Wagner *et al. 1990*). For a review of the current status of systematics of Miridae worldwide, see Schuh (1995).

Major funding for the publication of this monograph was made possible by generous donations by the Schlinger Foundation and Betsy Harrison Gagné, with additional contributions from George and Emma Cook, James D. Jacobi, Steven L. Montgomery, Charles and Lois O'Brien, Emily Schwab, and S.H. Sohmer.

Many Bishop Museum staff assisted in the compilation and editing of the manuscript, including Keith Arakaki, Karen Duffy, Francis Howarth, Karin Kami, G. Allan Samuelson, Tina Kuklenski, and Dan Polhemus. One of us (GMN) synthesized the various drafts of the manuscript after Wayne's death, and provided a consensus document. Another of us (NLE) coordinated figures with the text, edited, and typeset the final version for printing. Outside reviewers were Richard Froeschner (Smithsonian Institution), Tom Henry (U.S. Department of Agriculture), and Randall Schuh (American Museum of Natural History).

Gagné, W.C. 1988. Conservation priorities in Hawaiian natural systems. *BioScience* 38: 264–71.

Gagné, W.C. & L. Cuddihy. 1990. Vegetation, p. 45-114 *In:* Wagner, W.L., D.R. Herbst, & S.H. Sohmer. 1990. *Manual of the flowering plants of Hawai'i.* University of Hawaii Press & Bishop Museum Press, Honolulu. xviii + 1854 p.

Gagné, W.C. & L.T. Gill. 1989. Conservation education in Hawai'i: history, problems, and needs, p. 179-183 *In:* Stone, C.P. & D.B. Stone, eds., *Conservation biology in Hawai'i.* University of Hawaii Cooperative National Park Resources Studies Unit, Honolulu. 252 p.

Schuh, R.T. 1995. *Plant bugs of the world (Insecta: Heteroptera: Miridae).* New York Entomological Society, New York. xii + 1329 p.

Wagner, W.L. & V. Funk, eds. 1995. *Hawaiian biogeography: Evolution on a hot spot archipelago.* Smithsonian Institution Press, Washington, D.C. xvii + 467 p.

Wagner, W.L., D.R. Herbst, & S.H. Sohmer. 1990. *Manual of the flowering plants of Hawai'i.* 2 vols. University of Hawaii Press & Bishop Museum Press, Honolulu. xviii + 1854 p.

——Scott Miller, Gordon Nishida, and Neal Evenhuis
Department of Natural Sciences,
Bishop Museum, Honolulu

TABLE OF CONTENTS

ACKNOWLEDGMENTS

This study was made possible by the National Science Foundation Grant no. GB 23075. Part of the fieldwork was also supported by National Science Foundation Grant nos. 3105 and 5860 to Dr. P.D. Ashlock in 1968 and by National Science Foundation Grant no. GB 6761 to Dr. W.A. Steffan in 1970. This study is contribution no. 58 of Island Ecosystem IRP/IBP and Contribution No. 1997-011 to the Hawaii Biological Survey.

I am deeply indebted to the following persons for critical suggestions and comments during the course of this study and for critically reviewing the manuscript: Dr. E.I. Schlinger, Dr. L. Constance, and the late Dr. R.L. Usinger (who initially suggested this study, and who I profoundly regret did not live to see its conclusion) of the University of California, Berkeley, and the late Dr. J.L. Gressitt and Dr. F.G. Howarth of the Department of Entomology, Bishop Museum, Honolulu. I also wish to thank Mr. A. Hart and Mr. T. Parman, formerly of the Department of Entomology, Bishop Museum, for assistance with the figures and Mrs. N. Howarth and my wife Betsy for technical assistance. The late Mrs. I. Harrison typed the early drafts of the manuscript. The Bishop Museum provided the facilities and my superiors there, Drs. Roland Force, Edward Creutz, and W. Donald Duckworth, directors; Drs. J. Linsley Gressitt and Frank J. Radovsky, former chairmen, Department of Entomology; Dr. Scott Miller, present chairman, Department of Entomology. Dr. J.C.M. Carvalho, Museum Nacional, Rio de Janeiro, Brazil, kindly examined the holotype and a paratype of *Nesiomiris hawaiiensis* Kirkaldy during his visit to the British Museum during the initial phase of this work. I am grateful to the following persons and institutions for the loan of additional specimens: P.H. Arnaud, Jr., California Academy of Sciences, San Francisco; G. Funasaki, Hawaii State Department of Agriculture, Honolulu; D.E. Hardy, University of Hawai'i, Honolulu.

For photographic assistance I am indebted to William Mull, research associate, Department of Entomology, Bishop Museum, for his superb close-up photography, especially of the oviposition sequence of *N. timberlakei.* His work is an inspiration to all who wish to become insect photographers. I am indebted also to Eugene Drake, formerly assistant in entomology, Department of Entomology, Bishop Museum, for his assistance with color close-up photography of pinned specimens.

For field assistance I am especially indebted to Mr. John Obata, Dr. Steven Montgomery, and my wife Betsy who accompanied me on the many strenuous and sometimes hazardous hikes into the mountains and gulches, often getting soaked, frozen, or parboiled, depending on the terrain or weather.

I am also indebted to the many members of the Hawaiian *Drosophila* project at the University of Hawai'i, especially the codirectors, Drs. Hampton Carson and D. Elmo Hardy, who often provided company and sometimes furnished the logistics to enable me to get into and survive in often remote areas. The State Division of Forestry gave permission to enter watershed areas closed to the general public. The former superintendents of Hawai'i Volcanoes and Haleakalā National Parks, especially Bryan Harry and Russ Cahill, respectively, gave me permission to collect.

My wife, Betsy, provided other technical assistance and demonstrated great forbearance during my all too numerous tangential wanderings.

ABSTRACT

The formerly monotypic genus *Nesiomiris* Kirkaldy is revised and a cladistic classification is presented. Forty-nine new species are described from the Hawaiian Islands. A key to the species of *Nesiomiris* is presented. Both sexes, when available, as well as all associated instars of the nymphs, are described. Data are presented on the life cycles of some of the species of *Nesiomiris*. A detailed analysis of host-plant specificity is presented, which is correlated with the insular evolution and speciation of *Nesiomiris*. Methods of transoceanic dispersal and differing modes of insular speciation are hypothesized. It is inferred that contrasting modes of insular speciation are based on the relative specificity of founding species, which are either "generalist" or "specialist" phytophagous and predaceous species. Arguments are presented to support the claim that the founding species of *Nesiomiris* colonized Kaua'i and was relatively specialized to feed on Araliaceae at a time when this plant family was probably represented by only *Tetraplasandra* species.

INTRODUCTION

Prior to the present study, the genus *Nesiomiris* contained a single species, *N. hawaiiensis* Kirkaldy (1902) from the Hawaiian Islands. The adults and nymphs feed on practically all native Hawaiian species of Araliaceae and Aquifoliaceae (Swezey 1954). The placement of the genus in the Mirinae, tribe Stenodemini, was not known to be incorrect until the present study. This placement was seemingly based on erroneous assumptions that the arolia were fleshy and divergent and that the species are phenotypically similar in body form and coloration to the Holarctic genus *Teratocoris* Fieber. *Nesiomiris* is here removed to the Orthotylinae, Orthotylini, *Orthotylus* group (*sensu* Schuh 1974). No morphological study of the genus has been presented previously.

Timberlake (1921) and Zimmerman (1948b) stated that additional undescribed species of *Nesiomiris*, based on differences in male genitalia, were represented in collections they had made or examined, but they refrained from describing them. The external morphology of all available stages is described in this study, and 49 new species are described.

Previous references to the biology and ecology of *Nesiomiris* have consisted of little more than scattered host records, often made on the assumption that the species in question was *N. hawaiiensis*. This study includes new information on the biology of the species.

Most *Nesiomiris* species are single-island endemics (44 of 50 species), and the genus is distributed from the 5-million-year-old island of Kaua'i to the largest island in the chain, the island of Hawai'i, which is still volcanically active. Plesiomorphic species predominate at the older, Kaua'i end of the archipelago (Fig. 1). The inter-island founding

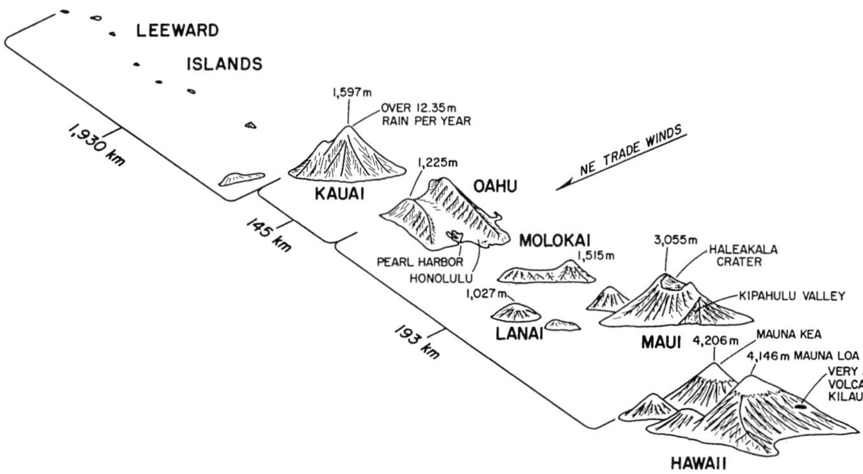

Fig. 1. Hawaiian Archipelago showing relative sizes and elevations, predominant wind direction, some major geologic features, and topography and comparative interisland distances.

events appear to have been predominantly southeastward or toward Hawai'i Island. The older islands are richer in species. This pattern appears to fit that of a genus that early colonized an older island in the archipelago and, with limited vagility, slowly moved eastward along the island chain. Several distinct lineages or species groups have formed on Kaua'i. Each lineage moved independently down the chain as a stock isolated from other lineages and speciated when it had successfully colonized another island.

Nesiomiris appears to have reached the islands, probably Kaua'i, preadapted to or previously associated with Araliaceae. Based on the feeding habits of the relatively plesiomorphic species of *Nesiomiris*, this colonization was probably on 1 or more of the pubescent-leaved species of *Tetraplasandra* (Section *Pterotropia*) (Sherff 1955). This study explores the insular and adaptive radiation of *Nesiomiris* on native Araliaceae and shows how an adaptive shift to another host-plant family (Aquifoliaceae) appears to have set the stage for another phase of insular radiation. Founder effects and the role of water barriers in interrupting gene flow are also discussed.

I have looked for *Nesiomiris* or its relatives on the araliaceous genera *Schefflera*, *Panax*, and *Borlaugiodendron* in Papua New Guinea, without avail.

Another logical place to look for these was the Marquesas Archipelago. The plant genera *Cheirodendron*, *Reynoldsia*, and relatives of *Tetraplasandra* are all present here. This archipelago also shares with Hawai'i several other floral and faunal elements such as *Bidens* (Asteraceae) (see Gillett 1972), and certain land mollusca (see Carlquist 1965). A search there in August 1977 was also negative.

For the 50 species of *Nesiomiris* covered in this study, I have endeavored to (1) determine and document host specificity and associate immature stages with adults; (2) determine inter- and intra-island distributions; (3) examine their insular evolution by examining isolating mechanisms, vagility, and morphological differences between species and conspecific populations on different islands; (4) compare seasonal population fluctuations; (5) delineate selective environmental and biotic factors driving speciation; and (6) integrate all of this information into a systematic treatment of the genus.

MATERIALS AND METHODS

The present study is based upon examination of several thousand specimens. Approximately 80% of the specimens were collected by me. A number of specimens, primarily immature stages, were preserved in 70% ethanol, while most of the adults were point mounted. The most effective collecting method for adults and nymphs proved to be sweeping foliage with a net. The large series collected by sweeping host plants shows that the comparative rarity of the insects in museum collections is largely due to lack of persistence or interest of previous collectors in this group of insects. Adults were rarely collected on other than host plants, and nymphs virtually never.

Eggs were discovered by searching for oviposition "scars" on petioles, rachises, and midribs of leaves or by watching for ovipositing females. The "scars" are brown and slightly raised (Fig. 2). Attempts at culture on potted, caged seedlings of known hosts failed. So did attempts to establish colonies on young trees in a botanical garden even when these were protected from ant predation by ringing the bases of the trunks with commercial Tanglefoot® (The Tanglefoot Company, Grand Rapids, MI).

The genitalia of the adults were dissected, cleared by heating gently in 10% KOH, and preserved in glycerol. They were drawn with the aid of an Olympus dissecting microscope and an ocular grid. As far as possible, the illustrations were prepared so that the corresponding figures represent exactly the same view for each species illustrated. In illustrations of male genitalia, particular views are indicated by standardization to the following format: (a) internal face of left clasper, (b) internal face of right clasper, (c) left lateral view of aedeagus, (d) left lateral view of spiculum, (e) left lateral view of pygofer (genital capsule) and claspers (parameres), and (f) terminal view of pygofer and claspers. Additional illustrations were prepared as required to provide either added perspective(s) on these structures, or to show intra-specific variation. These are labeled a–1, a–2, b–1, etc.

The holo- and allotypes of all new species are deposited in the Bishop Museum. The following acronyms are used in the text and in the species records to cite institutions in which some paratypic material was deposited:

AMNH	American Museum of Natural History, New York, USA
BMNH	The Natural History Museum, London, UK
BPBM	Bishop Museum, Honolulu, Hawai‘i, USA
CAS	California Academy of Sciences, San Francisco, California, USA
CNC	Canadian National Collection, Ottawa, Canada
HSDA	Hawaii State Department of Agriculture, Honolulu, Hawai‘i, USA
MNRJ	Museum Nacional, Rio de Janeiro, Brazil
NMNH	National Museum of Natural History, Washington, D.C., USA
UH	University of Hawai‘i at Mānoa, Honolulu, Hawai‘i, USA

Data for material examined is given as on the labels. Additional or clarifying information and conversions to metric equivalents that are not on the label are given in square brackets.

THE *NESIOMIRIS* FAUNA OF HAWAI'I

Although the Hemiptera fauna of Hawai'i contains both endemic and recently introduced elements, the endemics still predominate, at least in the native ecosystems. Zimmerman (1948a, 1948b) listed 728 named species of Hemiptera from the islands, approximately 75% of which are endemic. There are probably over 1,000 species of endemic Hemiptera. Zimmerman listed 43 endemic genera and 21 additional nonendemic genera of Hemiptera that contain endemic species. Additional undescribed endemic genera are known to exist.

The uneven representation of higher taxa in the Hawaiian biota is also reflected in the endemic Hemiptera. Many families that are important elements of continental faunas are completely absent or now represented only by recently adventive forms (Cicadidae, Aphididae, Aleyrodidae, Pyrrhocoridae, etc.). Other families contain tremendously developed species complexes derived from 1 or a few ancestral stocks (e.g., Miridae, Nabidae, and Lygaeidae).

Nearly all the endemic Hawaiian Hemiptera exhibit a high degree of host or ecosystem specificity. Phytophagous forms rarely attack more than 1 species or a few closely related species of native plants. Among the seed-feeding Lygaeidae of the genus *Nysius*, there are endemic species that can survive on introduced plants (Usinger 1942); some plagithmysines (Coleoptera: Cerambycidae) that attack introduced plants (Gressitt & Davis 1972); and a few native Lepidoptera that are of economic importance on sugar cane, coconut palms, and macadamia nuts. Host specificity is sometimes pronounced

Fig. 2. Oviposition scars on rachis of *Reynoldia sandwichensis.* Photo by W.C. Gagné.

even among predaceous forms. Certain of the endemic predaceous *Nabis* species are found only on a particular species of plant or are limited to a single type of plant community (Zimmerman 1948b).

Many Hawaiian Heteroptera now occur in unstable environments that are threatened with extinction, for example, those that occur on host plants on overgrazed slopes of the volcanoes on the island of Hawai'i, and species that occur on host plants that are now virtually extinct in the dry lowlands of O'ahu, Maui, and several of the other islands. These Heteroptera represent some of the finest examples of insular evolution and host specificity. The Orsillinae (Lygaeidae) (Ashlock 1966; Usinger 1942; Usinger & Ashlock 1966) with about 100 species are among described taxa by far the largest and most diverse example of this group found in any island archipelago or continent. The Hawaiian Orsillinae include endemic genera in the rain forests as well as the remarkably vigorous endemic species that can now also feed on introduced weeds. The works of Usinger and Ashlock have documented the Lygaeidae as the largest Heteroptera family in the islands. But, there is little question that the largest family of Heteroptera in the Hawaiian Islands is actually the family Miridae. Several hundred species, mostly still undescribed, have arisen from about 12 original colonizers. There are several mirid genera (e.g., "*Orthotylus*" and *Sarona*) that appear to be even more speciose than *Nesiomiris* and with apparently even more complex host relationships, as will be discussed in context with *Nesiomiris* host relationships later.

Although some ecological studies on the effect of introduced insects on the distribution and abundance of introduced range weeds have been made in Hawai'i (e.g., Davis 1974), virtually no one has evaluated the role of phytophagous insects in the ecology of Hawaiian forests. The present study also seeks to assess the impact of *Nesiomiris* bugs on their hosts, the native araliaceous and aquifoliaceous trees. Such information is important to our understanding of the factors responsible for the decline of certain of these native plants.

Host Specificity and Host Plants

Host relationships can sometimes provide valuable insights into the evolution of phytophagous mirids and indicate taxonomic affinities of the host plants as well. This evolutionary-ecological approach should encourage botanists and entomologists to collaborate, which could lead to new insights into the evolution of the biota as a whole.

Tables 1 and 2 show the geographical distribution and the plant hosts of *Nesiomiris* species. It is necessary to point out some apparent discrepancies in the data presented, especially with respect to the apparent absence of *Nesiomiris* on some potential host taxa. In many cases it is impossible to determine from Sherff's (1952, 1954, 1955) keys to which variety a given plant specimen belongs. More emphasis was placed on the identification of subspecific taxa of host plants in the earlier stages of the study than in the later stages. This neglect was compounded by the lack of any evidence that the insects distinguish among the varieties of host plants. At most, only distinct species of host plant (especially in *Tetraplasandra*) are isolating mechanisms to *Nesiomiris* species. Also, I was unable to find several potential host plants in the field or to identify them according

Table 1. Island distribution and host genera of *Nesiomiris* species[1]

TAXON	DISTRIBUTION							HOST GENUS				
	Kauai	Oahu	Molokai	Lanai	W. Maui	E. Maui	Hawaii	Cheirodendron	Reynoldsia	Tetraplasandra	Munroidendron	Ilex
GREEN AREOLE GROUP												
trifurcatus subgroup												
1. *trifurcatus*		x								+		
curvatus subgroup												
curvatus complex												
2. *curvatus*						x				+		
3. *pilosus*							x	+		+		
sinuatus complex												
4. *sinuatus*					x			+				
5. *parasinuatus*				x				+				
6. *swezeyi*						x		+				
7. *williamsi*							x	+				
reductus complex												
8. *reductus*					x			+				
elongatus subgroup												
elongatus complex												
9. *elongatus*	x									+	+	
curvipes complex												
10. *curvipes*			x		x					+		
11. *pallasatus*							x			+		
secularis complex												
12. *cheliferus*						x		+	(+)			
13. *spatulatus*			x					+				
14. *secularis*			x			x	x		+	+		
ohe complex												
15. *ohe*		x								+		
16. *trinesiotes*			x	x	x				+[2]			
17. *timberlakei*		x					x[3]	+	+	+		
18. *pallidus*					x	x				+		
RED AREOLE VEIN GROUP												
lihuensis subgroup												
19. lihuensis	x								+			
uncinatus subgroup												
cheirodendronae complex												
20. *cheirodendronae*	x							+				
21. *uncinatus*		x						+				
22. *limitatus*		x						+				
23. *beardsleyi*		x						+				
usingeri complex												
24. *usingeri*			x					+				
25. *kraussi*					x	x		+				
26. *rufonotus*						x		+				

Table 1. Island distribution and host genera of *Nesiomiris* species (continued)

	DISTRIBUTION							HOST GENUS				
TAXON	Kauai	Oahu	Molokai	Lanai	W. Maui	E. Maui	Hawaii	Cheirodendron	Reynoldsia	Tetraplasandra	Munroidendron	Ilex
fasciatus subgroup												
fasciatus complex												
27. *fasciatus*	x									+		
28. *maculatus*	x									+		
seminotatus complex												
29. *seminotatus*	x									+		
bifurcatus complex												
30. *bifurcatus*	x									+		
31. *tetraplasandrae*	x									+		
32. *waimeae*	x									+		
lineatus complex												
33. *lineatus*	x							+				
lapalapa subgroup												
lapalapa complex												
34. *lapalapa*	x							+				
35. *olapa*			x							+		
36. *ehu*	x							+				
37. *kauilamahu*	x							+				
38. *mahu*	x							+				
legnotus complex												
39. *legnotus*	x							+				
ILEX-ASSOCIATED GROUP												
hawaiiensis complex												
40. *lanaiensis*				x								+
41. *hawaiiensis*							x					+
42. *byroniae*						x						+
43. *mauiensis*					x							+
44. *kawau*					x	x						+
45. *oahuensis*		x										+
planatus complex												
46. *ilicis*		x										+
47. *ilicivorus*		x										+
48. *planatus*		x										+
49. *contortus*		x										+
50. *restrictus*		x										+

[1] All new species excepting *hawaiiensis* Kirkaldy.

[2] Possibly introduced by man

[3] Molokai only

to Sherff's revisions. The consensus among local botanists is that Sherff's treatments are excessively "split." Harold St. John (pers. comm.) believes *Reynoldsia* is represented in Hawai'i by only 1 species, *R. sandwicensis*.

Several *Nesiomiris* associations exemplify host specificity. A diverse assemblage of species is restricted to the genus *Tetraplasandra*. I interpret this as a demonstration of the founder effect (i.e., on "islands" of particular host trees in the rain forest, which are in turn on geographical islands) (Mayr 1963; Southwood 1960b). Two lineages of these bugs have speciated on these trees: the *trifurcatus* group on pubescent species (*Tetraplasandra*, sections *Pterotropia* and *Eutetraplasandra*), and the *lihuensis* group on glabrous species (see Sherff 1955) (sections *Nothotetraplasandra* and *Neotetraplasandra*). This situation may indicate the relative antiquity of the evolutionary bifurcation of the plant genus. Situations such as this seem to point strongly to an ancient association of insect and plant in the endemic biota. Conversely, on *Cheirodendron*, neither bugs nor plants conform to a discernable evolutionary pattern. A comparatively younger association, still in a state of evolutionary flux, seems to be implicated here. Such patterns provide a possible means of comparing both the sequence of Hawaiian colonization by host plant species and the colonization of host plants by the insects. One problem in this connection is whether botanists are correct in considering the morphologically aberrant araliad *Munroidendron* (a Kaua'i endemic) to be only a highly "advanced" segregate of *Tetraplasandra*. Here the entomologist may be helpful, since he can show that the same species of *Nesiomiris* feeds on both "genera", rather than being genus specific, as is the norm.

Host Specificity Patterns

The mechanics of host specificity have been reviewed by Dethier (1954) and Thorsteinson (1960). Both authors stress the importance of chemical stimuli for the elicitation or inhibition of oviposition and feeding responses. In the present discussion it is assumed that host limitation of monophagous and oligophagous *Nesiomiris* species is due to certain chemical stimuli present in all Hawaiian Araliaceae and possibly also with *Ilex anomala*. To a lesser extent, host limitation would appear to be due to the physical structure of feeding, oviposition, and resting sites (e.g., pubescent versus glabrous leaves, protected versus unprotected sites with respect to inclement weather, and cryptic habits and disruptive coloration due to visual predators).

The patterns of apparent host specificity in *Nesiomiris* are listed in Table 3 and are discussed under the characters examined for cladistic patterns. Absence of a species from a particular host is only negative evidence; the probability of its actual absence is in direct relationship to the intensity of sampling that the host has received.

The bugs were placed in the 3 host specificity categories based only on the presence or absence of nymphs, which I consider to be the definitive evidence of breeding on the host plant. Lack of nymphs indicated a probable perching record, not a true association with a host plant, but alternative feeding may occur on other Araliaceae that were not used for reproduction (see "Life Cycle").

The host specificity patterns of *Nesiomiris* species may be categorized as follows:

Table 2. Geographic Distribution of Host Plants and Associated *Nesiomiris* Species

Host Plant	Associated *Nesiomiris* Species
*Cheirodendron Dominii** (NW Kaua'i)	—[1]
Cheirodendron Fauriei	
var. unspecified (SE Kaua'i)	*olapa, cheirodendronae*
var.*Macdanielsii** (NE Kaua'i)	—
Cheirodendron "Gaudichaudii" (main islands)	*ehu, usingeri*
Cheirodendron Helleri	(see varieties)
var. unspecified	*lineatus, cheirodendronae*
var. *microcarpum** (NW Kaua'i)	—
var. *multiflorum* (NW Kaua'i)	*legnotus, ehu, mahu*
var. *sodalium** (south-central Kaua'i)	—
Cheirodendron kauaiense (Kaua'i)	*lineatus, lapalapa, ehu, kauilamahu, cheirodendronae*
var. *Forbesii* (E Kaua'i)	*lineatus, lapalapa, kauilamahu*
var. *keakuensis** (central Kaua'i)	—
Cheirodendron platyphyllum (Ko'olau and Wai'anae Mts, O'ahu) repeatedly sampled but apparently unsuitable host)	—
Cheirodendron trigynum (variety unspecified)	*pilosus, sinuatus, swezeyi, williamsi, reductus, cheliferus, timberlakei, limitatus, beardsleyi, kraussi, rufonotus, and usingeri*
var. *confertiflorum* (W Maui and E Maui)	*sinuatus, swezeyi, reductus*
var. *halawanum* (Oahu)	*uncinatus, beardsleyi*
var. *Hillebrandeii* (Oahu, Lāna'i)	*parasinuatus, beardsleyi*
var. *Rockii* (Lāna'i)	*parasinuatus*
var. *acuminatum* (NW Kaua'i)	*williamsi*
var. *molokaiensis* (E Moloka'i, W Maui)	*spatulatus*
[No species of *Nesiomiris* were recognized from the following varieties of *trigynum*: *subcordatum* (NW Hawai'i), *Fosbergii* (NE O'ahu), *mauiense* (E Maui), *oblongum* (E Maui), *osteostigma* (SE Molokai), *Skottsbergii* (Lāna'i, W Maui), *Ilicoides* (NW Hawai'i) and *Degeneri* (SE Hawai'i)]	
Cheirodendron "sp." (*kraussi* and *beardsleyi* probably on *C. trigynum*)	*mahu, kraussi, cheirodendronae, beardsleyi*
*Reynoldsia Degeneri** (SE Moloka'i)	—
Reynoldsia Hillebrandii (SE Hawai'i)	*secularis*
*Reynoldsia Hosakana** (NW O'ahu)	—
Reynoldsia huehuensis (NW Hawai'i)	*secularis*
var. *brevipes** (NW Hawai'i)	—
var. *intermedia** (NW Hawai'i)	—
Reynoldsia mauiense (E Maui)	*secularis*
*Reynoldsia oblongata** (southernmost Hawai'i)	—
Reynoldsia sandwicensis (variety unspecified) (O'ahu, E Maui and Hawai'i)	*secularis, timberlakei, trinesiotes*
var. *intercedens** (O'ahu)	—
var. *molokaiensis** (Moloka'i)	—

Table 2. Geographic Distribution of Host Plants and Associated *Nesiomiris* Species (continued)

Host Plant	Associated *Nesiomiris* Species
*Reynoldsia venusta** (NW O'ahu)	—
var. *lanaiensis* (E Lāna'i)	—
*Tetraplasandra bisattenuata** (SE Kaua'i)	—
Tetraplasandra gymnocarpa (variety unspecified) (Koolau Mts., Oahu)	*ohe, timberlakei*
var. *leptocarpa** (SE O'ahu)	—
var. *megalocarpa** (NE Ko'olau Mts., O'ahu)	—
var. *pupukeensis** (W Ko'olau Mts., O'ahu)	—
Tetraplasandra hawaiiensis (variety unspecified)	*timberlakei, pallidus, trinesiotes, curvipes, pallasatus*
var. *awiniensis** (NW Hawai'i)	—
var. *gracilis* (W Maui)	*trinesiotes, pallidus*
var. *micro* (Moloka'i, Maui, Lāna'i, NW Hawa'ii)	? *secularis, pallasatus*
Tetraplasandra kaalae (SW O'ahu)	(see varieties)
var. *kaalae*	*timberlakei*
var. *multiplex* (NW O'ahu)	*timberlakei*
Tetraplasandra kahanana (NE Ko'olau Mts., O'ahu)	*trifurcatus*
Tetraplasandra kavaiensis (variety unspecified) (Kaua'i)	*elongatus, maculatus*
var. *dipyrena*** (Lāna'i)	—
var. *grandis* (S Hawai'i)	*pilosus*
var. *intercedens** (E Maui)	—
var. *koloana* (Kaua'i)	*fasciatus, maculatus*
var. *nahikuensis*** (E Maui)	—
var. *occidua*** (W Maui)	—
var. *robustior*** (NW Maui)	—
*Tetraplasandra kohalae*** (N Hawai'i)	—
*Tetraplasandra lanaiensis*** (central Lāna'i)	—
Tetraplasandra lihuensis (variety unspecified) (Kaua'i)	*lihuensis*
var. *gracilipes** (SE Kaua'i)	—
*Tetraplasandra Lydgatei*** (SE O'ahu)	—
Tetraplasandra meiandra (variety unspecified)	*trifurcatus, curvatus, secularis, trinesiotes, seminotatus, bifurcatus*
var. *bisobtusa*** (NW Hawai'i)	—
var. *Bryanii** (N O'ahu)	—
var. *Degeneri** (NW Kaua'i)	—
var. *Hillebrandii** (Lāna'i)	—
var. *hiloensis** (NE Hawa'i)	—
var. *leptomera** (E Maui)	—
var. *longipedunculata** (Ko'olau Mts., O'ahu)	—
var. *makalehana** (NW O'ahu)	—
var. *mauiensis* (E Maui)	*secularis*
var. *molokaiensis** (Moloka'i)	—
var. *occidentalis** (W Maui)	—
var. *olowaluana*** (W Maui)	—
var. *ovalis* (E Maui)	*curvatus*
var. *polyantha** (E Maui)	—
var. *polystigmata** (E Maui)	—
var. *prolifica*** (NW Hawai'i)	—

Table 2. Geographic Distribution of Host Plants and Associated *Nesiomiris* Species (continued)

Host Plant	Associated *Nesiomiris* Species
Tetraplasandra meiandra (continued)	—
var. *prolificoldes*** (NW Hawai'i)	—
var. *ramosior* (W Maui)	*trinesiotes*
var. *rhynchocarpa** (SE Hawai'i)	—
var. *rhynchocarpoides** (SE Hawai'i)	—
var. *Rockii** (SE Hawai'i)	—
var. *simulans*** (NW Hawai'i)	—
var. *Skottsbergii** (E Moloka'i)	—
var. *tenuistylis** (SE Hawai'i)	—
var. *tenuistyloides* (SE Hawai'i)	*pilosus*
Tetraplasandra micrantha (Wai'anae Mts., O'ahu)	*timberlakei*
*Tetraplasandra Munroi*** (central Lāna'i)	—
*Tetraplasandra oahuensis** (SE O'ahu)	—
var. *eradiata** (SE O'ahu)	—
var. *Fauriei** (SE O'ahu)	—
var. *haillensis** (NW O'ahu)	—
var. *longipes** (SE O'ahu)	—
var. *pseudorhachis** (central O'ahu)	—
var. *pseudolongipes** (SE O'ahu)	—
var. *venulosior** (NE O'ahu)	—
*Tetraplasandra pupukeensis** (N O'ahu)	—
var. *decipiens* (N O'ahu)	—
var. *megalopoda* (N O'ahu)	—
var. *nitida* (NW O'ahu)	—
var. *venosa* (N O'ahu)	—
Tetraplasandra turbans (Wai'anae Mts, O'ahu)	*timberlakei*
*Tetraplasandra Waialeale*** (NE Kaua'i)	—
var. *acrior* (central Kaua'i)	—
var. *subsessilis* (NW Kaua'i)	—
var. *urceolata* (central Kaua'i)	—
var. *wahiawensis* (SW Kaua'i)	—
Tetraplasandra waimeae var. waimeae (NW Kaua'i)	*waimeae*
sp. (probably *waimeae)* (Kaua'i)	*tetraplasandrae*
Tetraplasandra waianaensis (var. unspecified) (S O'ahu)	*timberlakei*
var. *palehuana** (SW O'ahu)	—
Munroidendron racemosum (Kaua'i)	*elongatus*
Ilex anomala (all main islands)	*lanaiensis, ilicis, hawaiiensis, byroniae, mauiensis, kawau, ilicivorus, planatus, contortus, restrictus*

* According to Scherff's revisions (1952, 1954, 1955).
** Not found in the field during this study.
1. Indicates no *Nesiomiris* species; or host plant not found or determined thusly, when asterisked.

CATEGORY 1: Monophagous species with only 1 host (Total = 41 spp.; 82% of 50 spp.): *beardsleyi, bifurcatus, byroniae, cheliferus, contortus, curvatus, curvipes, fasciatus, hawaiiensis, ilicis, ilicivorus, kauilamahu, kawau, kraussi, lanaiensis, lapalapa, legnotus, lihuensis, limitatus, maculatus, mahu, mauiensis, oahuensis, ohe, olapa, pallasatus, pallidus, parasinuatus, planatus, reductus, restrictus, rufonotus, seminotatus, sinuatus, spatulatus, swezeyi, tetraplasandrae, uncinatus, usingeri, waimeae, williamsi*

CATEGORY 2: "Monophagous" species feeding on 2 or more species in the same genus, or in the same section of a genus (Total = 3 spp.; 6% of 50 spp.): *cheirodendronae, ehu, lineatus*

CATEGORY 3: Oligophagous species feeding on 2 or more sections within *Tetraplasandra* (Total = 1 sp.; 2% of 50 spp.): *trifurcatus*

CATEGORY 4: Oligophagous species feeding on 2 host genera (Total = 4 spp.; 8% of 50 spp.): *elongatus, pilosus, secularis, trinesiotes*

CATEGORY 5: Polyphagous species feeding on 3 genera of Araliaceae (Total = 1 sp.; 2% of 50 spp.): *timberlakei*

Monophagy

Some 39 species are associated exclusively with a single host species (Table 3). Two other species are also considered monophagous because they have been found only on 2 or more "species" within the same section of *Tetraplasandra.*

Of the 41 monophagous species, 18 species occur exclusively on 1 species of *Cheirodendron.* Eleven species occur on only 1 each of several species of *Tetraplasandra.* All 11 species that occur on *Ilex anomala* are monophagous. No species is restricted to either *R. sandwicensis* or *Munroidendron racemosum.* Three species, which occur on 2 or more hosts in this category, are associated with closely related species of *Cheirodendron* on Kaua'i.

Species that are considered advanced phylogenetically tend to be monophagous. Also, species that are so modified as to render them cryptic and inconspicuous on the host plant (*T. kavaiensis* and *T. hawaiiensis*) are monophagous. These are pallid species that often also have a pattern of green or black spots on the hemelytra. This suggests that this level of specificity has been enforced to some extent by visual predators such as insectivorous birds.

The level of specificity that might be reinforced by interspecific competition cannot be determined. Opler (1974a: 58) and Hering (1951: 167) found that some otherwise monophagous lepidopterous leaf miners of *Quercus* (Fagaceae) would breed on exotic *Quercus* species introduced into the natural range of the moths and their hosts. They considered this pattern to afford evidence for the enforcement of host specificity by interspecific competition by the moths. Within araliaceous hosts, monophagous bug species predominate on the older islands (Kaua'i and O'ahu). Host-plant transfers might be inhibited because of interspecific competition that might arise if hosts on the older islands were already "saturated" with bug species. But it should also be noted that 1 exotic araliaceous species from Australia (*Brassaia actinophylla*) has become extensively naturalized, at least in the lowland rain forests of O'ahu. No *Nesiomiris* species have yet been found to attack it.

TABLE 3. Degree Host Specificity in *Nesiomiris* Species

	HOST SPECIFICITY				
	Monophagous		Oligophagous		Polyphagous
***Nesiomitis* species**	1 host species	>1 Sp. in same genus or section	2 or more Sections within *Tetrapla-sandra*	2 host genera	3 genera of Araliaceae
1. *trifurcatus*			x		
2. *curvatus*	x				
3. *pilosus*				x	
4. *sinuatus*					
5. *parasinuatus*	x				
6. *swezeyi*	x				
7. *williamsi*	x				
8. *reductus*	x				
9. *spatulatus*	x				
10. *cheliferus*	x				
11. *elongatus*				x	
12. *ohe*	x				
13. *secularis*				x	
14. *timberlakei*					x
15. *trinesiotes*				x	
16. *pallidus*	x				
17. *curvipes*	x				
18. *pallasatus*	x				
19. *lihuensis*	x				
20. *cheirodendronae*		x			
21. *uncinatus*	x				
22. *limitatus*	x				
23. *beardsleyi*	x				
24. *usingeri*	x				
25. *kraussi*	x				
26. *rufonotus*	x				
27. *fasciatus*	x				
28. *maculatus*	x				
29. *seminotatus*	x				
30. *bifurcatus*	x				
31. *tetraplasandrae*	x				
32. *waimeae*	x				
33. *lineatus*		x			
34. *lapalapa*	x				
35. *olapa*	x				
36. *ehu*		x			
37. *kauilamahu*	x				
38. *mahu*	x				
39. *legnotus*	x				
40. *lanaiensis*	x				
41. *hawaiiensis*	x				
42. *byroniae*	x				
43. *mauiensis*	x				

TABLE 3. Degree Host Specificity in *Nesiomiris* Species (continued).

	HOST SPECIFICITY				
	Monophagous		**Oligophagous**		**Polyphagous**
***Nesiomiti* species**	1 host species	>1 Sp. in same genus or section	2 or more Sections within *Tetrapla-sandra*	2 host genera	3 genera of Araliaceae
44. *kawau*	x				
45. *oahuensis*	x				
46. *ilicis*	x				
47. *ilicivorus*	x				
48. *planatus*	x				
49. *contortus*	x				
50. *restrictus*	x				
TOTALS	41	3	1	4	1
% of total spp.	82	6	2	8	2
category total	44		5		1
% of total spp.	88		10		2

Oligophagy

About 5 species are considered to have oligophagous habits (Tables 3, 4) in that they feed either on 2 host genera of Araliaceae or on more than 1 section of *Tetraplasandra.* None was found to feed on taxa of more than 2 sections of the host genus even though these are or were available on all the main islands. This may be an artifact of man's impact, since the species of Section *Pterotropia*, especially, appear to have predominated in areas now extensively converted to agriculture.

The 2 oligophagous species in the geologically older islands, *N. elongatus* of Kaua'i and *N. trifurcatus* of O'ahu, are considered to be primarily oligophagous, relict species on *Tetraplasandra* and its close relative, *Munroidendron racemosum.* Alternatively, these 2 species might be sufficiently ancient to have coevolved with the speciation of the other hosts.

The 3 oligophagous species on the geologically younger islands (Moloka'i, Lāna'i, Maui, and Hawai'i) are thought to be utilizing previously unexploited or relatively unexploited hosts. Few other *Nesiomiris* species have been able to reach these younger islands. This might be considered evidence of the impact of interspecific interactions.

The small oligophagous species *N. secularis* and *N. trinesiotes* appear to be comparatively successful dispersers, as evidenced by their geographic distribution, and are able to exploit the seasonally available *R. sandwicensis.* Alternatively, they feed on a seasonal *Tetraplasandra* species at higher, wetter elevations.

Because the plesiomorphic relatives of *N. pilosus* of Hawai'i Island feed on *Tetra-*

TABLE 4. Geographic Distribution of Oligophagous (o) and Polyphagous (p) Species of *Nesiomiris*

ISLAND	GENERA AND/OR GENERIC SECTIONS OF SHARED HOSTS*	*NESIOMIRIS* SPECIES
Kaua'i	*Tetraplasandra, Munroidendron*	*elongatus* (o)
O'ahu	*Cheirodendron, Reynoldsia, Tetraplasandra* (sections *Pterotropia* and *Neotetraplasandra*	*timberlakei* (p)
O'ahu	*Tetraplasandra* (sections *Pterotropia* and *Neotetraplasandra)*	*trifurcatus* (o)
Moloka'i	*Reynoldsia* and *Tetraplasandra*	*trinesiotes* (o)
W Maui	*Tetraplasandra* (sections *Eutetraplasandra* and *Neotetraplasandra)*	*trinesiotes* (o)
E Maui	*Reynoldsia* and *Tetraplasandra*	*secularis* (o)
Hawai'i	*Reynoldsia* and *Tetraplasandra*	*secularis* (o)
Hawai'i	*Cheirodendron, Reynoldsia* and *Tetraplasandra*	*timberlakei* (p)

* Only one Section unless otherwise indicated.

plasandra species on the older islands, I consider that its exploitation of *Cheirodendron* indicates utilization of a relatively vacant niche. This assumption is strengthened by the fact that only 2 other species of *Nesiomiris*, the monophagous *N. williamsi* and polyphagous *N. timberlakei*, are found on *Cheirodendron* on Hawai'i Island.

Polyphagy

Only 1 species (*N. timberlakei*) is considered to be polyphagous (Tables 3, 4), that is, it feeds on all endemic genera of Araliaceae. This species has much in common with the oligophagous *N. secularis* and *N. trinesiotes*. It is relatively small and widely distributed (all over Hawai'i and O'ahu). I believe that it exploits *Reynoldsia* in the same manner as *N. secularis* and *N. trinesiotes*. *Nesiomiris timberlakei* additionally utilizes *Cheirodendron trigynum*. This may be due to the youthfulness of the association on Hawai'i, and it may have been introduced by humans into O'ahu, or vice versa. It may now be undergoing an exploitative phase. However, on Hawai'i *N. timberlakei* appears to be competitively excluded from *Reynoldsia* by *N. secularis*, but it is the only bug that attacks *Reynoldsia* on O'ahu. Both species build tremendous populations on this host during the onset of the summer dry season and the leaves are often completely chlorotic due to the heavy feeding.

Conclusions

Three basic patterns of host specificity are discernible: monophagy, oligophagy, and polyphagy.

Monophagous species are limited to a single host or to a closely related species pair of hosts. Although the basis of host selection by these bugs was not investigated, the occurrence of species occupying homologous niches on other Araliaceae (which might otherwise have been suitable hosts) indicates that interspecific displacement may be a primary reinforcing agent for monophagy. All the species associated with *Ilex anomala* are monophagous and allopatric. Phylogenetically advanced species and those associated with pubescent-leaved *Tetraplasandra* (sections *Eutetraplasandra* and *Pterotropia*) tend to be monophagous. The latter group is so specialized morphologically and behaviorally as to render its members cryptic on the underside of the host leaves. A greater number of monophagous species occur on *Cheirodendron* (18) than on *Tetraplasandra* (11). I consider that the former host genus is more recently arrived in the archipelago, since fewer distinct species and no sections are recognized by systematic botanists. Monophagous species predominate on the geologically older islands. This may be another aspect of interspecific competition in that the older ecosystems are more completely occupied.

Oligophagous species are restricted to 2 host genera or sections. Within Araliaceae, degree of specificity is due to a number of causes. Among those that are apparent are (a) the relict nature of some species that may have coevolved with the host plants on the older islands, (b) the exploitation of additional hosts by phylogenetically recent species on younger islands, and (c) the higher vagility of 2 species that are able to exploit hosts marginal to other *Nesiomiris* species because of seasonal availability.

Polyphagous species feed on all native genera of Araliaceae. Only 1 species (*N. timberlakei*) falls into this category, and it demonstrates all aspects of patterns (b) and (c) discussed above under oligophagous species. This species additionally attacks *Cheirodendron trigynum* and has a wider geographic distribution (O'ahu and Hawai'i) than any other species of *Nesiomiris*, although I cannot rule out the possibility that it may have been accidentally introduced onto one island or the other by humans.

Life Cycle

Adults and nymphs are found throughout the year on suitable host plants. Excepting the summer-deciduous *Reynoldsia* species, all potential host plants are always foliated. None of the 3 *Nesiomiris* species (*N. secularis*, *N. timberlakei*, *N. trinesiotes*) associated with *Reynoldsia sandwicensis* are specific to that host alone. If they were, there would be a period of several summer months when diapause would be required because of the lack of feeding sites. There was no evidence to indicate diapause, although it may exist. Several other possibilities are suggested: (1) these bugs are able to survive during the xeric months on the few remnant leaves that are almost invariably present on trees in larger groves of *R. sandwicensis*; and/or (2) the species can recolonize *R. sandwicensis* from other araliaceous hosts growing in adjacent wetter regions; and/or (3) they can feed on the inflorescences and green fruit produced during the summer dry season. During the summer the buds for the following season's growth are probably not available as a food source, since they have a thick waxlike coating presumably protecting them against desiccation and insect damage.

Oviposition was observed both during winter and summer months. Oviposition sites are on petioles, midribs, and rachises (Fig. 2). Feeding sites are usually on the underside of leaf blades. Nymphs and adults appeared to form loose aggregations on the same host plants and showed no evidence of preferring separate feeding sites.

No investigation was made of differences that may exist in sex ratio of populations during the year. Periodically, however, I would encounter a greater number of 1 sex, which did not appear to be due to chance. It is a common situation in mirids for males to mature faster and predominate for a brief period while females begin to mature. With the earlier death of the males, the females predominate before that proportion of a population resulting from eggs laid concurrently completes its life span.

Flight Activity and Dispersal

The inference can be made that most *Nesiomiris* species, at least those with an abundant supply of food plants, are usually sedentary and that there is little need or selection for dispersal capability or vagility.

From October 1970–May 1972, two Malaise traps were operated in the Kīlauea Forest Reserve, Hawai‘i Island, at an elevation of 1586 m. The contents were collected biweekly. Yet, only 14 specimens (6 males, 8 females) were intercepted, and all were *N. williamsi*. Most were caught following stormy, high-wind conditions. If the bugs were not avoiding the traps, their capture would appear to indicate low flight activity.

This forest is composed of a canopy of *koa* (*Acacia koa*) trees, a lower canopy of *‘ōhi‘a* (*Metrosideros polymorpha*) trees, and a still lower (10 m high) matrix of other species of trees, among them *‘ōlapa* (*Cheirodendron trigynum*), the host of *N. williamsi*, and *kāwa‘u* (*Ilex anomala*), the host of *N. hawaiiensis*. *Cheirodendron* is so abundant there that both Malaise traps were directly adjacent to a host tree.

On the other hand, species associated with summer-deciduous *Reynoldsia sandwicensis* appear to be active dispersers. Excepting the partly denuded island of Lāna‘i where there are only about a dozen *R. sandwicensis* trees left, even the most isolated host specimens each had a species of *Nesiomiris* attacking it. In fact, the island distribution of the bugs usually closely parallels the distribution of the host. These species of *Nesiomiris* tended to be the most widely distributed in the genus; *N. secularis* is found on both east Maui and Hawai‘i, and *N. timberlakei* is found on O‘ahu and Hawai‘i. Enough inter-island gene flow, for example, must be occurring to swamp tendencies toward speciation.

Body Size Correlations

Species at higher elevations are generally smaller than those at lower elevations. Examples of this phenomenon are species of the *rufonotus* group in the Alaka‘i Swamp of Kaua‘i, and species in the *sinuatus* complex and the *rufonotus* group at higher elevations on Haleakalā and Hawai‘i Island.

Species in relatively unstable environments are also smaller. This is particularly evident in species associated with the summer-deciduous *Reynoldsia sandwicensis* from O‘ahu to Hawai‘i. Smaller species also predominate on younger rather than on older islands.

Evolution and Speciation

Synopsis of Insular Evolution

First I shall attempt to categorize the broad modes of speciation of Hawaiian insects with reference to their evolutionary ecology. I am going to postulate the major evolutionary pathways that native insects may have taken by examining and comparing them with those species recently introduced either accidentally or deliberately and whose geographic spread and host associations in the islands are well documented.

My premise is that the phenomena we observe today with regard to introduced species are the same as those that accompanied natural colonizations. Of course, the number of candidates is now increased, and some strong, new selective forces (e.g., ants) are operative now that previously were not. I have paid special attention to the Heteroptera because of my greater familiarity with that group.

Some trends are apparent in the evolution of the species of the genus *Nesiomiris*. Among those species associated with araliads, the plesiomorphic species (*sensu* Hennig 1966) predominate at the northwest or Kaua'i end of the island chain, while the apomorphic species are more prevalent toward the southeast or Hawai'i end. In *Nesiomiris* I know of only 1 or possibly 2 instances in which an apomorphic species is found on an island northwest of its nearest, more primitive relative. In the highly speciated Hawaiian *Drosophila*, the most primitive species, *D. primaeva* Hardy & Kaneshiro, is endemic to Kaua'i (Carson & Stalker 1968). However, in the *hawaiiensis* group of *Nesiomiris*, which has speciated on *Ilex anomala*, the most plesiomorphic member is on Lāna'i near the center of the island chain, and the derived members extend in both directions along the chain. I believe that the adaptive shift from an araliad to *Ilex* took place on Lāna'i or Maui Nui and that *Ilex*-associated species are spreading northwest and southeast.

Possible Evolutionary Strategies

A comparative analysis of some possible evolutionary strategies of 4 genera of Heteroptera is detailed in Figures 3–7. Two of the genera analyzed, *Nabis* (Nabidae) and *Koanoa* (Miridae), are wholly or mostly predaceous; the other 2, *Sarona* (Miridae) and *Nesiomiris*, are phytophagous.

I have assumed that any species that successfully colonizes the Hawaiian Archipelago may fall into 1 of 2 classes: (a) relatively generalized or (b) specialized in its ecological requirements. While a "generalist" would seem to have an advantage, a relative "specialist" might colonize successfully provided suitable hosts were already present. These 2 colonizing classes fall into 4 evolutionary strategies.

Strategy I: Speciation of a "generalist" phytophagous species into "specialist" species (example: *Sarona*, Fig. 3). The distribution and host genera associations indicate that a "generalist" colonizing species of *Sarona* spread quickly throughout the main islands of the archipelago, attacking a variety of unrelated host plants. This strategy is seen today in several genera of Miridae that appear to have arrived relatively recently (e.g, *Hyalopeplus pellucidus* Stål) or have recently been accidentally introduced by humans (e.g., *Spanagonicus albofasciatus* Reuter). After an initial explosive population phase, selective factors begin to operate and isolating mechanisms increasingly affect diverging populations. The combination of insular isolation and increasing host-plant specialization may eventually result in a large number of species. From the Bishop Museum collections of *Sarona* I have tentatively sorted out about 30 species, all but 1 undescribed. These are associated with species of 14 genera of host plants in 11 different plant families (Table 5). My preliminary observations indicate also that an evolutionary history similar to that postulated for *Sarona* may have occurred in several other genera of phytophagous Hawaiian Miridae (e.g., "*Orthotylus*" and *Sulamita*).

Strategy II: Speciation of "specialist" phytophagous species to "generalist" species (example: *Nesiomiris*, Fig. 4). The precursor species of *Nesiomiris* appears to have been relatively specialized, apparently preadapted to *Tetraplasandra* (Araliaceae). It appears to have reached Kaua'i first and did not successfully colonize a younger island until several lineages had evolved in Kaua'i on *Tetraplasandra* and then *Cheirodendron*. Successful host shifts to *Cheirodendron* species, for example, appear to have occurred repeatedly, both on Kaua'i and on other islands of the chain. A new association developed with *Reynoldsia*, a genus of Araliaceae absent from Kaua'i. This portion of the evolution of *Nesiomiris*, here termed Phase I, appears to correspond to the present stage of the mirid genus *Kalania*, which attacks only *Urera* (Urticaceae). In a rich vascular flora such as that of Hawai'i, with about 1,800 endemic species and varieties in 93 families (Fosberg 1948) there was wide opportunity for fortuitous and successful adaptive shifts to other host plant families here termed Phase II. In *Nesiomiris*, Phase II occurred when an adaptive shift was accomplished onto *Ilex anomala* (Aquifoliaceae) (Fig. 5), which was followed by another phase of explosive speciation as other islands were successfully colonized. This adaptive shift apparently took place at the center of the main island chain, since the most plesiomorphic *Ilex*-associated species (*N. lanaiensis*) and its closest relatives occur in that vicinity. Phase III, representing adaptive shifts to additional plant families, has not yet occurred in *Nesiomiris*. However, it does appear to have occurred in *Cyrtopeltis* (Miridae) (Gagné 1969), which demonstrates geographical patterns similar to those of *Nesiomiris* but attacks a greater variety of unrelated host-plant families (Table 6). *Cyrtopeltis* is at least diphyletic here, and the major monophyletic lineage attacks 4 families of host plants.

Strategy III: Speciation of a "generalist" predator species to "specialist" species (example: *Koanoa*, Fig. 6). Only 2 species, *K. hawaiiensis* Kirkaldy and *K. williamsi* Usinger, have been described. I have seen several additional species in local collections, but it appears that the genus will remain comparatively small. I have some familiarity with *K. hawaiiensis* through work on the insect ecology of *Metrosideros polymorpha* and *Acacia koa* on Hawai'i Island and from incidental collecting on other islands. *Koanoa* species appear to be facultative predators, taking both insect prey and plant material. They are small and occur throughout the main islands in many habitats and on many plants. Consequently, there would appear to be frequent gene exchange between and within the islands of the archipelago. Speciation appears to have resulted from increased specificity to particular host plants, presumably because of the uniform availability of associated prey. Examples of "host" plants are *Freycinetia* for *K. williamsi* and *Acacia koa* for an undescribed *Koanoa* species.

Strategy IV: Speciation of a "specialist" predator species to "generalist" species (ex-

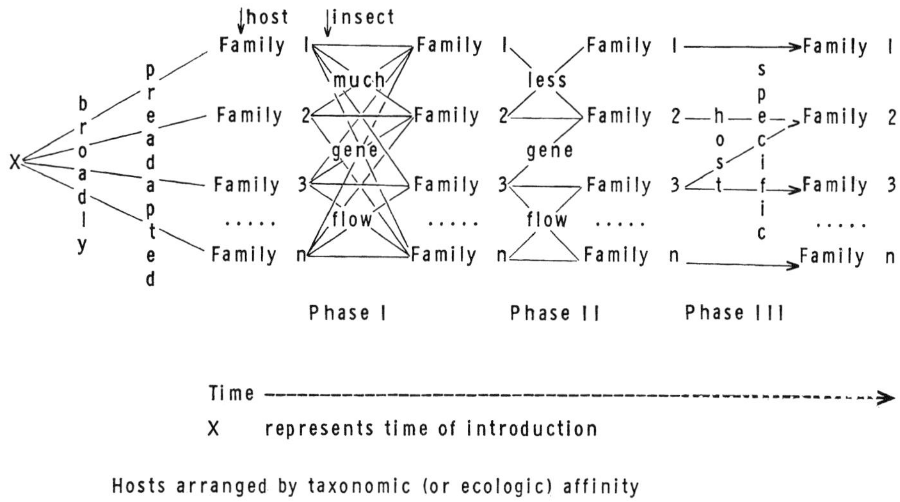

Fig. 3. Strategy I: Phytophage, "generalist" to "specialist". Example: *Sarona* (Miridae).

ample: *Nabis*, Fig. 7). *Nabis* now contains 26 described species in Hawai'i, but there are a number of undescribed species. Zimmerman (1948b) estimated there were probably over 50 species in the archipelago and postulated that these had originated from 1 colonizing species. The features that have allowed isolating phenomena to proceed in this group are (1) the evolution of flightlessness, (2) prey associated host-plant preferences, (3) habitat preferences (ground dwelling, riparian, and tree dwelling), and (4) isolation conferred by separate islands.

The most primitive species was thought to be *N. blackburni* White, which Zimmerman (1948b: 140) said was widespread and had "been reported from many kinds of plants."

Potential Modes of Insular Speciation

Of the 12 potential modes of species formation proposed by Mayr (1963: 428; table 15–1), those involving "gradual speciation" mechanisms seem to especially lend themselves to analysis here. The genetic basis for insular speciation was not investigated in *Nesiomiris* because their chromosomes do not lend themselves to this type of analysis.

My analysis of speciation will center, then, on such speciation mechanisms as geographic barriers between (i.e., instant speciation) and within (i.e., gradual speciation) islands and ecological barriers (host-plant associations). These barriers can be enumerated as follows: (1) inter-island founder events, (2) intra-island geographic isolation, (3) intergeneric host-plant transfers, and (4) multiple invasions of the same host plant.

Intra-island geographic isolation requires some additional explanation. While some

TABLE 5. Host Associations and Distribution of *Sarona* Species

HOSTS	ISLANDS						
	Ka	O	Mo	La	WM	EM	H
Aquifoliaceae							
Ilex		x		x			
Asteraceae							
Dubautia						x*	
Euphorbiaceae							
Claoxylon	x						
Euphorbia	x						
Loranthaceae							
Korthalsella							x
Myoporaceae							
Myoporum	x						
Myrtaceae							
Metrosideros			x				
Eugenia		x					
Oleaceae							
Osmanthus				x		x	
Pittosporaceae							
Pittosporum						x	x
Rutaceae							
Pelea	x*	x					x*
Saxifragaceae							
Broussaisia		x					
Urticaceae							
Pipturus			x		x	x	x

* Several *Sarona* spp. present. Abbreviations: Ka = Kauai; O = Oahu; Mo = Molokai; La = Lanai; WM = W Maui; EM = E Maui; H = Hawaii.

islands, notably O'ahu and Maui, are now single geographic units, each in actuality owes its origin to the merging of 2 separate volcanoes by continued volcanic activity (Stearns 1966). The connecting landscapes of O'ahu and Maui with their xeric lowland conditions prevailing now and for probably a considerable period into the past are unsuitable for any potential host plants other than *R. sandwicensis*.

Inter-island Founding

There would appear to be a competition gradient of greater severity for island founding toward the northwest of the island chain, where already established *Nesiomiris* species occupy the requisite niches. For species insufficiently isolated reproductively, there would be the danger of genetic swamping by ancestral populations, since the number of potential founders would be infinitesimally smaller than the established populations, at least on a given host genus. Hence, the history of the genus should indicate a paucity of successful island founding westward (i.e., toward Kaua'i) in the island chain. The *lanaiensis*

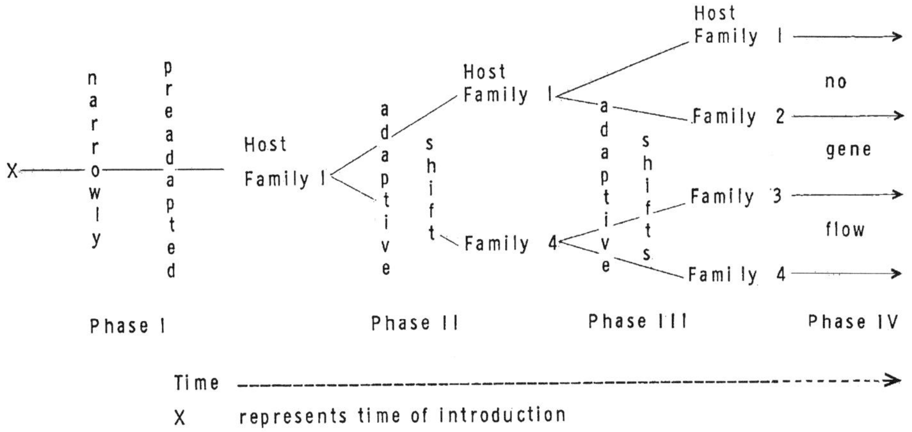

Fig. 4. Strategy II: Phytophage, "specialist" to "generalist". Example: *Nesiomiris* (Miridae).

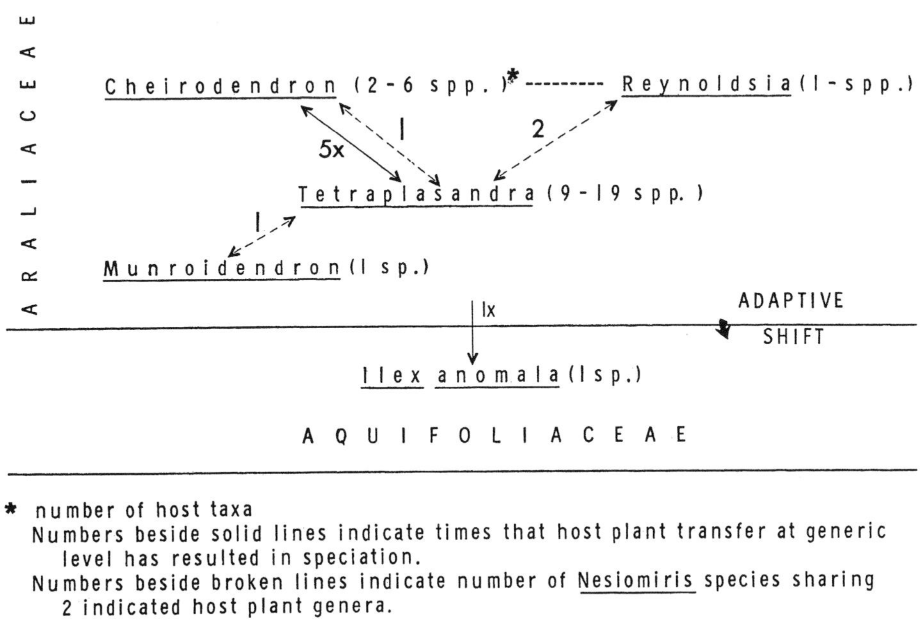

Fig. 5. Adaptive radiation of *Nesiomiris* spp. on host-plant genera.

TABLE 6. Host Association and Distribution of Hawaiian *Cyrtopeltis* spp.[1]

Species[2]	Host Plant	Host Family	Distribution
confuse Perkins	*Cyrtandra cordifolia*	Gesneriaceae	O'ahu
cyrtandrae Gagné	*Cytandra* sp.	Gesneriaceae	O'ahu
hawaiiensis Kirkaldy	*Dubautia* sp.	Asteraceae	Maui, Moloka'i, Hawai'i
lysimachia C. & U.	*Lysimachia* sp.	Primulaceae	Kaua'i
perplexa Gagné	*Raillardia* sp.	Compositae	Maui
pyllostegiae C. & U.	*Phyllostegia* sp.	Labiatae	O'ahu
sidae Gagné	*Sida* sp.	Malvaceae	Maui
n. sp. 1	*Sida* sp.	Malvaceae	Maui
n. sp. 2	*Sida* sp.	Malvaceae	Lana'i
n. sp. 3	*Abutilon sandwicensis*	Malvaceae	O'ahu
terminalis Gagné	*Cyrtandra* sp.	Gesneriaceae	O'ahu
n. sp. 4	*Sicyos cucumerinus*	Cucurbitaceae	Hawai'i
n. sp. 5	*Wilkesia hobdyi*	Asteraceae	Kaua'i

1. Adapted from Gagné, 1969
2. *hawaiiensis, perplexa* and n. sp. 4 appear to have evolved from a different progenitor than the remainder of the species; that is, *Cyrtopeltis* appears to be diphyletic in the Hawaiian Islands.

group feeding on *Ilex anomala* is an exception in that the initial adaptive shift to this host appears to have taken place on Maui Nui (Maui, Moloka'i, Lāna'i) (Figs. 8–14).

Since climate, topography, and the hosts are relatively the same, particularly in the rain forests of each island, the physical environment would appear to be less important in speciation than genetic phenomena (founder effect, genetic drift, loss mutations, etc.). Because nearly identical environmental conditions occur somewhere on each of the main islands, a gravid female blown to a neighboring uncolonized island would have a chance of establishing an independent population. With at least 3 independently evolving species groups involved, the stage was set for explosive speciation.

In sympatric taxa in the same species group, or even the same species complex, characteristically 1 member of the pair will have a more developed male clasper and a less complex aedeagus than the other member.

A summary of the 13 postulated inter-island founder events is presented in Figure 13. Founding was primarily from northwest to southeast, that is, from the older to the younger islands. The source of the presumed progenitor species is called the donor island; that which it colonized is called the recipient island.

It is evident that Kaua'i has been colonized from outside the archipelago or from now extinct populations to the northwest. Although the presently most plesiomorphic species in the genus (*N. trifurcatus*) is on O'ahu, I argue that this species was derived from a still more plesiomorphic founder from Kaua'i, whose relatives are now extinct. Another factor would appear to be the relative isolation of Kaua'i from the other islands (Fig. 1): Kaua'i is 122 km from O'ahu. The greatest distance between other major islands is 50 km

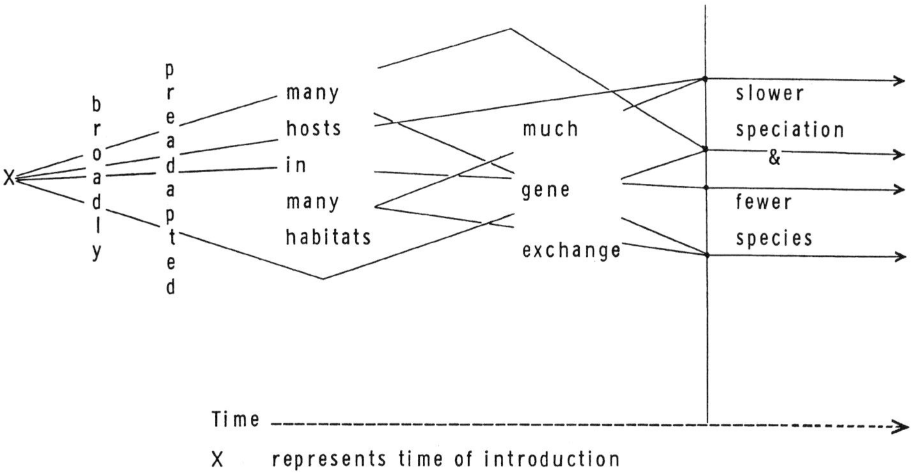

Fig. 6. Strategy III: Speciation of a "generalist" predator. Example: *Koanoa* (Miridae).

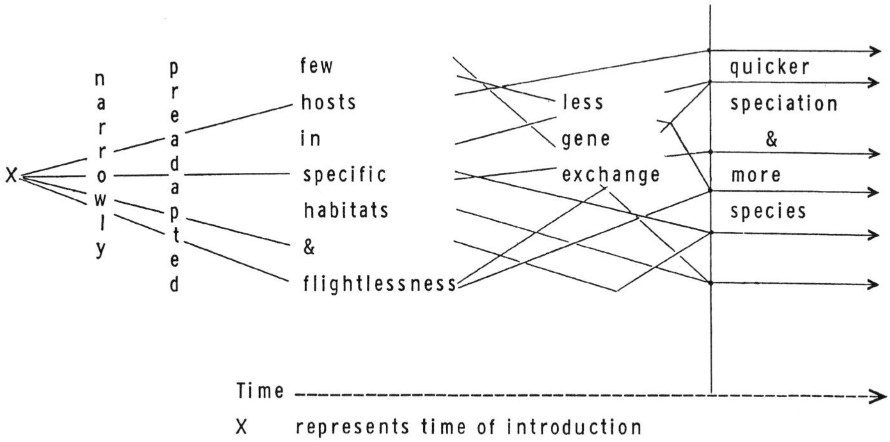

Fig. 7. Strategy IV: Speciation of a "specialist" predator. Example: *Nabis* (Miridae).

between Maui and Hawai'i. The isolation of Kaua'i is also reflected in the fact that all of its 16 species of *Nesiomiris* are monoinsular endemics (Fig. 14).

About 26% of the speciation in *Nesiomiris* can be attributed to inter-island founder events. Inter-island founding might be considered a form of "instantaneous speciation" (see Mayr 1963: 432) in that new species populations are normally initiated by single gravid females or by a few individuals that are immediately isolated geographically from the parental stock.

While I have relied on the occurrence of the morphologically closest relative on the same host genus on another island to try to authenticate the occurrence of a founder event, Carson (1970) has supported morphological data with chromosomal structure. He clearly demonstrated the origins of several species of *Drosophila* on Hawai'i Island from adjacent Maui on the basis of the occurrence and distribution of shared complex chromosomal inversions.

A founder would be successful in establishing a new population only if its host plants on the donor island already occurred on the recipient island. Hence, I sought founder progenitors on donor islands among those species attacking the same host genus on the recipient island. On the newly colonized island, the host would be less likely to harbor previously arrived and potentially competitive *Nesiomiris* species. This might be another factor explaining the bulk of founding events from older to newer islands. Other genera of potential host plants on the newly colonized island would also be less likely to harbor potentially competitive *Nesiomiris* species. Thus host-plant genus transfers and polyphagy are more likely on the younger islands (Fig. 14, Table 4). In other words, the older the island, the greater the likelihood that its potential host plants will already be more saturated with *Nesiomiris* species.

Intra-island Founding

This category is actually inseparable from the inter-island founding just discussed in that several of the present islands in the archipelago, such as O'ahu and Maui, were each once composed of 2 islands. Conversely, Lāna'i, Moloka'i, and Maui were once a single island (Stearns 1966). Even the youngest island, Hawai'i, has been built up by 5 volcanoes of different ages that have anastomosed. It thus becomes impossible to say with certainty whether a particular founding event was inter-island or intra-island in character. It seems likely, however, that more of the founding events here classified as intra-island were in actuality inter-island than vice versa. Regardless of sea level fluctuation during interglacials, each of the volcanoes that constructed what are now O'ahu, Maui, and Hawai'i Islands. would have risen independently above the ocean surface as separate islands

Figs. 8–9. 8, The Hawaiian Archipelago. Islands become progressively younger geologically from left to right. Islands from Kure (30 my) to Nihoa (10 my) are atolls and basaltic pinnacles now devoid of potential host plants of *Nesiomiris,* which is now distributed on islands in the main chain: Kaua'i (5 my) toHawai'i (0.7 my). **9,** Origin of *Nesiomiris* from a founding event into the Hawaiian Archipelago, presumably to Kaua'i, and the subsequent history of *trifurcatus* (sp. no. 1) and *curvatus* (spp. nos. 2–8). Subgroups: existing species, numbered circles; open circles, hypothetical populations. solid lines join species on same genus of host plant; dashed lines join species or populations that are on different host plants; arrows indicate directionality of inter- and intra-island founding.

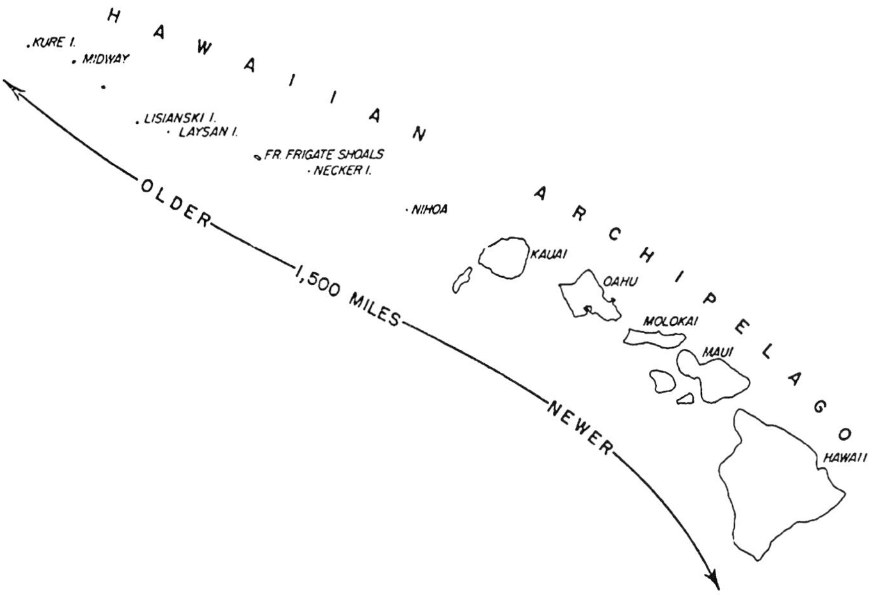
HAWAIIAN ARCHIPELAGO
KURE I.
MIDWAY
LISIANSKI I.
LAYSAN I.
FR. FRIGATE SHOALS
NECKER I.
NIHOA
KAUAI
OAHU
MOLOKAI
MAUI
HAWAII
OLDER
1,500 MILES
NEWER

8

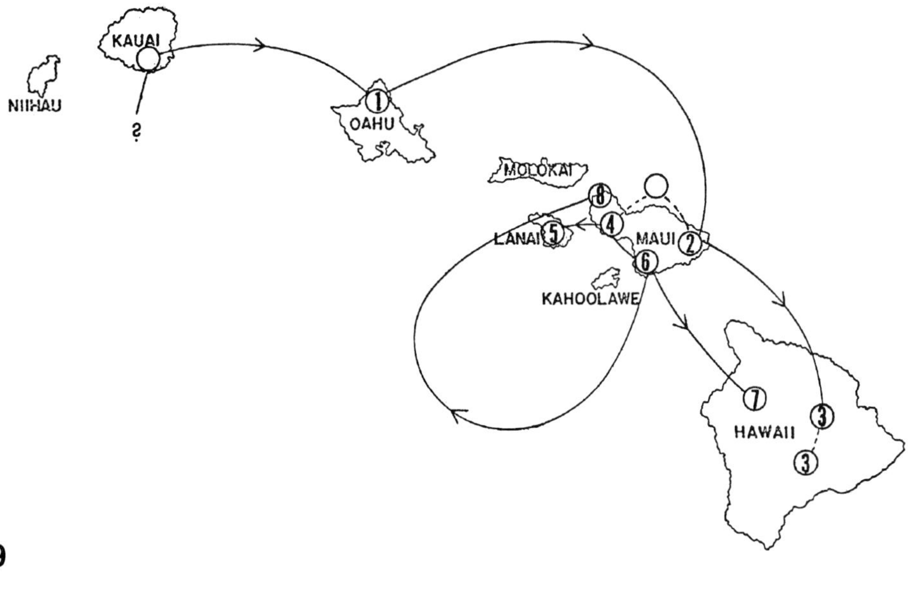
KAUAI
NIIHAU
?
OAHU
MOLOKAI
LANAI
MAUI
KAHOOLAWE
HAWAII

9

until sustained vulcanism on adjacent peaks converged to form single islands. Also, if the lowland vegetation were virtually devoid of suitable host plants, the result was effectively insular. If the intra-island and inter-island founding events are combined, about 50% of the speciation in *Nesiomiris* can be attributed to them.

Host Plant Shifts

Between Host Plant Genera

The most succinct host plant transfers are those between different genera. A quantum jump as dramatic as the successful colonization of *Ilex anomala* from an araliaceous species is termed an adaptive shift (Fig. 4). The transfers occurring between araliaceous genera are **not** considered to be adaptive shifts. I assume that the first *Nesiomiris* to colonize the archipelago was preadapted to attack *all* genera of Araliaceae present on Kaua'i. It presumably arrived when only the genus *Tetraplasandra* occurred there. When *Cheirodendron* appeared later, probably after considerable speciation of *Nesiomiris* on various species of *Tetraplasandra*, several species of *Nesiomiris* transferred to *Cheirodendron* and speciated.

The question arises as to whether *Nesiomiris* had managed to colonize *Tetraplasandra* on any or all of the other main islands by the time *Cheirodendron* arrived in the archipelago. On which island did *Cheirodendron* first appear? If speciation is any measure of antiquity, then *Cheirodendron* has apparently been longest on Kaua'i, since 4 of its 6 species are endemic (Sherff 1954). O'ahu has 2 species of *Cheirodendron*, 1 of which is endemic and shares *C. trigynum* with the islands to the southeast.

Two processes may be occurring with respect to the shifts in host between *Tetraplasandra* and *Cheirodendron*: (1) the movement of *Cheirodendron* southeastward to Hawai'i Island into the same habitats as *Tetraplasandra* species already attacked by *Nesiomiris* species giving these insects the opportunity to attack *Cheirodendron* if not excessively host specific; (2) the movement of *Nesiomiris* species associated with *Cheirodendron* on older islands southeastward down the chain to noninfested *Cheirodendron* hosts, speciating as a result of isolation from parental populations on older islands.

Where does *Reynoldsia* fit into this scheme? This genus appears to be an even more recent arrival in the archipelago than *Cheirodendron* and *Tetraplasandra*. *Reynoldsia* has not yet reached Kaua'i and occurs only from O'ahu to Hawai'i. Only 3 species of *Nesiomiris* (*N. secularis, N. timberlakei,* and *N. trinesiotes*) attack *Reynoldsia*, but none is host specific. These 3 species are the only ones that are not monoinsular endemics.

It seems more plausible that shifts between host genera would occur as intra-island events rather than as inter-island events. This fits with Opler's (1974b) findings on "guilds" of leaf-mining microlepidoptera of oaks (Fagaceae) in the "California Oak region," which he considered to be an "evolutionary island." He found that successful colonization of host plant islands often leads to the evolution of new species. "In the case of California oaks . . . the source areas for colonists are sympatric oaks. At first, the colonists are conspecific with their congeners from the source oak. Eventually, they become adapted to the special life requirements forced upon them by . . . the new host" (p. 71–72).

In island rain forests, *Tetraplasandra* and *Cheirodendron* grow in close association.

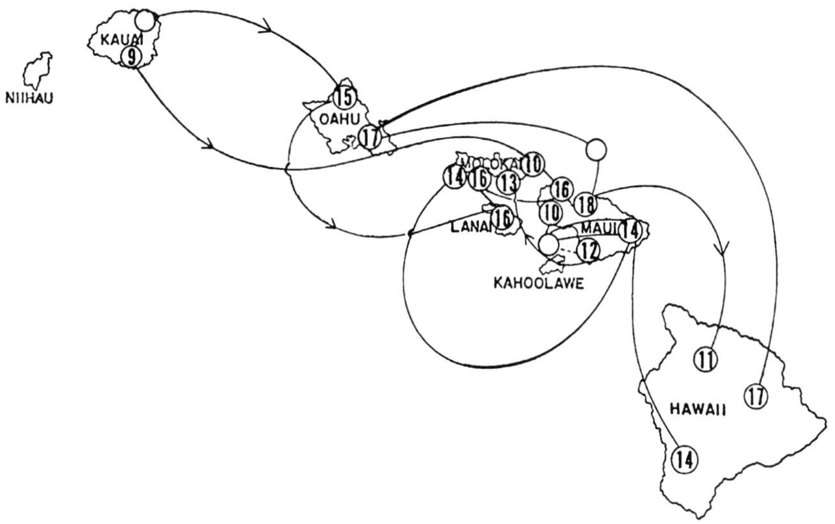

Fig. 10. Origin of *elongatus* subgroup on Kaua'i and its subsequent history. The transfers of *curvipes* (no. 10), *secularis* (no. 14), *trinesiotes* (no. 16), and *timberlakei* (no. 17) are considered to represent simple inter-island colonizations without speciation. Subgroups: existing species, numbered circles; open circles, hypothetical populations. Solid lines join species on same genus of host plant, broken lines join species or populations that are on different genera of host plants, arrows indicate directionality of inter- and intra-island founding.

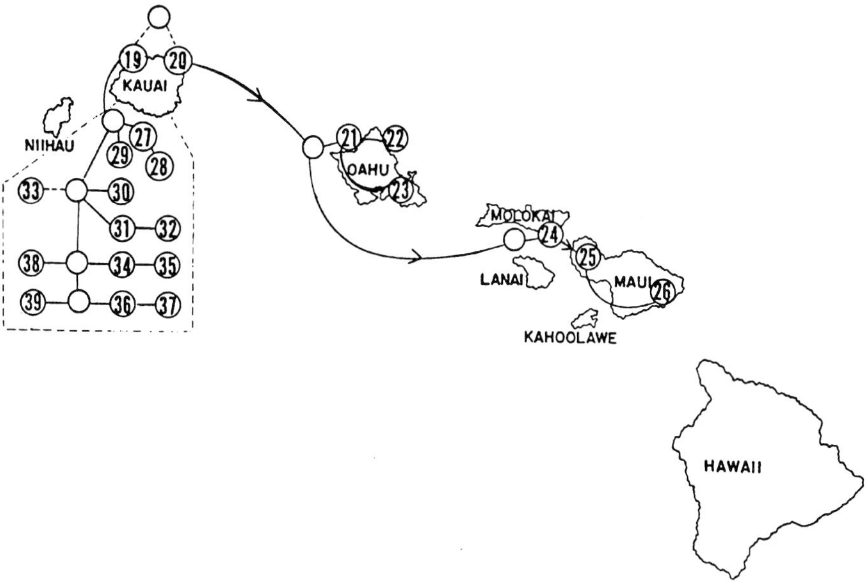

Fig. 11. Origin of *lihuensis* group on Kaua'i and its subsequent history. Subgroups: existing species, numbered circles; open circles, hypothetical populations. Solid lines join species on same genus of host plant, broken lines join species or populations on different genera of host plants, arrows indicate directionality of inter- and intra-island founding.

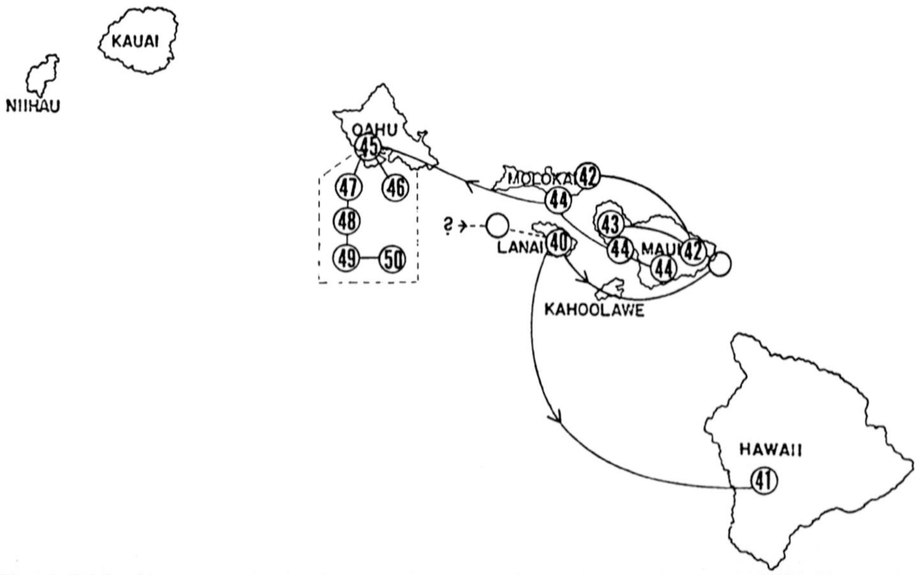

Fig. 12. Origin of *lanaiensis* group on Lāna'i (possibly from a Kaua'i Araliaceae-associated founder) and its subsequent history. Subgroups: existing species, numbered circles; open circles, hypothetical populations. Solid lines join species on same genus of host plant, broken lines join species or populations on different genera of host plants, arrows indicate directionality of inter- and intra-island founding.

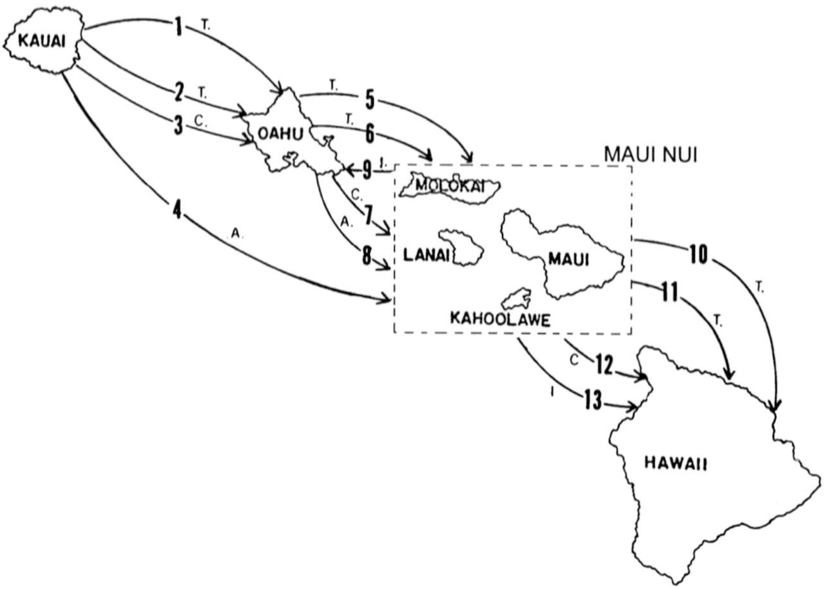

Fig. 13. Summary of the proposed minimum number (13) of inter-island founders in the genus *Nesiomiris*. A = species associated with unknown araliaceous genus; C = species associated with *Cheirodendron*; I = species associated with *Ilex*; T = species associated with *Tetraplansadra.* Maui, Moloka'i, and Lāna'i are considered as a single unit (Maui Nui), since these islands have been joined at various times (Stearns 1966).

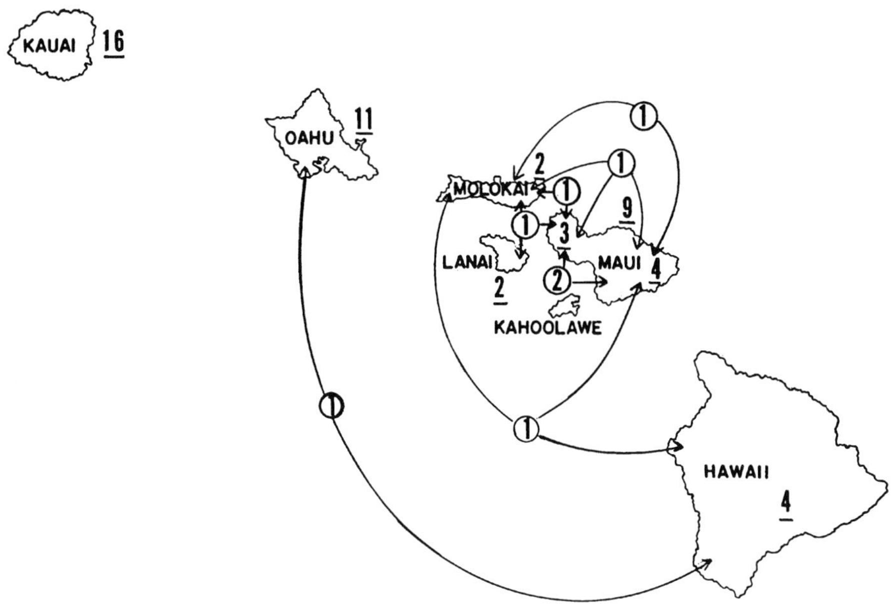

Fig. 14. Insular endemicity of *Nesiomiris* spp. Lines join islands with shared species; number of shared species indicated inside circles. Underlined number within and adjacent to islands indicates number of monoinsular endemic species for each island and for E and W Maui.

Hence, there would appear to be more likelihood of repeated and eventually successful inter-generic colonization in this situation than between genera on different islands. Hence, I have restricted my analysis of host plant transfers to the intra-island level.

Between Host Plant Species

I consider this situation to apply only between species of *Tetraplasandra*. The tomentose species of sections *Eutetraplasandra* and *Pterotropia* and predominantly glabrous species of sections *Neotetraplasandra* and *Neothotetraplasandra* harbor different species of *Nesiomiris*. These species are usually host specific at the sectional level.

Figure 15 delineates the hypothetical speciation of *Nesiomiris* as a result of colonization and subsequent isolation on different species of *Tetraplasandra*. The following restrictions were placed on this assumed speciation to arrive most parsimoniously at the result figured: (1) different *Tetraplasandra* sections, or a probably recently extinct population, would have to exist on the same island; and (2) the closest plesiomorphic relative from which a bug species had presumably arisen would have to be on a different *Tetraplasandra*, or else on an extinct hypothetical one on the same island. Given the above restrictions, 8 transfers to different plant hosts resulting in *Nesiomiris* speciation appear to have occurred, or about 16% of the speciation.

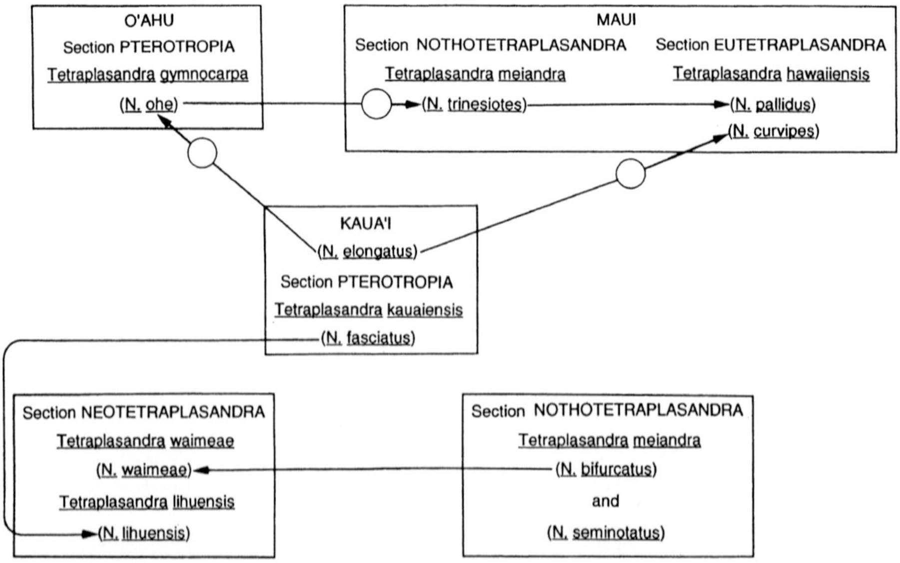

Fig. 15. *Nesiomiris* speciation resulting from transfers between named host species of *Tetraplasandra*.

MULTIPLE INVASIONS OF HOST PLANT

In this category fall those speciating events that most plausibly occurred when a host plant already occupied by 1 or more species of *Nesiomiris* was colonized by yet another from the same island. In this case the "newcomer" must have been sufficiently distinct from those species already present so that it was not immediately swamped out by the latter, but at the same time, it had to be sufficiently competitive to induce speciation, for example, through divergent character displacement.

My definition of character displacement follows that given by Grant (1972): "Character displacement is the process by which a morphological character state of a species changes under natural selection arising from the presence, in the same environment, of 1 or more species similar to it ecologically and/or reproductively." Grant cited the "best documented" examples of character displacement in the literature and concluded that "the evidence for the ecological aspect of character displacement is weak." I suggest that the major character displacement in *Nesiomiris* lies in the structure of the male genitalia, particularly the size and conformation of the claspers. In holophyletic pairs of species considered to belong to the same species complex, it frequently happens that one has larger, longer, more robust claspers than the other.

The insects chosen to analyze the occurrence of speciation from multiple host plant invasions were selected from sympatric members of the species complexes listed in Table 7. Only the species in the *N. fasciatus*, *N. usingeri* and *N. lapalapa* complexes were clearly sympatric and of synchronous occurrence on their respective hosts. The remainder,

TABLE 7. Analysis of Species Complexes Occurring on the Same Host on the Same Island

Species Complex	Island Distribution	Contained Species	Host Plant	Geographical Relationship
uncinatus	O'ahu	*uncinatus, limitatus, beardsleyi*	*Cheirodendron trigynum*	Narrowly Allopatnc
planatus	O'ahu	*ilicis, ilicivorus, planatus, contortus, restrictus*	*Ilex anomala*	Narrowly Allopatric
usingeri	E Maui	*kraussi, rufonotus*	*Cheirodendron trigynum*	Sympatric
fasciatus	Kaua'i	*fasciatus, maculatus*	*Tetraplasandra kauaiensis*	Sympatric
lapalapa	Kaua'i (Makaleha Mts)	*lapalapa, kauilamahu*	*Cheirodendron kauaiense*	Sympatric
lapalapa	Kaua'i (Alaka'i Swamp)	*olapa, ehu, mahu*	*Cheirodendron spp.*	Sympatric
hawaiiensis	E Maui, Moloka'i	*byroniae, kawau*	*Ilex anomala*	Narrowly Sympatric
hawaiiensis	W Maui	*mauiensis, kawau*	*Ilex anomala*	Narrowly Sympatric

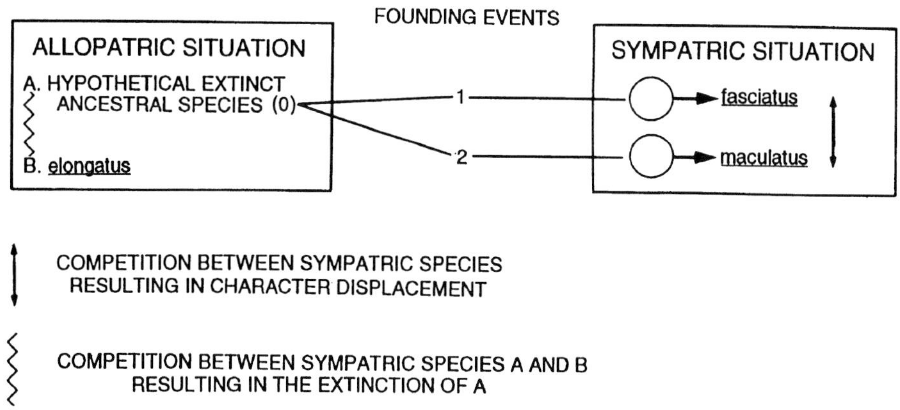

Fig. 16. Speciation of the *fasciatus* complex on Kaua'i via multiple host invasions on *Tetraplasandra kavaiensis* and their subsequent character displacement.

while narrowly sympatric, were for the most part altitudinally and/or geographically allopatric. The situation for *N. kawau* and *N. byroniae* cannot be resolved without additional collecting, but presently they appear to be narrowly allopatric. If multiple host invasions and character displacement are responsible for all or part of the speciation, which are the likely candidates? The 3 species complexes, *fasciatus*, *usingeri*, and *lapalapa*, in addition to having the attributes outlined above, have the most immediate plesiomorphic ancestor of each species distributed allopatrically on the same host (Figs. 16–18).

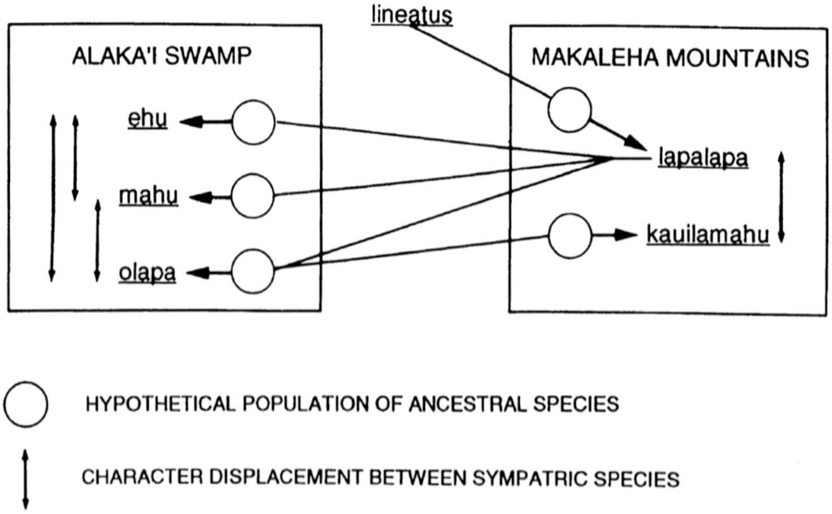

Fig. 17. Speciation of the *lapalapa* complex on Kaua'i via multiple host invasions on *Cheirodendron* spp. and their subsequent character displacement.

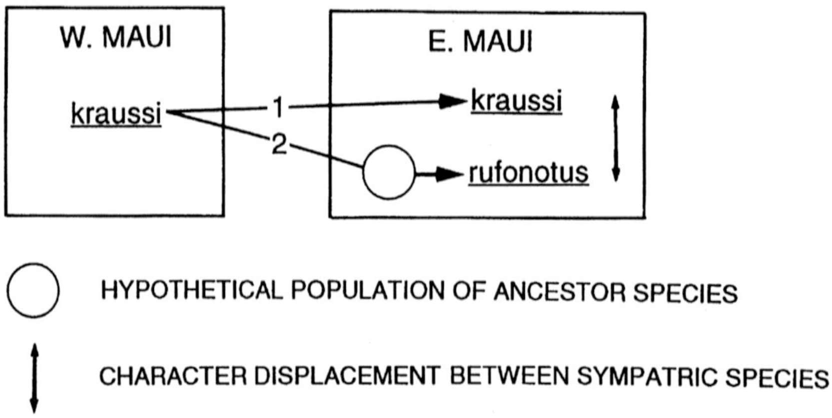

Fig. 18. Speciation of the *usingeri* complex on Kaua'i via multiple host invasions of *N. kraussi* from W Maui onto *Cheirodendron trigynum* on E Maui and subsequent character displacement.

The phylogenetic sequence of the *lapalapa* complex might be somewhat as follows: a *lapalapa*-like ancestor might have colonized the central Alaka'i Swamp massif from the Makaleha Mountains and then speciated in isolation, resulting in *N. olapa*. The same process could have resulted in the evolution of *N. ehu*, that is, a recolonization of the Alaka'i Swamp massif from the Makaleha Mountains by a *N. lapalapa*-like ancestor, which then

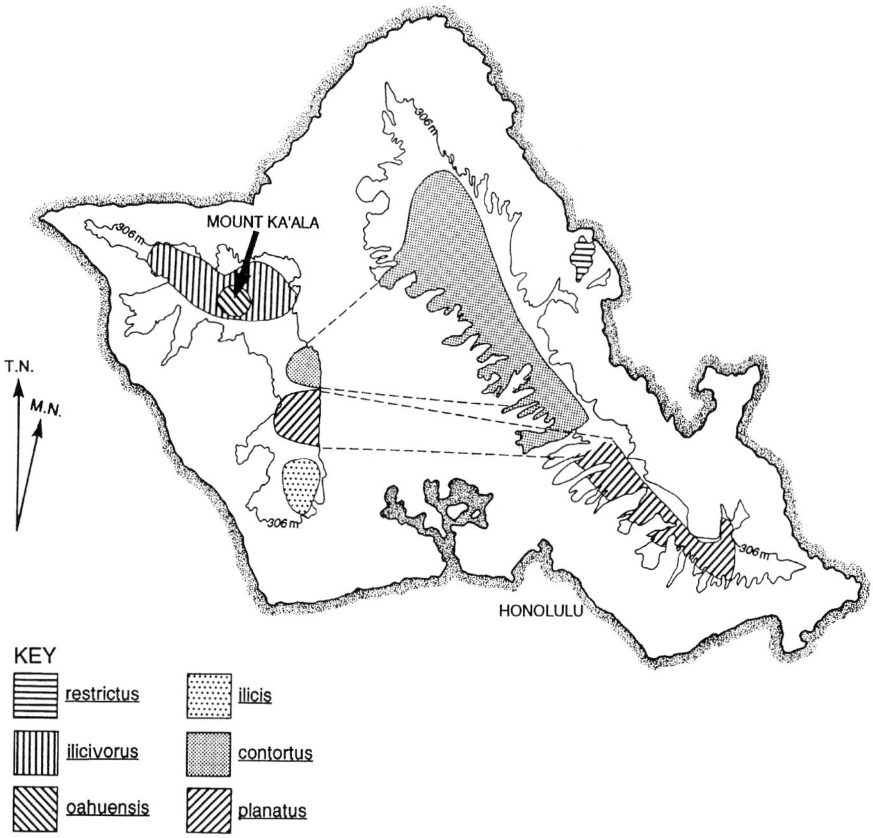

Fig. 19. Distribution of the *Ilex anomala*-associated species of *Nesiomiris* on O'ahu. The 300 m elevation contour indicates present approximate lower limits of *I. anomala*. Broken lines join populations of *N. contortus,* n. sp. (above) and *N. planatus,* n. sp. (below), which occur in both Wai'anae (left) and Ko'olau (right) Mountain

encountered there the reproductively isolated *N. olapa* or *olapa*-like species. The differences would be reinforced and magnified by character displacement, in this case in the male genitalia, to minimize the possibility of hybridization. Bock (1970) has made a cogent argument for the possible occurrence of this process in psittirostrine Hawaiian honeycreepers, based upon analysis of differences in bill shape and body size in the subfamily. Grant (1972) however, in an analysis of Bock's and other data, has just as cogently argued that there is as yet no well-grounded explanation for the evolution of closely related sympatric species complexes on oceanic islands.

An analysis of Figures 16–18 indicates that multiple host invasions might have generated 4 species in the *lapalapa* complex and 2 each in the *fasciatus* and *usingeri* complexes. Thus, 8 (ca. 16%) of the species of *Nesiomiris* could be attributed to multiple host invasions with subsequent character displacement, particularly of the male clasper.

The remainder of the species complexes, *N. planatus* and the Maui and Moloka'i representatives of the *hawaiiensis* complex, do not appear to conform to the pattern outlined above. Neither do they conform to the pattern of clinally distributed subspecies or populations as outlined by Mayr (1963). If conformation of the male claspers is indicative, there is no "smoothing out" of existing differences, nor is there increased variability in the narrow zones of contact, and there apparently is no gene flow between adjacent populations. These discrete populations are then considered full species. In *N. mauiensis* and *N. kawau*, which are narrowly allopatric members of the *lanaiensis* complex on west Maui, *N. kawau* (Fig. 88) has much-reduced claspers as compared with *N. mauiensis* (Fig. 87). This could be an example of displacement of a genital character in a sympatric invasion of the same host. Once differentiated, these 2 populations were selected for different ecological advantages on *Ilex* at different altitudes. Now *N. mauiensis* predominate at the higher, wetter, and colder climatic regimes rather than *N. kawau*. The same situation might also be applicable to *N. kawau* and *N. byroniae* on east Maui, where the latter predominates on the wetter windward areas of the island. If these transpecific phenomena can be added to speciation from multiple host invasions (now with subsequent allopatry), then the total for speciation from this factor would rise to about 20%.

Intra-island Complex Speciation Pattern

The *planatus* Complex

This group presents a much more complex geographical pattern (Fig. 19) than those outlined above. *Nesiomiris restrictus* at least, and possibly *N. ilicis*, are presently geographical isolates that appear to have speciated allopatrically, the former from a *N. contortus*-like ancestor and the latter from an *N. oahuensis*-like ancestor.

N. oahuensis is postulated to have arisen as a result of island founding from Moloka'i. If *N. oahuensis* reached the Wai'anae Mountains of O'ahu and an *N. ilicis*-like species had speciated before the Ko'olau Mountains had substantially risen above the surface of the ocean, then the following pattern of geographical speciation of the *planatus* complex might be postulated, using the geographical genesis of O'ahu as outlined by Stearns (1966: pls. 3–6) as a guide. This could be the basis of the *Nesiomiris* speciation sequence. Alternatively one could postulate a strictly allopatric (meaning only on Ko'olau and Wai'anae Mountains) sequence of speciation. Males of the *planatus* complex, especially in *N. ilicivorous*, *N. planatus*, *N. contortus*, and *N. restrictus*, have the left clasper and tooth on the pygofer relatively constant in shape.

Returning to the geographical building and subsequent erosion of O'ahu following Stearns' (1966) reconstructions (see his pls. 3–6), the pattern is potentially ever more complicated. Not only are there differences in the ages of the 2 mountain ranges as can be concluded from an inspection of these figures, but there were numerous changes in sea level, climate and undoubtedly in host plant distribution on O'ahu. Presently *Ilex anomala* does not occur below about 300 m [1,000 ft] (Fig. 19), and the region between the mountain ranges is now virtually completely converted to agriculture. This intermountain area recently might have been too low for *Ilex*. But, undoubtedly in the past this area would have been widely occupied by this species either because the climate was cooler and wetter during the glacial epochs, or because this intermountain region was up to 60 m higher in elevation (Macdonald & Abbott 1970) due to lower sea levels during these glacial epochs. Thus, I postulate that much of the speciation in the *planatus* complex took place during periods when there was an ocean barrier between the Wai'anae and Ko'olau Mountains, and/or periods when the inter-mountain area was not suitable for *Ilex anomala*. Otherwise, I would be confronted with a possible continuum of suitable host plants between the 2 mountain ranges. If my hypothesis of the phylogeny of the *planatus* complex is correct, the species might have evolved in roughly the following chronological sequence: (1) *N. ilicis* in the southern Wai'anae Mountains from an *N. oahuensis*-like ancestor, which was adapted to higher, more montane climate; (2) *N. ilicivorous* at the north end of the Wai'anae Mountains in the same fashion as *N. ilicis*; (3) a *N. ilicivorous*-like species founded in the southern Ko'olau Mountains and evolved into *N. planatus*; (4) *N. planatus* recolonized the Wai'anae Mountains, possibly during a time when extensive gene exchange between these 2 ranges was possible; (5) *N. contortus* speciates in the northern Ko'olau Mountains when a *N. planatus*-like population is isolated from *N. planatus*; (6) a *N. contortus*-like species also inhabited the isolated Turnover Peak area and evolved into *N. restrictus* by a combination of small population size and geographical isolation from the Ko'olau Mountains.

Another factor to take into account is the age of Maui Nui (Lāna'i, Moloka'i, east Maui, and west Maui) from which the precursor of *N. oahuensis* presumably came. Macdonald and Abbott (1970) indicated that the oldest of this group of islands is the western half of Moloka'i, which is 1.8 million years old. This would indicate that the evolution of *N. oahuensis* and subsequently the *planatus* complex cannot exceed 1.8 million years and, likely, is hundreds of thousands of years younger than this, since the western half of Moloka'i, presently about 400 m elevation, presumably did not much exceed this in the past. This would mean that the western end was too low to induce rain-producing orographic clouds. This, coupled with the partial shielding of the higher mountains of the east end (Macdonald & Abbott 1970), would have made the western end quite dry and marginal for *Ilex anomala* throughout most or much of its geologic history.

The *uncinatus* Complex

The O'ahu members of this complex could have speciated thus: (1) a *N. cheirodendronae*-like species founded onto the Ko'olau Mountains from Kaua'i and speciated to form *N. uncinatus*; (2) part of the *N. uncinatus* population is then isolated on the Turnover Peak massif and evolves into *N. limitatus*; (3) a *N. uncinatus*-like population reached the Wai'anae Mountains and speciates in isolation to form *N. beardsleyi*; and (4) *N. beardsleyi* reaches the Ko'olau Mountains recently, or sufficiently frequently, so that gene exchange between the 2 mountain ranges is still relatively frequent, and *N. beardsleyi* spreads throughout the Ko'olau Mountains, displacing all but remnant populations of *N. uncinatus* at the southern end of this range.

If I am correct in postulating that intra-island, allopatric (geographic) speciation ac-

Fig. 20. *Nesiomiris pallidus*, n. sp., male left lateral view. Photo by Wayne C. Gagné.

Fig. 21. *Nesiomiris maculatus*, n. sp., left lateral view. Maculate pattern gives cryptic coloration effect on underside of leaflets of *Tetraplasandra kavaiensis*. Photo by Wayne C. Gagné.

counts for the taxa in the *planatus* and *uncinatus* complexes, then a total of 7 species, for which 5 in the former and 2 in the latter complex, could be accounted.

Other Evolutionary Phenomena

Observations of *Nesiomiris* species collected on *Tetraplasandra hawaiiensis* indicate that ecological separation of sympatric *Nesiomiris* species on the same host plant may be under way. The underside of the mature leaves of *T. hawaiiensis* normally have a dense white tomentum. There invariably are, however, sucker or "water shoots" on which the leaves are green and glabrous on the underside. On Hawai'i Island, tomentose leaves are attacked by *N. pallasatus* while glabrous leaves often are occupied by *N. timberlakei*. A correlation may be seen in the body color of these 2 insects. *Nesiomiris pallasatus* is pallid with a spotted pattern on the hemelytra rendering it extremely difficult to see on the underside of tomentose leaves, which is its normal feeding and resting site. Conversely, *N. timberlakei* has a green body and is most abundant on the underside of green sucker growth, where it is likewise cryptic. By contrast, both species are quite conspicuous to the human eye when *N. pallasatus* is placed on a green background, or when *N. timberlakei* is placed on a tomentose substrate. This suggests that some visual predator, probably birds, is taking advantage of this difference. I have frequently observed honeycreepers, especially the *'amakihi* (*Loxops virens virens*), searching for insects on the underside of leaves of *T. hawaiiensis* on Hawai'i Island. A similar situation appears to apply to 4 species of *Nesiomiris* that attack *T. hawaiiensis* on Maui and Moloka'i. Two of these, *N. curvipes* and *N. pallidus* (Fig. 20), are pale-bodied and are closely associated with tomentose leaves; the other 2, *N. secularis* and *N. trinesiotes*, are green-bodied and usually attack glabrous, green leaves.

Three species of *Nesiomiris* occur on *T. kavaiensis* on Kaua'i. The leaves of the *T. kavaiensis* are scruffy tomentose and green mottled on both surfaces. Two of these insects, *N. maculatus* (Fig. 21) and *N. fasciatus*, are green spotted, while the 3rd, *N. elongatus*, is predominantly greenish. The latter is sympatric with *N. maculatus* and *N. fasciatus*, while these 2 are narrowly sympatric.

Predation and Behavior

I agree with the idea advanced by Spieth (1966, 1968) that predation has been an important factor in the behavioral evolution of many Hawaiian Drosophilidae. The inference for several of these species was of avian predation.

Some *Nesiomiris* species, for example those on *Cheirodendron*, are "startled" by slight motions of the hand, and one can play "hide and seek" with adults and nymphs from 1 side of a leaf to the other. Crypsis behavior is evident in other *Nesiomiris* species such as *N. curvipes*, *N. pallasatus*, and *N. pallidus* on *Tetraplasandra hawaiiensis*, which characteristically "freeze" on the underside of the leaves when disturbed. *Nesiomiris maculatus* and *N. fasciatus* also show crypsis.

In addition to insectivorous birds, a number of potential invertebrate predators inhabit host plants of *Nesiomiris*. Some of the groups most frequently encountered are spiders (Araneida), larval Chrysopidae and Hemerobiidae (Neuroptera), and Reduviidae and

Nabidae (Heteroptera). At lower elevations, especially below 600 m in the drier, leeward climatic regions, ants (Hymenoptera: Formicidae) undoubtedly take a toll. An adult of *Zelus renardii* (Kolenati) (Heteroptera: Reduviidae) was observed to feed on an adult of *N. lanaiensis* on *Ilex anomala* on Lāna'i.

No evidence of parasitism or disease was encountered during the study; however, close attention was not paid to this aspect.

SYSTEMATICS

Historical Review

Nesiomiris essentially, has not attracted attention since the beginning of this century, when Kirkaldy (1902) established the genus for *N. hawaiiensis*, in the *Fauna Hawaiiensis*. Since then, several authors (e.g., Timberlake 1921; Zimmerman 1948b; Swezey 1954) have indicated that numerous species were present in material that they and others had collected.

In the 1960s, J.C.M. Carvalho and R.L. Usinger undertook a preliminary investigation of the genus and set out some 20 presumed undescribed species in the material before them. Carvalho selected, examined, and illustrated the type of *N. hawaiiensis* (Fig. 84) during a visit to BMNH. For various reasons, primarily the recurring illness of Usinger, the study was postponed. Much of the material available to them was old, faded, and often without any indication of the host plant. Early in 1968, Usinger suggested to me that the genus might provide an interesting doctoral thesis. When P.D. Ashlock, then of the Bishop Museum, offered funds for fieldwork, study was undertaken.

Placement of the genus in the higher classification of the family has never been critically attempted. In his original description, Kirkaldy (1902) assigned *Nesiomiris* to the subfamily Mirinae, Division Miraria (= Tribe Mirini). His only comment was "allied to *Teratocoris* Fieber." Later, Kirkaldy (1906a) and Reuter (1910) reaffirmed this classification, although Reuter did so with some reservation, since it was 1 of the genera he had not seen. The genus remained in Miraria until Carvalho (1952) reassigned it to the Stenodemini in the same subfamily, without comment. Eyles and Carvalho (1975) keyed out *Nesiomiris* in the Stenodemini, also without comment.

Genus *Nesiomiris* Kirkaldy

Nesiomiris Kirkaldy, 1902, *Fauna Hawaiiensis* 3: 145, pl. 5, fig. 50 (n. gen.). Kirkaldy, 1906b, *Trans. Am. Entomol. Soc.* 32: 144 (cat.). Zimmerman, 1948a, *Insects of Hawaii* 1: 74; Zimmerman, 1948b, *Insects of Hawaii* 3: 205, 216, 217 (note, comparison). Carvalho, 1952, *An. Acad. Bras. Cienc.* 24(1): 84 (cat.). Carvalho, 1955, *Bol. Mus. Goeldi* 11(2): 104 (key). Carvalho, 1959, *Arq. Mus. Nac.* (Rio J.) 4: 295 (cat.). Eyles & Carvalho, 1975, *J. Nat. Hist.* 9: 267 (key).

Kirkaldy's original description is inadequate by present standards to distinguish the genus, so it is hereby redescribed.

Redescription. Orthotylinae, Orthotylini, *Orthotylus* group, *sensu* Schuh (1974). Slender, elongate species, mostly greenish or maculated with green when alive, fading to yellows and pale browns with age. Characterized by the pronotum with an anterior constriction dividing it into an anterior impunctate lower and narrower, laterally convex lobe, and a longer, higher convex, rugose-punctate posterior lobe anteriorly declivous, and a narrow rounded pronotal collar (Fig. 22). *Head*. Smooth, slightly exerted, carinate across base between eyes, wider than anterior lobe of pronotum; eyes not in contact with pronotum, separated by less than the width of the eye; eyes when viewed from front, rounded anteriorly, face declivous; dorsum and tylus usually darker than remainder.

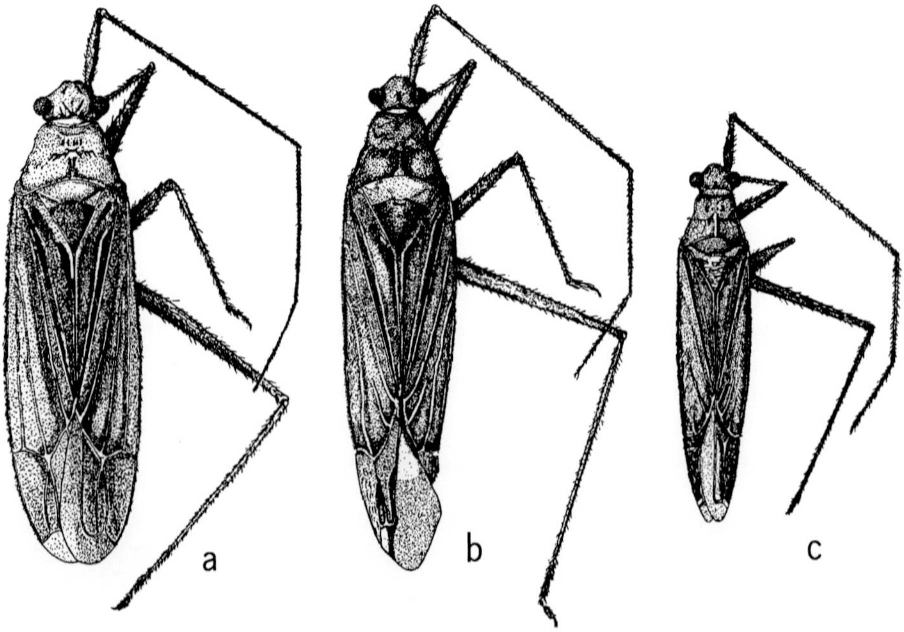

Fig. 22. Dorsal view of *Nesiomiris* species. **a**, *contortus*, n. sp.; **b**, *hawaiiensis* Kirkaldy; **c**, *timberlakei*, n. sp.

Antennae linear, segment II thicker than III or IV; segment I thickest, medially thickened, curving slightly exteriorly; inserted in front of eye; 1.5 × or more longer than body length; I separated from eye by a space less than its diameter; pubescence equalling or less than diameter of segments. Rostrum reaching procoxae and beyond. *Pronotum*. Sparsely pubescent, dull, anterior and posterior margins concave, humeral angles declivous, rounded; collar distinct, nonattenuated; deeply impressed transversely in the middle; mesoscutum exposed, glabrous, declivous posteriorly; scutellum smooth, flat, pubescent apex, acute. *Hemelytra*. Opaque, weakly rugose, pubescent; embolium narrow; clavus, corium, and cuneus distinct; apex of clavus with a darkened sensory (?) pit; membrane biareolate; hind wing usually infumate along 1A and posterior to anal ridge (Figs. 34, 52, 83). Ostiolar peritreme whitish to creamy, forelobe covering canal. *Legs*. Pubescent, impunctate, elongate, femora normal, not noticeably enlarged nor incrassate (excepting *N. curvipes* and *N. pallasatus*); 1st tarsal segment shorter than 2nd; last tarsal segment darker than I and II; claws of the orthotyline type, with convergent parempodia.

MALE. *Genitalia*. Pygofer variously infuscate dorsally, claspers highly irregular in form; in general left clasper bifurcate, right clasper trifurcate, bifurcate or uniramous; bases of clasper arms externally setaceous; rim of gonopore horseshoe shaped; ductus seminis a flexible cylindrical tube; process of vesica a single apically sinuate spiculum, which is subapically flabellate; aedeagus or phallotheca a sclerotized cylinder, often apically appendiculate, enclosing spiculum when at rest.

FEMALE. *Genitalia*. (Terminology follows Slater 1950.) Posterior wall of bursa copulatrix oriented on a dorso-ventral plane. Sclerotized rings (Figs. 23, 24) and posterior wall (Figs. 23, 25,

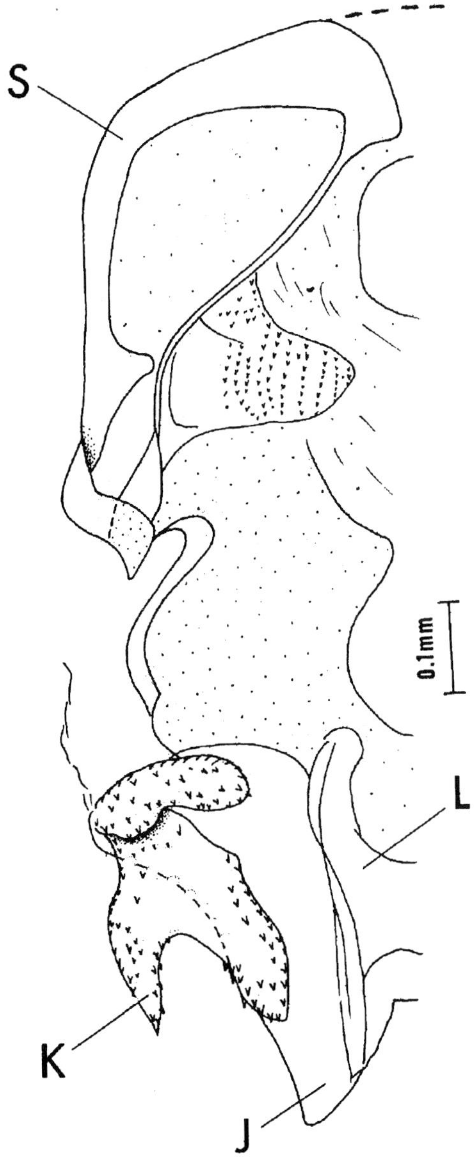

Fig. 23. *Nesiomiris curvatus,* n. sp., female sclerotized rings and anterior view of right posterior wall. J = *J* structure; K = *K* structure; L = *L* structure; S = sclerotized rings

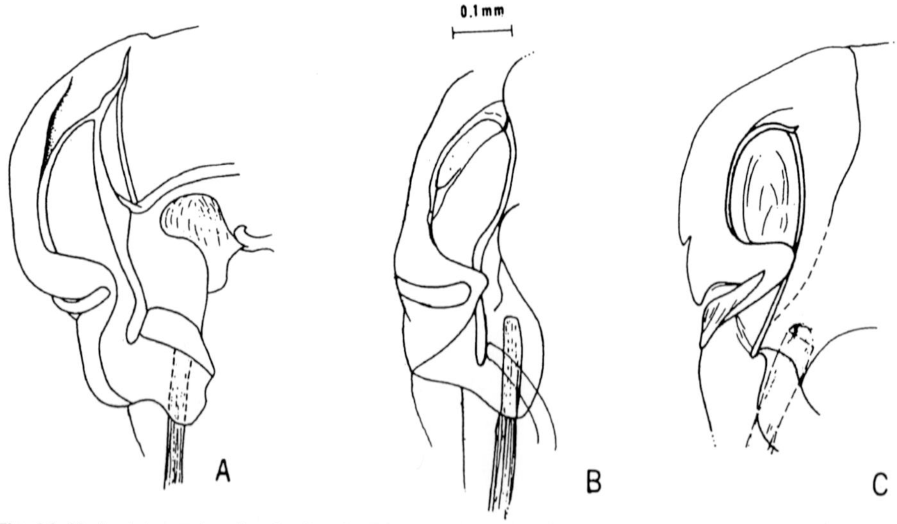

Fig. 24. *Nesiomiris* species, female, dorsal view of sclerotized rings. a, *secularis,* n. sp.; b, *hawaiiensis* Kirkaldy; c, *beardsleyi,* n. sp.

26) (see *N. curvatus*, especially) typical highly infolded orthotyline type. Anterior face of posterior wall with paired lobes (*J* structure); membranous flange arising from dorsal margin of each *J* structure and extending ventrad, with 2 apical, spiculate, fingerlike projections (*K* structures), and another smaller, usually spiculate evaginate lobe arising from the base of the *K* structure. A median lobe, the *L* structure between the *J* structures. A sclerotized hoodlike flap or pocket that is attached dorsally and laterally on 1 side (see Fig. 30) anterior to the sclerotized rings and between these and bases of the 1 valifers. Ovipositor just surpasses apex of abdomen.

NYMPHS (Fig. 27). Dorsal abdominal scent gland opens on suture between abdominal tergites 3 and 4; this is a simple oval opening with a sclerotized bar above it ("Type 2," *sensu* Akinbohungbe 1974). Head and antennae similar in shape and proportions as in adult. Dorsum and appendages variously pubescent; setae on borders of pronotum and on wing buds of later instars more prominent.

EGG (Fig. 28, from gravid *N. trifurcatus* ♀). Shape closest to "*Dicyphus*-type" (*sensu* Cobben 1968). Length: 1.30–1.45 mm; width: 0.25–0.27 mm, operculum with 2 opercular respiratory horns, these equal in length (0.30 mm), inwardly directed at their apices, aeropylar opening apparently on apex; surface of chorion smooth when viewed at 60 ×; posterior pole rounded; spongelike substance between the arms of the respiratory horns (serosal plug?).

Type of genus. *Nesiomiris hawaiiensis* Kirkaldy (monobasic).

Classification and Character Comparison

Carvalho (1952) in his world catalog to the genera of Miridae placed *Nesiomiris* in the tribe Stenodemini, subfamily Mirinae. In his keys to the world genera of the Miridae (Carvalho 1955), *Nesiomiris* keys out in couplet 12 (p. 104), containing the Holarctic *Lep-*

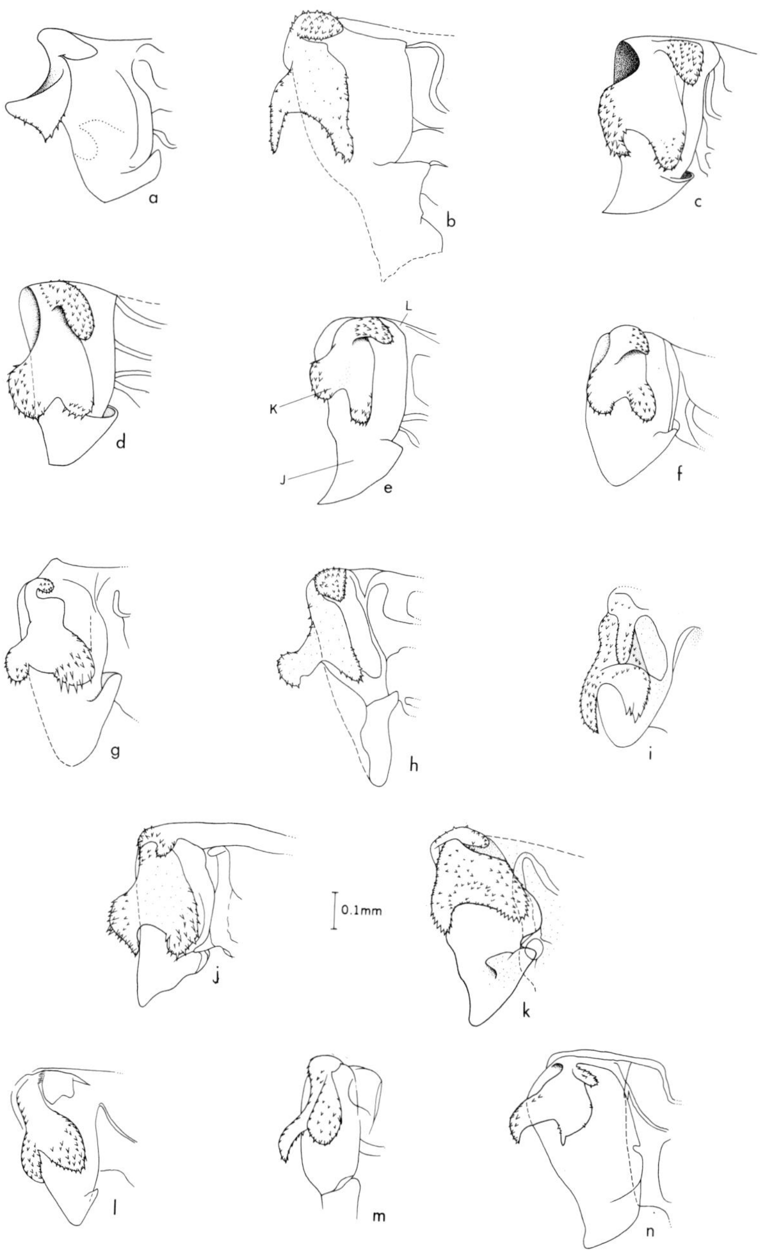

Fig. 25. *Nesiomiris,* n. spp., females, anterior view of right posterior wall. **a**, *trifurcatus;* **b**, *pilosus;* **c**, *sinuatus;* **d**, *parasinuatus;* **e**, *swezeyi;* **f**, *williamsi;* **g**, *spatulatus;* **h**, *cheliferus;* **i**, *elongatus;* **j**, *secularis* (Hawai'i Island population); **k**, *secularis* (E Maui population); **l**, *trinesiotes;* **m**, *pallidus;* **n**, *curvipes.* J = *J* structure; K = K structure; L = *L* structure. all drawn to same scale.

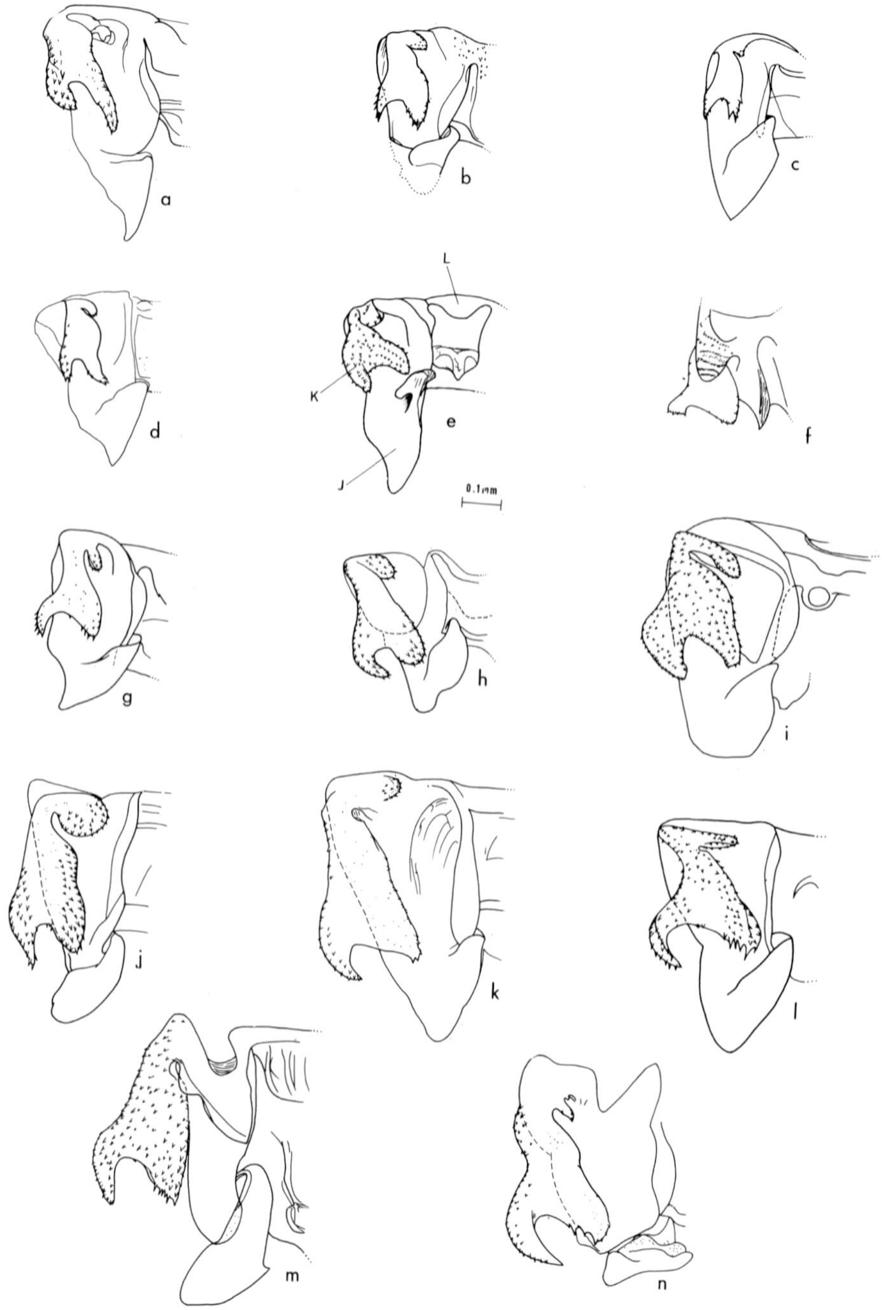

Fig. 26. *Nesiomiris* species, female, anterior view of right posterior wall: **a**, *beardsleyi*, n. sp.; **b**, *usingeri*, n. sp.; **c**, *kawau*, n. sp.; **d**, *rufonotus*, n. sp.; **e**, *olapa*, n. sp.; **f**, *cheirodendronae*, n. sp.; **g**, *ehu*, n. sp.; **h**, *mahu*, n. sp.; **i**, *lanaiensis*, n. sp.; **j**, *hawaiiensis* Kirkaldy; **k**, *byroniae*, n. sp.; **1**, *oahuensis*, n. sp.; **m**, *ilicivorus*, n. sp.; **n**, *planatus*, n. sp.. J = *J* structure; K = *K* structure; L = *L* structure. All drawn to same scale.

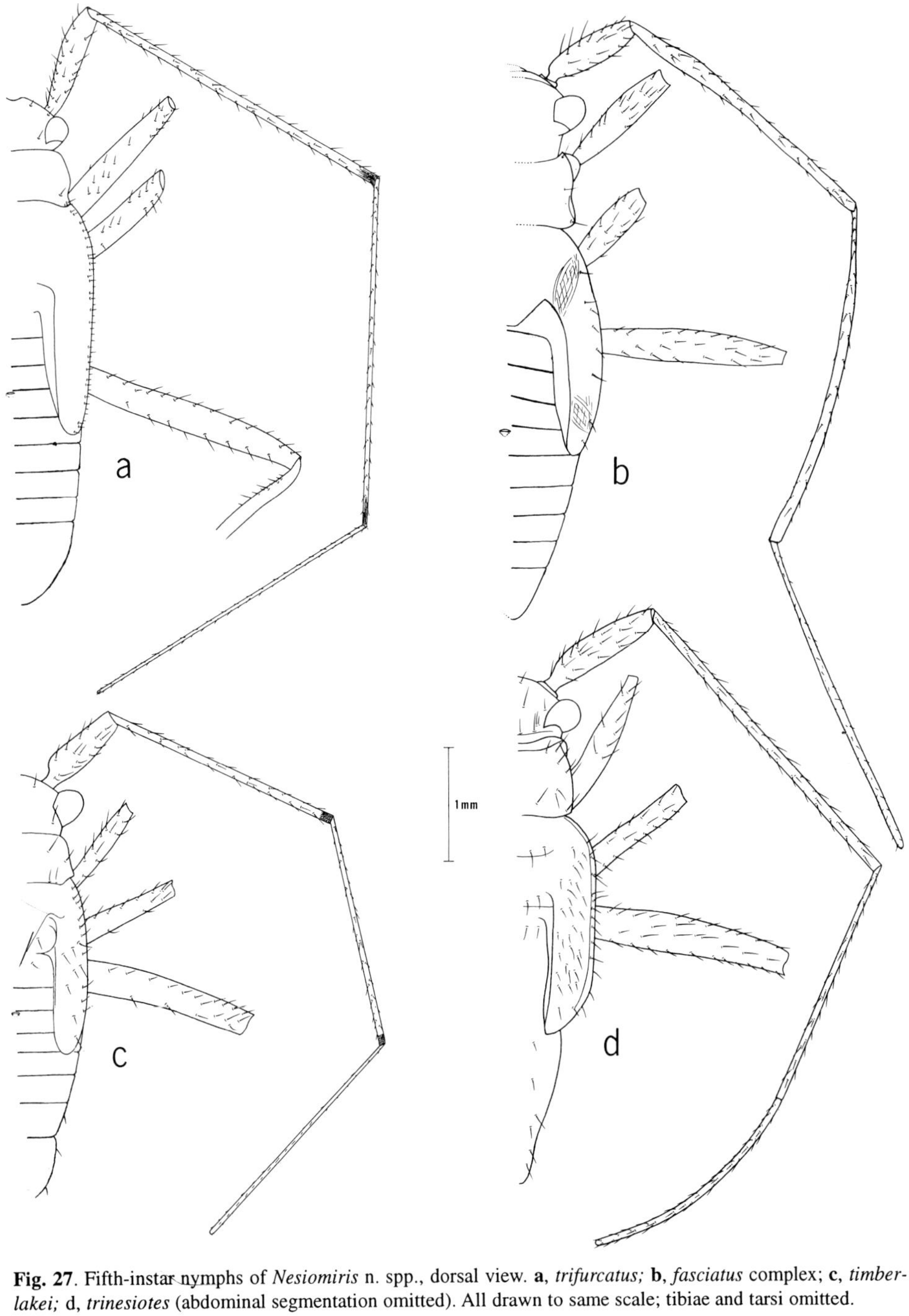

Fig. 27. Fifth-instar nymphs of *Nesiomiris* n. spp., dorsal view. **a**, *trifurcatus;* **b**, *fasciatus* complex; **c**, *timberlakei;* d, *trinesiotes* (abdominal segmentation omitted). All drawn to same scale; tibiae and tarsi omitted.

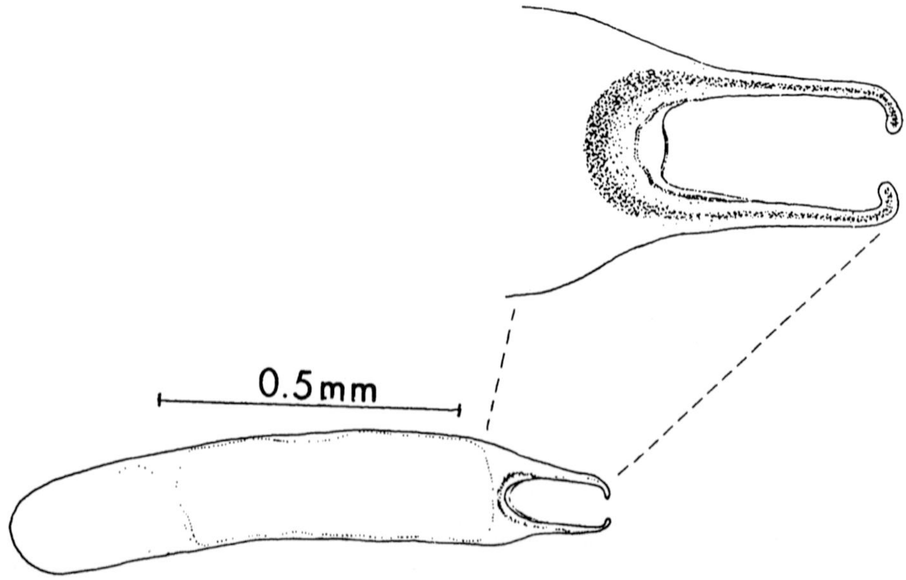

Fig. 28. *Nesiomiris trifurcatus*, n. sp., egg, lateral; operculum detail at right.

topterna Fieber. Eyles and Carvalho (1975) keyed *Nesiomiris* out between *Lasiomiris* Reuter and *Notostira* Fieber, both also in the Stenodemini.

With the combination of characters presented above, *Nesiomiris* has to be moved to Orthotylinae: Orthotylini: *Orthotylus* group, *sensu* Schuh 1974.

The closest relatives of *Nesiomiris* are unknown. Schuh (1974) placed *Solanocoris* Carvalho in another generic grouping of the Orthotylini, the "*Falconia* group." Nothing remotely resembling *Nesiomiris* has been discovered in the extensive Bishop Museum collections from Oceania, New Guinea, and adjacent areas.

Kirkaldy (1902) and Zimmerman (1948b) considered *Nesiomiris* to be "allied to *Teratocoris* Fieber", which is close in general coloration and habitus to *Nesiomiris*, but the similarity ends there. *Teratocoris* is a member of Mirinae and is associated with Holarctic sedges. In the tribe Orthotylini, *Nesiomiris* resembles the Palearctic genus *Cyllecoris* Hahn, as both possess a pronotal collar and pronotum with a median constriction (when viewed dorsally and laterally) and the eyes are placed similarly (see Carvalho 1955, pl. 3, figs. 34, 36).

Certain features of the male genitalia of *Nesiomiris* resemble those of *Cyllecoris* and *Pseudoxenetus* Reuter. Kelton's (1959) study of male genitalia showed that the combination of characters described above support assignment of *Nesiomiris* to the tribe Orthotylini. However, in most of the species that Kelton analyzed, the right clasper is larger than the left. In the species that I consider most plesiomorphic in *Nesiomiris* (e.g., *trifurcatus*, *curvatus*, *pilosus*, etc.), the left clasper, although less robust, is longer and more

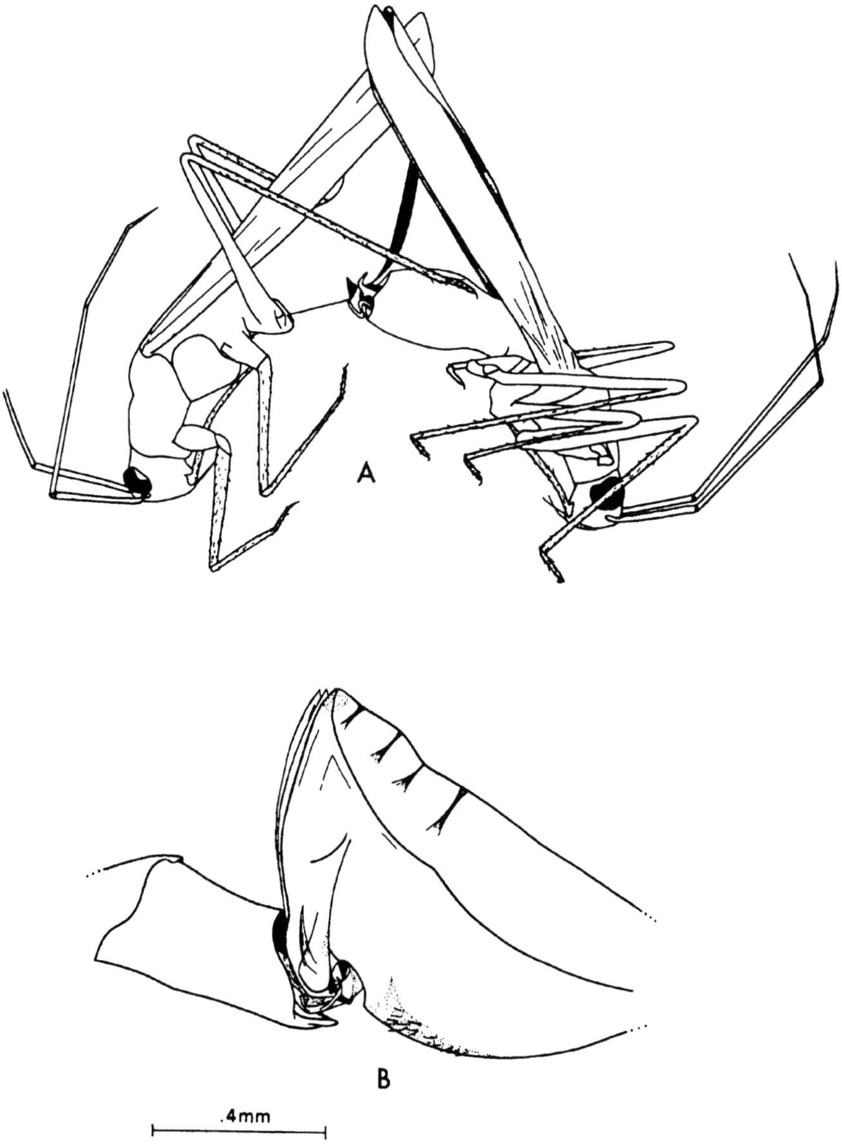

Fig. 29. *Nesiomiris* species *in copula.* **a**, *williamsi,* female left, male right; **b**, *ohe,* male left (pygofer and claspers), female right (abdomen).

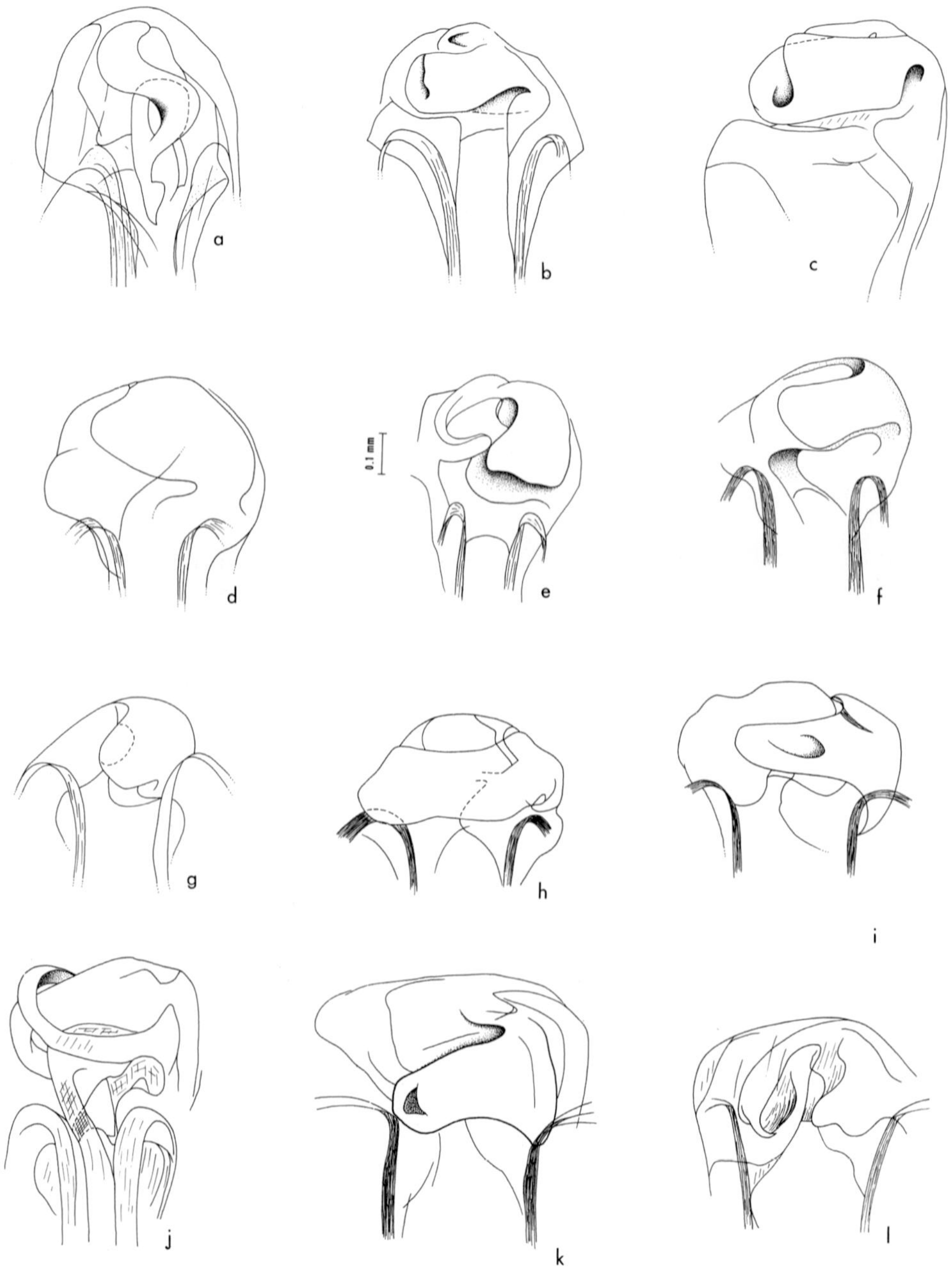

Fig. 30. *Nesiomiris* n. spp., females, sclerotized "pocket" anterior to bases of 1st valvifers, ventral view: **a**, *trifurcatus*; **b**, *curvatus*; **c**, *pilosus*; **d**, *sinuatus*; **e**, *parasinuatus*; **f**, *swezeyi*; **g**, *williamsi*; **h**, *spatulatus*; **i**, *cheliferus*; **j**, *elongatus*; **k**, *trinesiotes*; **1**, *pallidus*. All drawn to same scale.

slender than the right. My analysis suggests that the left clasper is more stable in the Orthotylinae, both in form and size between genera and species, and that there is an evolutionary tendency for the reduction in size and complexity of the right clasper. This situation seems to be prevalent in *Nesiomiris* species.

Pronotal conformation and the structure of the female genitalia suggest that *Cyllecoris* and *Pseudoxenetus* may form a natural group. It may also be noted that the left clasper of *P. scutellatus*[1] (Uhler) (see Kelton 1959: 67, fig. 95a) has the general conformation of that of *Nesiomiris*, but the vesicae are quite different. The species belong to a group having 1 rather than 3 spiculi of the vesica. Schuh (1974) placed the ant-mimicking genera *Pseudoxenetus* and *Cyllecoris* within the *Orthotylus* group of Orthotylini.

Of the figures of female genitalia presented by Slater (1950), the basic conformation of the sclerotized rings and the anterior face of the posterior wall of *Nesiomiris* most closely conform with those of the Nearctic *Orthotylus modestus* Van Duzee (see Slater's pl. 8, fig. 2; pl. 6, fig. 27) and *Pseudoxenetus scutellatus* [see Slater's pl. 7, fig. 3; pl. 6, fig. 17; as *regalis* (Uhler)]. These 2 species have the *K* structures covered with short, stout setae. Slater (1950) considered *Pseudoxenetus* to be "very closely related to the European species of the [Palearctic] genus *Cyllocoris* [= *Cyllecoris*], if not congeneric with them." He also noted the resemblance of the sclerotized rings of *P. regalis* to those illustrated by Kullenberg (1947) for *C. flavoquadrimaculatus* (De Geer) and commented that the posterior wall is "of the same type although exhibiting some specific difference." The pronota of *Nesiomiris* and *Cyllecoris* species are also of the same basic type, so there may be a natural group containing *Nesiomiris*, *Cyllecoris*, and *Pseudoxenetus*.

To my knowledge, the heavily sclerotized pocket or hoodlike flap between the 1st valvifers, which are grasped by the claspers of the male (Fig. 29), has not been examined for its potential systematic value in Miridae. I illustrate (Fig. 30) these structures for 11 *Nesiomiris* species. On the basis of this structure, *N. trifurcatus* is in 1 subgroup; *N. curvipes*, *N. pilosus*, *N. sinuatus*, *N. parasinuatus* *N. swezeyi*, and *N. williamsi* are in a second; and the remaining species are in a third subgroup. Pending studies of these structures in other genera, I will not speculate further on their systematic value, if any, except to call attention to their potential value.

Key to Species of Hawaiian *Nesiomiris*
(based largely on males)

1. Areolar veins greenish, often fading to yellows and browns after death; areolar cells partially to wholly coriaceous (Fig. 32a); head and antennal segment I usually yellowish. On Araliaceae and Aquifoliaceae; *trifurcatus* and *lanaiensis* groups. All main islands...... **2**

1′. Areolar veins reddish (except greenish in *bifurcatus*, which has upper arm of left clasper biramous), not fading after death; areolar cells completely membranous (Figs. 49b, 82b); head and antennal segment I usually reddish. On Araliaceae; *lihuensis* group. All main islands except Hawai'i **20**

2(1). Adults not sexually dichromatic; antennal segment I yellow to reddish; apex of venter of male without a tuft of short, dense setae; propleural stripe usually present (Fig. 78); spiculum usually subapically flabellate (Fig. 32d). Nymphs with apices of antennal seg-

1. Henry (1985: 1128) synonymized *Pseudoxenetus scutellatus* under *P. regalis.*

ments I and III reddish (pallid in *cheliferus*); propleural stripe present or absent. On Araliaceae; *trifurcatus* group. All main islands **3**

2′. Adults sexually dichromatic; antennal segment I yellowish; apex of venter of male with a tuft of short, dense setae (Fig. 81e); propleural stripe usually absent (Fig. 80) (except male of *N. lanaiensis*); spiculum not subapically flabellate (Fig. 81d). Nymphs with apices of antennal segments II and III sometimes infuscate but never reddish; propleural stripe always absent; *lanaiensis* group. All main islands except Kaua'i **40**

3(2). Propleural stripe extending from collar onto hind lobe of pronotum (Fig. 78); head, antennal segment I, apex of hind femur and base of hind tibia rufescent; pygofer with a dorsal tooth (Fig. 32c); right clasper 3-pronged (Fig. 32b). O'ahu .... 1. **trifurcatus, n. sp.**

3′. Propleural stripe usually not extending onto hind lobe of pronotum but ending at mesal constriction (Fig. 20); head, antennal segment I, apex of hind femur, and base of hind tibia greenish, fading to yellowish; pygofer not toothed dorsally; right clasper at most 2-pronged **4**

4(3′). Upper arm of left clasper without teeth (Fig. 35a); right clasper not reduced externally beyond 2-pronged condition (Fig. 35b). Nymphs lacking propleural stripe; apex of hind femur rufescent. On *Cheirodendron trigynum* and *Tetraplasandra meiandra*; *curvatus* subgroup. Lāna'i, Maui, and Hawai'i **5**

4′. Upper arm of left clasper sometimes with teeth; right clasper reduced externally; 1-pronged condition with other arm internally facing and not externally visible (Fig. 44f). Nymph usually with or without propleural stripe; apex of hind femur usually pallid. On Araliaceous genera; *elongatus* subgroup. All main islands **11**

5(4). Upper arm of right clasper long and hooked (Fig. 35b); upper arm of left clasper long and sinuate (Fig. 35a); aedeagus sometimes thickened near left apex (Fig. 35c–1). On *Cheirodendron trigynum* and *Tetraplasandra* spp.; *curvatus* and *sinuatus* complexes. Lāna'i, Maui, and Hawai'i **6**

5′. Upper arm of right clasper at least reduced to knoblike structure and not reaching more than 1/2 the distance to dorsum of pygofer; upper arm of left clasper evenly, inwardly hooked, not sinuate or S-shaped (Fig. 42a); aedeagus with a flange or process on the left side apically (Fig. 42c–1). On *Cheirodendron trigynum*; *reductus* complex. W Maui .... 8. **reductus, n. sp.**

6(5). Arms of right clasper elongate (Fig. 35b); upper arm of left clasper long and sinuate but not S-shaped (Fig. 35a). On *Cheirodendron trigynum* and *Tetraplasandra* spp.; *curvatus* complex. Maui and Hawai'i **7**

6′. Lower arm of right clasper reduced to short lobe (Fig. 38b,f); upper arm of left clasper long and S-shaped (Fig. 38a, e). On *Cheirodendron trigynum*; *sinuatus* complex. Lāna'i, Maui, and Hawai'i **8**

7(6). Vestiture of hemelytra short, reclined; setae on hind femur mostly reclined; pronotum 2/3 or less as long as basal width; rostrum reaching apex of mesosternum; scutellum brown in male. On *Tetraplasandra meiandra*. E Maui .... 2. **curvatus, n. sp.**

7′. Vestiture of hemelytra long, suberect; setae on hind femur mostly erect; pronotum 3/4 as long as basal width; rostrum just surpassing apex of mesosternum; scutellum green in male. On *Cheirodendron trigynum* and *Tetraplasandra meiandra*. Hawai'i .... 3. **pilosus, n. sp.**

8(6′). Lower arm of right clasper consisting of a fingerlike projection (Fig. 38b, f); upper arm of left clasper aligned with edge of pygofer (Fig. 38a, e); propleuron infuscate apically anterior to coxal cleft **9**

8′. Lower arm of right clasper usually truncated (Fig. 40b, f) or consisting of a very short fingerlike projection (Fig. 40b–1, f–1); upper arm of left clasper not aligned with edge of pygofer (Fig. 40e, e–1); propleuron not infuscate apically anterior to coxal cleft **10**

9(8). Upper arm of right clasper surpassing dorsum of pygofer (Fig. 38f); smaller areolar cell partially coriaceous (Fig. 31c) ♀ outer lobe of *K* structure thinner than inner lobe, hind femur weakly reddish near apex. W Maui .... 4. **sinuatus, n. sp.**

9′. Upper arm of right clasper not reaching dorsum of pygofer (Fig. 39b); smaller areolar cell totally coriaceous; ♀ outer lobe of *K* structure thicker than inner lobe, hind femur not reddish near apex. Lāna'i .... 5. **parasinuatus, n. sp.**

10(8′). Lower arm of right clasper bluntly rounded (Fig. 40b, f); basal carina of head extending to inner aspect of eye (Fig. 33e); rostrum just surpassing middle of mesosternum. E Maui .... 6. **swezeyi, n. sp.**

10′. Lower arm of right clasper composed of a fingerlike projection (Fig. 41f) basal carina of head extending only to a line drawn through inner aspect of antennal fossa perpendicular to carina; rostrum reaching apex of mesosternum. Hawai'i .... 7. **williamsi, n. sp.**

11(4′). Hind femora weakly to strongly incrassate (Fig. 46b); apex of aedeagus not twisted to face left clasper (Fig. 44f); propleural stripe present (*elongatus*), reduced to spot at mesal constriction, or absent. On *Tetraplasandra* (sections *Pterotropia* and *Eutetraplasandra*); *munroidendron*, *elongatus*, and *curvipes* complexes. Kaua'i, Moloka'i, Maui, and Hawai'i **12**

11′. Hind femora normal, straight, not incrassate (Fig. 46a); apex of aedeagus twisted such that atrium faces left clasper (Fig. 53f); propleural stripe present. On *Cheirodendron trigynum*, *Reynoldsia* and *Tetraplasandra* spp. (all sections except *Neotetraplasandra*); *secularis* and *ohe* complexes. O'ahu, Moloka'i, Lāna'i, Maui, and Hawai'i **14**

12(11). Dark greenish in life, fading to paler greens after death; left apex of aedeagus with prominent tooth (Fig. 44c, d); spiculum with long subapical flabellum (Fig. 44d); propleural stripe not extending posteriorly beyond mesal constriction. Nymphs and adults with apex of hind femur and base of hind tibia sometimes reddish annulate, neither incrassate; nymphs usually with a propleural stripe. On *Tetraplasandra kavaiensis* and *Munroidendron racemosum*; *elongatus* complex. Kaua'i .... 9. **elongatus, n. sp.**

12′. Pale yellowish species with faint greenish areas on dorsum in life; left apex of aedeagus, flanged (Fig. 47c); spiculum without subapical flabellum (Figs. 47d, 48d); propleural stripe reduced to a spot on the mesal constriction; hind femur and tibia strongly incrassate and pallid (Fig. 46b). Nymphs lacking propleural stripe. On *Tetraplasandra hawaiiensis*; *curvipes* complex. Moloka'i, Maui, and Hawai'i **13**

13(12′). Setae on hemelytra not subtended by brown spots; dorsal abdominal setae of late instar nymphs not set on raised papillae. Moloka'i and W Maui .... 10. **curvipes, n. sp.**

13′. Setae on hemelytra subtended by brown spots (Fig. 52b); dorsal abdominal setae of late instar nymphs set on raised papillae. Hawai'i .... 11. **pallasatus, n. sp.**

14(11′). Right clasper reduced to 2 small knobs or arms, 1 of which is internally directed (Fig. 50b); apex of aedeagus usually extending beyond atrium of pygofer (except *secularis*) (Fig. 50e, f); *secularis* complex. E Maui and Hawai'i **15**

14′. Right clasper reduced to small lobes (Figs. 54b, 55b, 56b, 57b), lacking an internally directed arm; apex of aedeagus not extending beyond atrium of pygofer (Fig. 54e); *ohe* complex. O'ahu, Lāna'i, Moloka'i, Maui, and Hawai'i **17**

15(14). Right clasper with a thumblike (Fig. 51b) or biramous (Fig. 50b) inner-directed arm; process on apex of aedeagus extending beyond atrium of pygofer (Fig. 51e). On *Cheiro-*

dendron trigynum. Moloka'i and E Maui **16**

15′. Right clasper with a bladelike, excavate inner arm (Fig. 53b); process on aedeagus not extending beyond atrium of pygofer (Fig. 53e). On *Reynoldsia* and *Tetraplasandra* spp. Moloka'i, E Maui, and Hawai'i .... 14. **secularis, n. sp.**

16(15). Apex of aedeagus chelate (Fig. 50c); apex of head rounded (male) (Fig. 33d), to weakly bifid (female); antennal segment I subequal to head width (ratio 8: 8) in male. Nymphs with apex of antennal segments I and II pallid; IV infuscate. E Maui .... 12. **cheliferus, n. sp.**

16′. Process on apex of aedeagus spatulate (Fig. 51c, e); apex of head slightly bifid (male) to straight (female), not striolate; antennal segment I longer than head width (ratio 8: 7) in male. Nymphs with apex of antennal segments I and II reddish; segment IV pallid. Moloka'i .... 13. **spatulatus, n. sp.**

17(14′). Larger species, body length greater than 6 mm; inner face of upper arm of left clasper with lumps (Fig. 54a, e) and/or denticles (Fig. 55a, e). On *Reynoldsia* and *Tetraplasandra* spp. O'ahu, Lāna'i, Moloka'i, and W Maui **18**

17′. Smaller species, body length less than 6 mm; inner face of upper arm of left clasper smooth (Fig. 56a). On *Cheirodendron trigynum*, *Reynoldsia* and *Tetraplasandra* spp. O'ahu, Maui, and Hawai'i **19**

18(17). Pygofer with a knoblike protuberance on atrium above base of right clasper (Fig. 54f); spiculum without a long subapical flabellum (Fig. 54d); propleural stripe extending onto hind lobe of pronotum (see Fig. 78); upper arm of left clasper at most lumpy, lacking denticles (Fig. 54a, e). Nymphs lacking a propleural stripe. On *Tetraplasandra gymnocarpa*. O'ahu .... 15. **ohe, n. sp.**

18′. Pygofer unarmed (Fig. 55e, f); spiculum with a long subapical flabellum (Fig. 55d); propleural stripe ending on mesal constriction; upper arm of left clasper armed with teeth and denticles on inner face (Fig. 55a, e, f). Nymphs with propleural stripe. On *Reynoldsia* and *Tetraplasandra* spp. Moloka'i, Lāna'i, and W Maui .... 16. **trinesiotes, n. sp.**

19(17′). Body with at least posterior lobe of pronotum, and hemelytra, greenish; length of antenna I less than head width (male); propleural stripe extending uninterrupted to mesal constriction; dorsal setae short, recumbent, inconspicuous. On *Cheirodendron trigynum*, *Reynoldsia* and *Tetraplasandra* spp. O'ahu and Hawai'i .... 17. **timberlakei, n. sp.**

19′. Body totally pallid (Fig. 20); portions of hemelytra only faintly greenish when alive; length of antennal segment I greater than head width (male); propleural stripe interrupted above coxal cleft (Fig. 20); dorsal setae long, suberect, conspicuous. On *Tetraplasandra hawaiiensis*. Maui .... 18. **pallidus, n. sp.**

20(1′). Spiculum flabellate and/or uncinate (Figs. 58d, 59d); upper arm of left clasper uniramous; pygofer dorsally or left laterally toothed or with sclerotized lobe (Figs. 58c, 59c). On *Tetraplasandra lihuensis* and *Cheirodendron* spp.; *lihuensis* and *uncinatus* subgroups. Kaua'i, O'ahu, Moloka'i, and Maui **21**

20′. Spiculum sinuate but not apically flabellate or uncinate (Fig. 66d); upper arm of left clasper biramous or terminally toothed (Fig. 66a); uniramous in *N. seminotatus* (Fig. 68a) and in *lapalapa* subgroup; pygofer, when toothed, only so above base of left clasper (Fig. 66e). On *Tetraplasandra* and *Cheirodendron* spp.; *bifurcatus*, *fasciatus*, and *lapalapa* subgroups. Kaua'i **28**

21(20). Spiculum subapically flabellate but not uncinate (Fig. 58d); pubescence on dorsum erect, dense, golden brown; propleural stripe absent. On *Tetraplasandra lihuensis* .... 19. **lihuensis, n. sp.**

21′. Spiculum subapically flabellate and/or uncinate (Fig. 59d, e); pubescence on dorsum recumbent, sparse, dark brown; propleural stripe present or absent. On *Cheirodendron* spp.; *uncinatus* subgroup. Kaua'i, O'ahu, Moloka'i, and Maui 22

22(21′). Spiculum subapically flabellate and uncinate (Fig. 59d, e); area about gonopore usually denticulate (Fig. 60d–1); right clasper strongly biramous and chelate (Fig. 59b); apex of head bifid when viewed from above; pygofer dorsally toothed (Fig. 60e, f) or subdorsally lobate (Fig. 59e); *uncinatus* complex. Kaua'i and O'ahu 23

22′. Spiculum apically uncinate but not flabellate (Fig. 63c, d); area about gonopore not denticulate (Fig. 63d); right clasper uniramous (Fig. 63b) or simply lobed (Fig. 65b); apex of head rounded when viewed from above. On *Cheirodendron trigynum*; *usingeri* complex. Moloka'i and Maui 26

23(22). Fleshly lobe beneath gonopore denticulate (Fig. 60d–1); arms of right clasper subequal in length (Fig. 60b, f); an incipient vein at apex of large areolar vein; aedeagus with a terminal "lip" (Fig. 60c); apex of clavus rufescent. On *Cheirodendron trigynum.* O'ahu 24

23′. Fleshly lobe beneath gonopore not denticulate (Fig. 59d); upper arm of right clasper much shorter than arm (Fig. 59b); apex of large areolar vein rounded, incipient vein absent; apex of clavus green or yellowish green. On *Cheirodendron Fauriei*, *C. kavaiensis*, and *C. Helleri.* Kaua'i .... 20. **cheirodendronae, n. sp.**

24(23). A weakly defined propleural stripe extending from pronotal collar to posterior humeral angle; apex of upper arm of left clasper downwardly directed, not facing dorsal tooth on pygofer (Fig. 60e, f); tooth on pygofer simple (Fig. 60e); right clasper with blunt lobe opposite bases of arms (Fig. 60b). SW Ko'olau Mts, O'ahu .... 12. **uncinatus, n. sp.**

24′. Propleuron without any indication of propleural stripe; apex of upper arm of left clasper dorsally directed and facing dorsal tooth on pygofer (Fig. 61e, f); tooth on pygofer subtended by teeth and denticles (Figs. 61e, 62e); right clasper rounded opposite bases of arms (Fig. 63b) 25

25(24′). Upper arm of left clasper not denticulate (Fig. 61a, e); apical "lip" of aedeagus produced into hornlike process (Fig. 61c); tooth on pygofer with a few small denticles on right base (Fig. 61e). NW Ko'olau Mts, O'ahu .... 22. **limitatus, n. sp.**

25′. Upper arm of left clasper denticulate (Fig. 62a, e, f); apical "lip" of aedeagus not produced (Fig. 62c); tooth on pygofer with large subsidiary teeth on both bases (Fig. 62e, f). Ko'olau and Wai'anae Mts, O'ahu .... 23. **beardsleyi, n. sp.**

26(22′). Arms of right clasper reduced to 2 acute lobes (Fig. 63b); tooth on pygofer acute, sclerotized (Fig. 63e, f); rostrum reaching apex of mesosternum. Maui 27

26′. Right clasper simply lobate (Fig. 65b); tooth on pygofer rounded, fleshly (Fig. 65e, f); rostrum reaching apex of procoxae. E Maui .... 26. **rufonotus, n. sp.**

27(26). Apical arm of right clasper longer than ventral arm (Fig. 63b); ventral arm of left clasper apically excavate (Fig. 63a). Moloka'i .... 24. **usingeri, n. sp.**

27′. Apical arm of tight clasper shorter than ventral arm (Fig. 64b); ventral arm of left clasper not apically excavate (Fig. 64a). Maui .... 25. **kraussi, n. sp.**

28(20′). Propleural stripe present, but reduced (Fig. 21); antennal segments I and II yellowish; upper arm of left clasper uniramous, sometimes apically toothed. On *Tetraplasandra* spp.; *fasciatus* subgroup 29

28′. Propleural stripe usually absent; when present, extends from collar to base of pronotum (Fig. 78); antennal segments I and II yellowish to rufescent; upper arm of left clasper biramous (Figs. 66a, 73a) or uniramous (Fig. 74a), but never terminally toothed. On *Tet-*

raplasandra and *Cheirodendron* spp.; *bifurcatus* and *lapalapa* subgroups **31**

29(28). Pale yellow species with dorsal pattern of green spots excepting head, of adults and nymphs (Fig. 21); pygofer toothed above base of left clasper (Fig. 66e); upper arm of left clasper toothed at apex (Fig. 66a). On *Tetraplasandra kavaiensis*; *fasciatus* complex **30**

29′. Uniformly greenish species; pygofer unarmed (Fig. 68e); upper arm of left clasper uniramous, not toothed at apex (Fig. 68a, e). On *Tetraplasandra meiandra* .... 29. **seminotatus, n. sp.**

30(29). Right clasper longer and more robust than left clasper (Fig. 66b); antennal segment I yellowish in male; tooth on pygofer acute when viewed laterally (Fig. 66e) .... 27. **fasciatus, n. sp.**

30′. Right clasper shorter and smaller than left clasper (Fig. 67b); antennal segment I reddish (male), yellowish (female), shorter than head width; tooth on pygofer rounded when viewed laterally (Fig. 67e) .... 28. **maculatus, n. sp.**

31(28′). Apex of head bifid; upper arm of left clasper strongly bifurcate (Fig. 69a); pygofer toothed above base of left clasper (Fig. 69e); large species, body length greater than 5.5 mm. On *Cheirodendron* and *Tetraplasandra* spp.; *bifurcatus* subgroup **32**

31′. Apex of head usually rounded; upper arm of left clasper uniramous (weakly, apically bifurcate in *lapalapa*); pygofer toothed or not, above base of left clasper (Fig. 73e); smaller species, body length less than 5.5 mm. On *Cheirodendron* spp.; *lapalapa* subgroup **35**

32(31). Head, antennal segments I and II yellowish; propleural stripe, if present, not extending beyond mesal constriction; right clasper uniramous (Fig. 69b, f); area about gonopore not denticulate (Fig. 69c, d). On *Tetraplasandra* spp.; *bifurcatus* complex **33**

32′. Head rufescent, antennal segment I brownish, II rufescent; propleural stripe present, extending from base of head to base of pronotum; area ventral to gonopore denticulate (Fig. 72d–1). On *Cheirodendron Helleri* and *C. kavaiense*; *lineatus* complex .... 33. **lineatus, n. sp.**

33(32). Areolar veins greenish, fading to yellow after death; rostrum reaching apex of mesocoxae; propleural stripe extending from collar to mesal constriction. On *Tetraplasandra meiandra* .... 30. **bifurcatus, n. sp.**

33′. Areolar veins reddish in life and after death; rostrum not passing middle of mesocoxae; propleural stripe obsolescent in male, or absent **34**

34(33′). Antennal segment I shorter than head width; rostrum reaching middle of mesocoxa; propleural stripe obsolescent (male), extending from collar to mesal constriction (female); femora and antennal segment I not pilose; upper arm of left clasper apically bifid (Fig. 70d) .... 31. **tetraplasandrae, n. sp.**

34′. Antennal segment I longer than head width; rostrum reaching middle of mesosternum; propleural stripe absent (male and female); pronotum flat in profile; femora and antennal segment I pilose; upper arm of left clasper subapically bifid (Fig. 71a, e). On *Tetraplasandra waimeae* .... 32. **waimeae, n. sp.**

35(31′). Pronotum with strongly developed propleural stripe (Fig. 78); right clasper uniramous, consisting of a single apically hooked arm (Fig. 79b, e); *legnotus* complex .... 39. **legnotus, n. sp.**

35′. Pronotum lacking propleural stripe; right clasper with an apical hook and a subapical, blunt arm (Fig. 73b); *lapalapa* complex **36**

36(35′). Pygofer with tooth above base of left clasper (Fig. 79e); nymphs sparsely setose **37**

36´. Pygofer without tooth above base of left clasper; nymphs densely setose .... 38. **mahu, n. sp.**

37(36). Tooth on pygofer ventrally rounded (Fig. 73e); upper arm of left clasper apically bifurcate (Fig. 73a, e); right clasper larger and longer than left clasper (Fig. 73e); rostrum reaching apex of mesosternum. On *Cheirodendron kauaiense* .... 34. **lapalapa, n. sp.**

37´. Tooth on pygofer becoming ventrally acute (Fig. 74e), or immediately acute (Fig. 75e, f); upper arm of left clasper uniramous (figs. 75a, e, f); right clasper smaller and shorter than left clasper (Fig. 75b, f); rostrum reaching apex of mesocoxae. On *Cheirodendron kauaiense* and other *Cheirodendron* spp. **38**

38(37´). Tooth on pygofer wider than basal diameter of left clasper (Fig. 74e), ventrally acute; area about gonopore denticulate (Fig. 74c, d). On *Cheirodendron fauriei* .... 35.**olapa, n. sp.**

38´. Tooth on pygofer much narrower than basal diameter of left clasper (Fig. 75e). On other species of *Cheirodendron* **39**

39(38´). Tooth on pygofer directly situated on pygofer (Fig. 75e, f); upper arm of left clasper long, upright, almost reaching dorsum of pygofer (Fig. 75e); head and tylus rufescent; antennal segment I almost as long as head width (ratio 8: 8.5). On *Cheirodendron Helleri* and *C. kauaiense* .... 36. **ehu, n. sp.**

39´. Tooth on pygofer situated on a subsidiary tooth (Fig. 76e); upper arm of left clasper short, apex downwardly directed, not reaching dorsum of pygofer (Fig. 76e); antennal segment I about 3/4 of head width (ratio 4.75: 7). On *Cheirodendron kauaiense* .... 37. **kauilamahu, n. sp.**

40(2´). Body widest across pronotum (Fig. 22c); legs and antennal segment I not pilose; setae inclined, brown; male dark green, dorsum polished, female pale green; membrane of forewing not reduced, exceeding apex of cuneus by a length greater than length of small areolar cell; infuscation on membrane extending into large areolar cell; dorsum of pygofer blackish; aedeagus, when toothed, tooth located apically on venter; spiculum not abruptly elbowed apical to gonopore (Fig. 86c); appendages of nymphs not pilose; *lanaiensis* complex. O'ahu, Moloka'i, Lāna'i, Maui, and Hawai'i **41**

40´. Body widest across hemelytra (Fig. 22a); legs and antennal segment I pilose, setae, erect, golden brown; male darker green, but dorsum not polished; membrane of forewing reduced, exceeding apex of cuneus by a length less than length of small areolar cell (Fig. 82e, f, g); infuscation on membrane not extending into large areolar cell; dorsum of pygofer pallid; spiculum abruptly elbowed apical to gonopore (Fig. 90c); appendages of nymphs pilose; *planatus* complex. O'ahu **46**

41(40). Propleural stripe sometimes present in male, always absent in female; lower arm of right clasper apically blunt and uncinate (Fig. 81b, f); dorsal tooth on pygofer extending laterally toward base of left clasper (Fig. 81e, f). Lāna'i .... 40. **lanaiensis, n. sp.**

41´. Propleural stripe always absent; lower arm of right clasper apically acute (Fig. 85b); dorsal tooth on pygofer, when present, posteriorly directed (Fig. 85e). O'ahu, Maui, and Hawai'i **42**

42(41´). Arms of right clasper chelate, not anastomosed (Fig. 85b, f); about 2/3 of large areole coriaceous (Fig. 82d). Hawai'i and Maui **43**

42´. Arms of right clasper not chelate, arms anastomosed (Fig. 88b, f); about 1/3 of large areole coriaceous. Maui, Moloka'i, and O'ahu **45**

43(42). Ventral apex of aedeagus without a tooth (Fig. 84c); dorsum of pygofer with 2 distinct, blunt teeth (Fig. 85e, f); scutellum light brown; large areole coriaceous at most on inner 1/2 (Fig. 82c). Hawai'i .... 41. **hawaiiensis** Kirkaldy

43´. Ventral apex of aedeagus with a tooth (Fig. 86c); dorsum of pygofer with 2 blunt, indistinct teeth (Fig. 86e, f) or lacking teeth (Fig. 87e, f); scutellum yellowish green; large areole coriaceous at least on inner 2/3. Maui and Moloka'i **44**

44(43´). Pygofer with 2 indistinct blunt teeth (Fig. 86e, f); large areole almost completely coriaceous (male); completely coriaceous (female); head not striolate; rostrum reaching middle of mesosternum. E Maui and Moloka'i .... 42. **byroniae, n. sp.**

44´. Pygofer unarmed (Fig. 87e, f); large areole coriaceous on inner 2/3 (male and female); apex of head striolate; rostrum reaching apex of mesosternum. W Maui .... 43. **mauiensis, n. sp.**

45(42´). Antennal segment I longer than head width; pygofer unarmed (Fig. 88e); upper arm of left clasper reduced, not reaching more than 1/2 distance to dorsum of pygofer when viewed laterally (Fig. 88e, f). Maui .... 44. **kawau, n. sp.**

45´. Antennal segment I subequal to head width; pygofer with large tooth above base of left clasper (Fig. 89e, f); upper arm of left clasper reaching dorsum of pygofer when viewed laterally (Fig. 89e). O'ahu .... 45. **oahuensis, n. sp.**

46(40´). Tooth on pygofer wider than base of left clasper, apically rounded (Fig. 90e, f); tooth on aedeagus incipient (Fig. 90c). SE Wai'anae Mts, O'ahu .... 46. **ilicis, n. sp.**

46´. Tooth on pygofer narrower than base of left clasper when viewed laterally, apically acute (Fig. 91e); tooth on aedeagus pronounced (Fig. 91c). Wai'anae and Ko'olau Mts, O'ahu **47**

47(46´). Lower arm of right clasper once uncinate (Fig. 92b, f); tooth on pygofer situated above left claspers by a distance equal or greater than basal thickness of left clasper when viewed laterally (Fig. 92e) **48**

47´. Lower arm of right clasper at least 2 × uncinate (Fig. 93b, f); tooth on pygofer resting against base of left clasper or at a distance above left clasper less than basal thickness of left clasper when viewed laterally (Fig. 93e) **49**

48(47). Large areole completely coriaceous (Fig. 82f); apex of lower arm of right clasper pointing at atrium of pygofer (Fig. 91f); tooth on pygofer not dorsally serrate (Fig. 91e, f); lateral border of mesonotal wing bud of 5th-instar nymph linear and anterior humeral angle of pronotum pallid. Northern 1/2 of Wai'anae Mts, O'ahu .... 47. **ilicivorus, n. sp.**

48´. Large areole membranous along 1/2 of apical border with areolar vein; apex of lower arm of right clasper pointing at upper arm of right clasper (Fig. 92f); tooth on pygofer dorsally serrate (Fig. 92f); lateral border of mesonotal wing bud of 5th-instar nymph convex and anterior humeral angle of pronotum infuscate. Middle of Wai'anae Mts. and southern 1/2 of Ko'olau Mts, O'ahu .... 48. **planatus, n. sp.**

49(47´). Lower arm of right clasper 2 × uncinate, that is, S-shaped (Fig. 93b, f); apex of head not striolate; rostrum reaching apex of mesosternum in 5th-instar nymph and lateral border of mesonotal wing bud evenly convex. Central Wai'anae Mts and NW Ko'olau Mts, O'ahu .... 49. **contortus, n. sp.**

49´. Lower arm of right clasper 3 × uncinate, that is, almost double S-shaped (Fig. 94b, f); apex of head weakly striolate; rostrum reaching middle of mesosternum in 5th-instar nymph and lateral border of mesonotal wing bud sublinear. Extreme NW Ko'olau Mts, O'ahu .... 50. **restrictus, n. sp.**

Species Descriptions

Since the species are primarily diagnosed on the basis of males, males have been used for all holotypes. If the male of a species was unknown, the species was not described.

Descriptions

Adult descriptions are based on the holotype or allotype. Size variations of paratypes are included in parentheses behind each measurement on the respective holotype or allotype. Other morphological variation considered significant is contained in a section on "variation," which follows the description of the types.

The description of each nymphal instar for a species is a composite, incorporating morphological and size variation for the respective stadia. Subsequent nymphal instars are described only to the extent they differ from the previous instar.

Measurements

All measurements are in millimeters. Specimens were oriented such that the long axis was in a horizontal plane. The head was measured across the eyes to their edges; the pronotum was measured along its greatest length, from the edge of the collar to the posterior humeral angle; the pronotal width was measured across the posterior lobe to the edges of the humeral angles. The body length is from the apex of the head to the tip of the membrane of the wing. The length of the rostrum was taken from the base of the tylus to the apex of the rostrum.

Illustrations

The apparently nondiagnostic base of the aedeagus (phallobase, ductus seminis, and dorsal attachments) is shown somewhat schematically. Attention to detail was paid only to the apical region of the spiculum.

The setae on the pygofer are usually omitted. This is because, apart from being of little diagnostic value for differentiating species, if included, they tend to obscure left lateral views of the apex of the pygofer and the claspers.

The claspers were drawn from their inner aspects to show that view of the greatest longitudinal axis, that is, the clasper as essentially lying on its exterior surface.

The spiculum was drawn separate from the body of the aedeagus if found in an everted position when the pygofer was dissected. If it was found in "normal" or uneverted position, it was drawn as found resting within the phallus, with the aedeagus lying on its right side. This, or the spiculum, if separate, was sometimes rotated to best show the entire length of the subapical flabellum on the spiculum (if present), the atrium of the phallus, and the apex of the ductus seminis, if denticles were present on the latter.

The *trifurcatus* Group

***Description*. ADULTS.** Not sexually dimorphic for color; areolar veins greenish, areolar veins greenish, areolar cells partially coriaceous, membrane usually infuscate, not reduced; antennal segment I yellow to reddish; apex of venter of male without tuft of short, dense setae; propleural stripe usually present; spiculum subapically flabellate.

NYMPHS. Propleural stripe present or absent; apex of antennae II and III reddish.

Distribution. All main islands.

Host. Species of *Cheirodendron*, *Reynoldsia*, *Munroidendron*, and *Tetraplasandra*.

Included species. Nos. 1–18.[2]

The *trifurcatus* Subgroup

Description. **ADULTS**. Hind lobe of pronotum and sometimes collar greenish across hind lobe of pronotum; apex of head weakly bifid; head, antennal segment I, apex of hind femur, bases of hind tibia rufescent; pygofer with dorsal teeth; right clasper 3-pronged; upper arm of left clasper with prominent teeth; female with outer lobe of *K* structure much shorter than inner lobe; and weakly differentiated from inner lobe mesodorsal evagination of *K* structure minutely spiculate (Fig. 25a).

NYMPHS. Propleural stripe absent; apices of hind femora red.

Distribution. O'ahu I (Ko'olau Mts).

Host. *Tetraplasandra kahanana*, *T. meiandra*.

Included species. No. 1.

Discussion. The unreduced state of the claspers is considered plesiomorphic, as all other species have the right clasper at most with 2 prongs and a much reduced stub in place of the 3rd prong.

The color of the antennae, head, areolar vein, and the reddish apices of the femora in the adults and nymphs, the presence of the propleural stripe, the elongate form, and the association with *Tetraplasandra* species on an older island (in this case O'ahu) appear to be plesiomorphic character traits that place this complex near the base of the genus.

1. *Nesiomiris trifurcatus* Gagné, **new species** — Figs. 25a, 27, 28, 31a, 32

The unreduced, 3-pronged condition of the right clasper at once distinguishes *trifurcatus*.

Description. **MALE** (n = 9). Body: 7.11 (6.51–7.35) long; 1.35 (1.24–1.39) wide. Head: 1.01 (0.99–1.01) wide; apex slightly bifid; rufescent on frons and dorsally, excepting on shining callosities of vertex; forms reddish striate; carinate across base, tylus yellowish brown, remainder of head yellowish; vestiture sparse, brownish at inner aspects of eyes, pale below. Antennal segments lengths: I, 1.25 (1.10–1.29), reddish becoming yellowish on inner basal aspect; longer than head width; vestiture uniform in size and length, shorter than thickness of antennal segments; II, 3.02 (3.02–3.70), reddish on basal 3/4, brownish for next 1/4, then blackish at apex; IV, 1.40 (1.40–2.96); blackish. Rostrum: 1.51 (1.41–1.55), reaching middle of mesosternum. Pronotum: 1.19 (1.10–1.19) long; 1.35 (1.24–1.39) wide; greenish behind collar and becoming obsolescent beyond middle of posterior lobe; subglabrous; a depression between the weakly developed calli. Mesoscutellum and scutellum yellow, greenish at apex. Hemelytra: greenish, more strongly so on veins; areoles and membrane coriaceous as indicated (Fig. 31a) and greenish, remainder of membrane smoky. Legs: apices of femora red, base and apical 1/2 of tibiae yellow; dorsum of genital atrium brown; atrial tooth dorsal, margin serrate (Fig. 32e, f); claspers as illustrated (Fig. 32a, b, e, f).

---

2. Numbers indicate numerical listing of species, which roughly parallels the phylogenetic sequence (see Table 1).

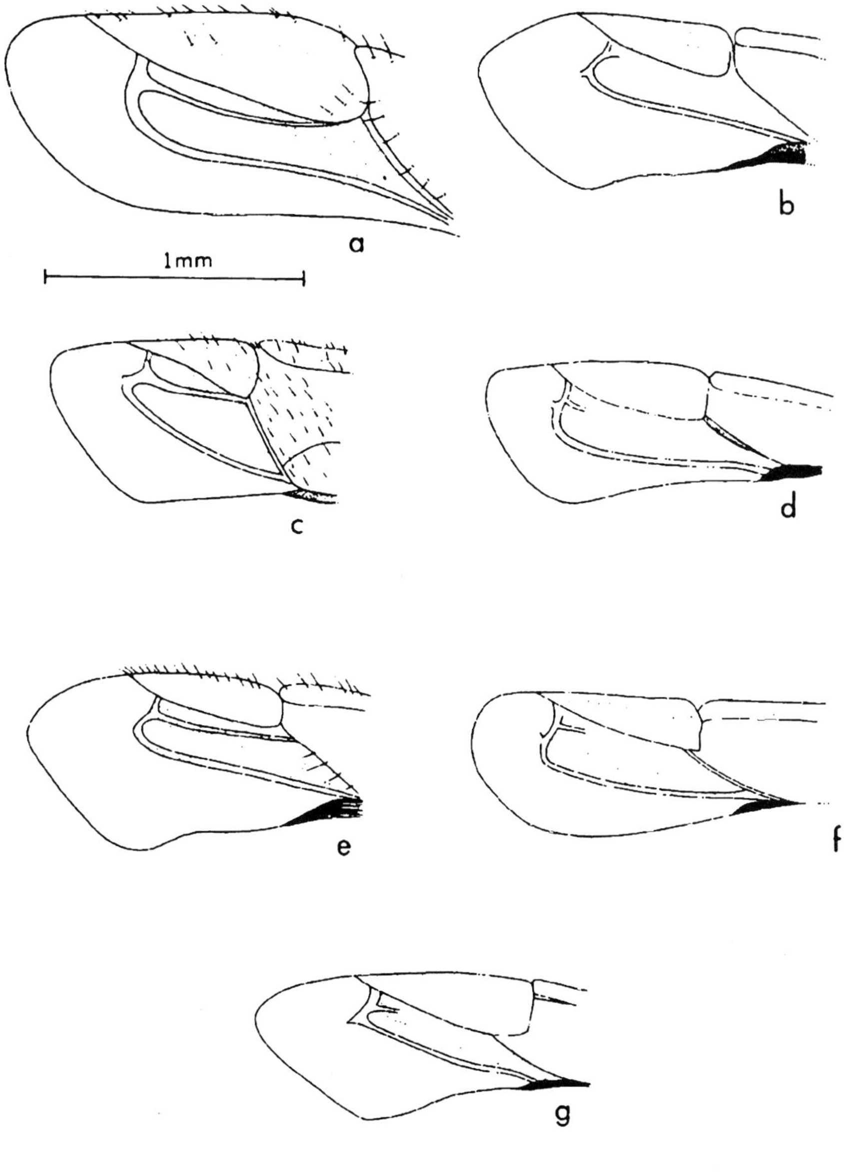

Fig. 31. Apex of hemelytron and membrane of *Nesiomiris* n. spp. **a**, *trifurcatus*; **b**, *curvatus*; **c**, *sinuatus*; **d**, *swezeyi*; **e**, *spatulatus*; **f**, *cheliferus*; **g**, *pallasatus*. Setae omitted on b, d, f, g.

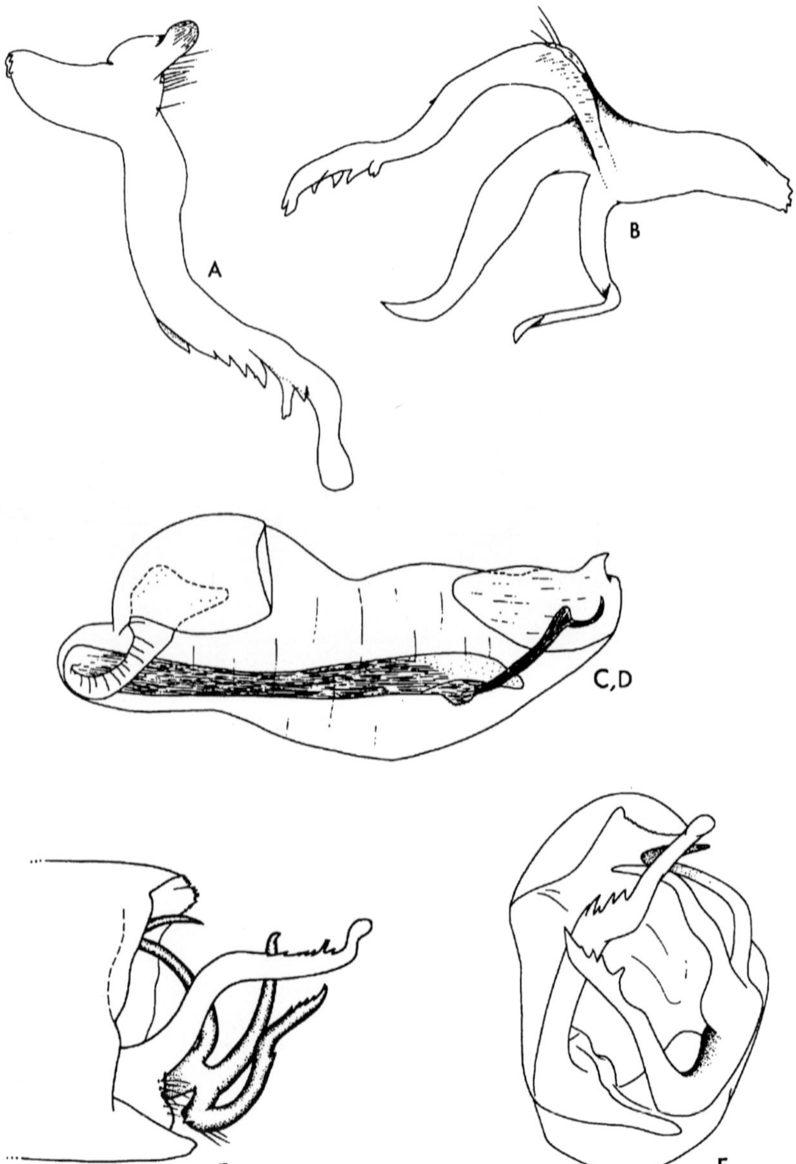

Fig. 32. *Nesiomiris trifurcatus*, n. sp., male genitalia. **a**, left clasper, internal; **b**, right clasper, internal; **c**, **d**, aedeagus and spiculum, lateral; **e**, pygofer, left lateral; **f**, pygofer, terminal.

FEMALE (n = 5). Body: 7.11 (6.60–7.46) long; 1.39 (1.30–1.39) wide. Antennal segments lengths: I, 1.25 (1.21–1.25); II, 3.52 (3.25–3.65); III, 3.20 (2.59–3.20); IV, 1.33 (1.33–1.60). Rostrum: 1.50 (1.49–1.50) long. Pronotum: 1.21 (1.14–1.21) long; 1.39 (1.30–1.39) wide. Differing from male as follows: reddish in same areas but less pronounced; apex of vertex and areas between reddish striae becoming pulverulent; clavus darker green than remainder of hemelytra; clavus darker on apical 1/3, remainder yellow; *K* and *L* structure (Fig. 25a).

NYMPHS. *1st and 2nd instars*. Unknown. *3rd instar* (n = 1). Body: 2.58 long; 0.57 wide. Head: apex broadly rounded; lacking basal carina; chaetotaxy: a prominent bristle between eye and base, and 2 pairs of lesser setae, 1 at apex of vertex, and the other pair above and farther apart, row of 4 evenly spaced bristles across head and base. Antennal segment lengths: I, 0.53, with a few irregularly distributed setae that are about 1/2 as long as width of segment; II, 1.32, reddish tinged at apex, with scattered, more reclined bristles becoming shorter than those on I; III, 1.75, immaculate; chaetotaxy still shorter and denser. Rostrum: 1.07 long; reaching apices of hind coxae. Pronotum: 0.35 long; 0.57 wide; propleural stripe absent; slightly convex anterior, and slightly concave posterior margins, lateral margin convex; a few scattered bristles with 1 at anterior humeral angles and a pair at posterior angles, not divided into anterior and a posterior lobe, but shallowly impressed at middle. Mesonotum laterally concave with an anterolateral constriction, a prominent bristle at the constriction, slightly longer at middle than pronotal length, shallowly impressed in middle. Metanotum laterally convex, shallowly and evenly concave anteriorly, abruptly concave at middle posteriorly; about 1/3 length of mesonotum; a prominent bristle mesolaterally, another at the posterior humeral angle. Legs: all surfaces of femur with randomly placed bristles, which are subequal in length to the greatest thickness of the femur; femoral bristles same length as those comprising 2 posterior ill-defined rows on the tibia, the latter being interspersed with shorter, more reclined hairs, especially on the hind tibia (both types on pro- and mesotibia are shorter and less conspicuous); chaetotaxy on tarsus mostly as short as the shortest hairs on tibia, and with a longer pair on upper apex of last segment. Abdomen: chaetotaxy consisting of a pair of bristles on either side of midline on 1st 3 segments, scattered hairs ventrally; shape generally elongate acuminate, lateral margin evenly concave through 7th segment, then becoming abruptly tubular. *4th instar* (n = 2). Body: 2.95–3.70 long; 0.66–0.70 wide. Head slightly indented at middle of apex; chaetotaxy same as 3rd instar. Antennal segment I with scattered erect bristles subequal to greatest thickness of segment with more recumbent, slightly shorter hairs; antennal segments II and III reddish annulate near apex, chaetotaxy of these and antennal segment IV similar to 3rd instar; ratio of lengths of segments I and IV as 0.45: 1.35: 2: 1.55. Rostrum: 1.25–1.45 long; reaching apices of hind coxae. Pronotum: 0.38–0.45 long; 0.66–0.70 wide; posterior lobe (see adult description) evident as a slight lateral invagination near humeral angle, and a prominent bristle; otherwise shape and chaetotaxy similar to 3rd instar. Mesonotal wing pad reaching apex of 1st visible abdominal tergite; chaetotaxy consisting of a pair of bristles at middle, and a row along the exterior edge; anterior margin convex; metanotum slightly longer along middle than pronotum; metanotal wing bud attaining middle of 2nd visible abdominal tergite. Legs: similar to 3rd instar except tarsi have 3 to 4 pairs of prominent dorsal bristles. Abdomen: as described for 3rd instar. *5th instar* (n = 8). (Fig. 27a). Body: 3.24–4.60 long; 0.80–0.92 wide. Head: 0.85–0.92 wide. Antennal segment lengths: I, 0.92–1.18; II, 2.38–2.95; III, 1.79–2.60; IV, 1.65–2.50. Rostrum: 1.30–1.60 long. Pronotum: 0.45–0.76 long; 0.67–0.92 wide. Structure and chaetotaxy as in 4th instar, but mesonotal wing pads now extending to middle of 6th visible abdominal tergite. Apices of last tarsal segments now dark. Last 2 abdominal segments tubular (apparently male nymph only). Setae much paler than adult. Ratio of antennal segments I–IV as 4: 11.5: 12: 11.

All instars generally green in life, but green bleaches to pallid in alcohol.

Distribution. O'ahu I (Ko'olau Mts).

Host. *Tetraplasandra meiandra*, *T. kahanana*.

Etymology. Named for the 3-armed right clasper of the male.

Type data. Holotype ♂ (BPBM 14,826), allotype ♀, paratypes, 8♂, 2♀, 10N, **O'AHU**: Ko'olau Mts, Castle Trail, 2000 ft. [605 m], 15.III.1970, on *Tetraplasandra meiandra*, W.C. Gagné. Additional paratypes with data differing from holotype only as follows: 2♂, 3♀, 6N, 666 m, 13.X.1968, on *Tetraplasandra*; 6N, 666 m, 12.X.1968, on *Tetraplasandra*; 3♂, 3♀, Waiāhole Forest Reserve, Turnover Peak, 2000 ft. [605 m], 31.V.1970, on *T. kahanana*; 5♂, 3♀, 3N Waimano Trail, 725 m, 23.IX.1972, on *T. meiandra* (BPBM).

Other material examined. 1♀, same data as holotype, except: Hālawa Trail, 457 m, 20.VI.1971; 1♀, same data as holotype, except Hālawa Ridge Trail, 540 m, 20.VI.1971 (BPBM).

Discussion. Without associated diagnostic males, the unassociated females and nymphs in "other material examined" above have not been designated as paratypes.

The *curvatus* Subgroup

Description. **ADULTS**. Antennal segment I and apices of hind femora yellow to yellowish brown; apex of head rounded; hind lobe of pronotum and sometimes collar greenish in fresh specimens; propleural stripe extending from collar and ending abruptly at mesal constriction; pygofer unarmed; right clasper 2-pronged, 3rd prong much reduced or absent; upper arm of left clasper without teeth; female with outer lobe of *K* structure about the same length as the inner lobe and well differentiated from inner lobe; mesodorsal evagination of *K* structure usually minutely spiculate (Fig. 25b-f).

NYMPHS. Propleural stripe absent; apices of hind femora rufescent.

Distribution. O'ahu, Lāna'i, Maui, and Hawai'i Is.

Host. *Cheirodendron trigynum* and *Tetraplasandra* spp.

Included species. Nos. 2–18.

The *curvatus* Complex

Description. **ADULTS**. Apex of head weakly incised; outer arms of right clasper unreduced, inner or 3rd arm of this clasper much reduced, represented by a short knob; upper arm of the left clasper long and sinuate, but not S-shaped; left apex of aedeagus with blunt flange.

Distribution. Maui and Hawai'i Is.

Host. *Cheirodendron trigynum*, *Tetraplasandra meiandra*, *T. kavaiensis*.

Included species. Nos. 2–3.

2. *Nesiomiris curvatus* Gagné, new species — Figs. 23, 31b, 33a, 34a, 35, 36c

The male genitalia in particular indicate a close relationship to *N. pilosus*, with which this species forms a species complex.

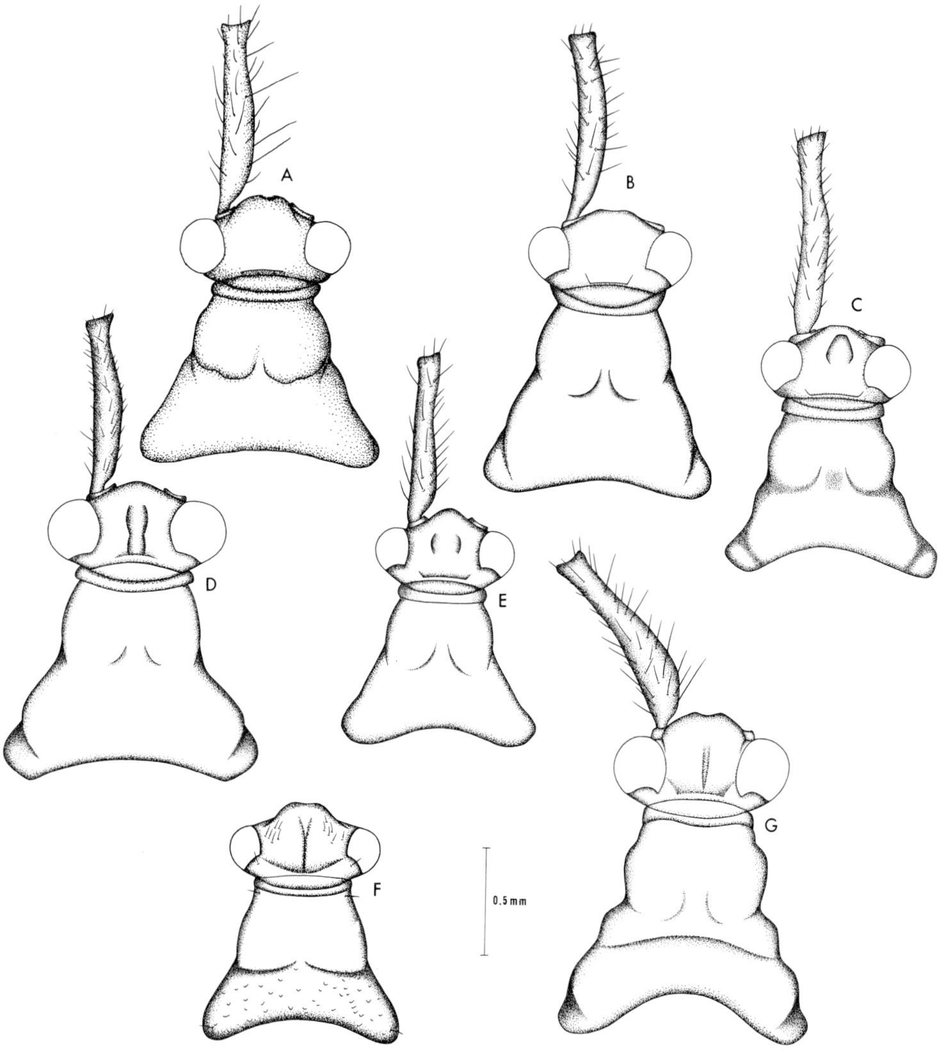

Fig. 33. *Nesiomiris* n. spp., antennal segment I, head, and pronotum, dorsal view. **a**, *curvatus;* **b**, *pilosus;* **c**, *sinuatus;* **d**, *cheliferus;* **e**, *swezeyi;* **f**, *ohe* (antenna omitted); **g**, *elongatus.*

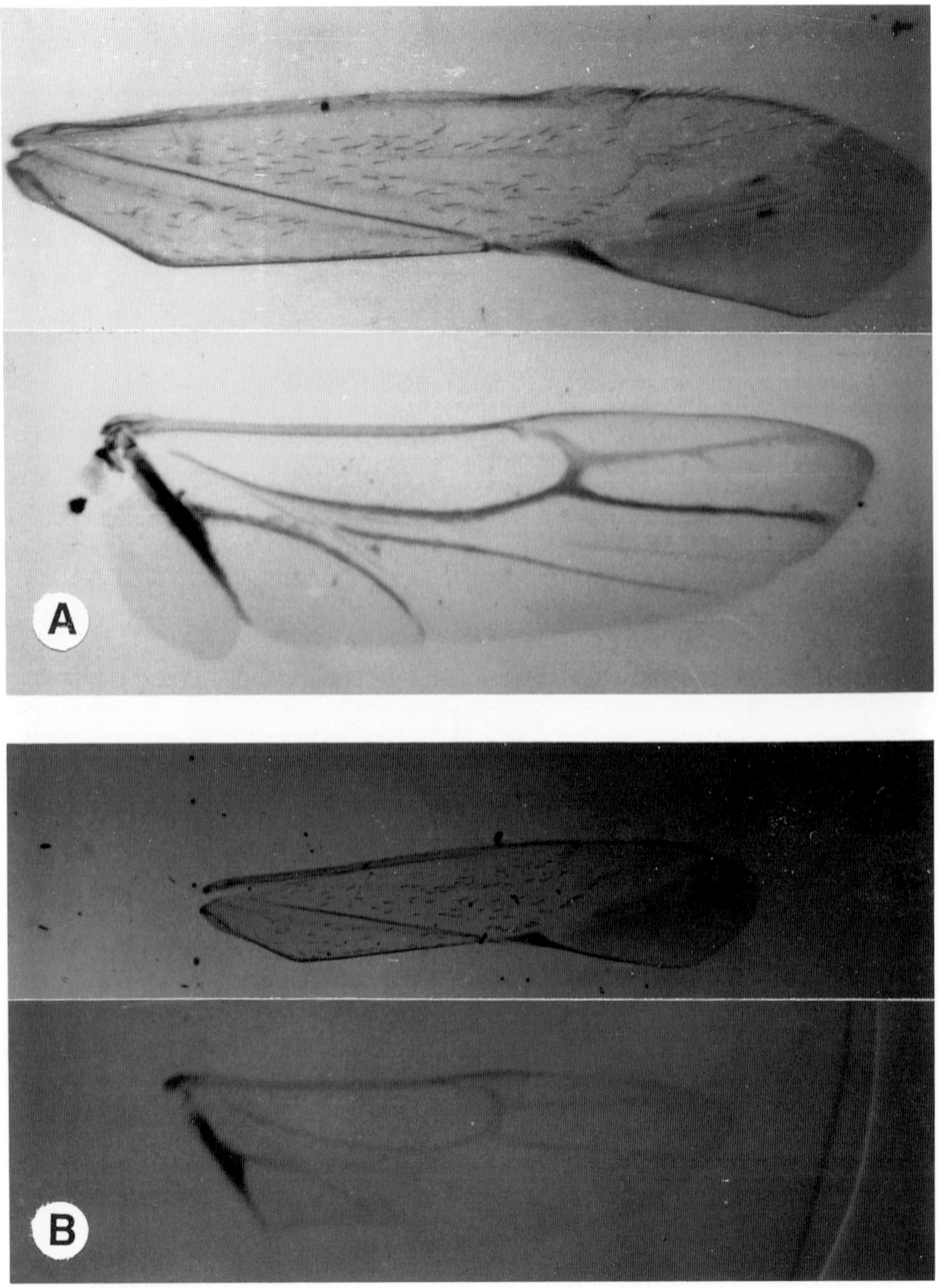

Fig. 34. *Nesiomiris* n. spp., right wings, dorsal view. **a**, *curvatus,* hemelytron above, hind wing below; **b**, *williamsi,* hemelytron above, hind wing below.

The narrow elongate form, 2 elongate arms of the right clasper, the structure of the spiculum and phallus of the male, as well as the rufus apices of the 2nd and 3rd antennal segments of the nymphs relate this species with the *trifurcatus* subgroup. *Nesiomiris curvatus* differs from *N. pilosus* of Hawai'i as noted under the latter.

Description. **MALE** (n = 7). Body: 7.05 (6.81–7.17) long; 1.48 (1.35–1.48) wide. Head: 1.01 (0.99–1.01) wide; apex weakly incised (Fig. 33a); yellowish brown on frons and dorsally paler below; callosities shining, with a shallow impression between them vestiture sparse, consisting of a few scattered recumbent, brown hairs, these absent on callosities weakly carinate across base. Antennal lengths: I, 1.17 (1.11–1.17), yellowish, longer than head width, ratio 9: 8, vestiture brownish, semierect hairs more prevalent on inner aspects (Fig. 33a); II, 3.36 (3.20–3.95), yellowish with a blackish apex; III, 2.52 (2.50–2.65), yellowish at base, remainder blackish; IV, 2.15 (1.75–2.31), blackish. Rostrum: 1.52 (1.45–1.55), reaching apex of mesosternum. Pronotum: 1.11 (.90–1.15) long; 1.48 (1.35–1.48) wide, shape as illustrated (Fig. 33a); hind lobe pale greenish, remainder yellow; mesoscutellum and scutellum with scattered brown, suberect setae on disc. Large areole coriaceous as illustrated (Figs. 31b, 34a); vestiture and infumate membrane as for *N. trifurcatus*, membrane clear adjacent to the dark last tarsal segments; vestiture inclined, about as long as tibial width, brownish on dorsal surfaces; paler on ventral surfaces. Abdomen: atrium (Fig. 35e) lacking a dorsal tooth; claspers as illustrated (Fig. 35).

FEMALE (n = 7). Body: 6.86 (6.86–7.45) long: 1.47 (1.40–1.50) wide. head 0.92 (0.92–1.03) wide. Antennal lengths: I, 1.15 (1.15–1.21); II, 3.52 (3.52–3.95); III, (2.58–2.85); IV, 1.95 (1.40–1.45). Rostrum: 1.51 (1.45–1.56) long. Pronotum: 1.15 (1.03–1.17) long; 1.47 (1.40–1.50) wide. Differing from male as follows: ratio of antennal segment I to head width as 3.5: 3, shining callosities of head darker; antennal segment I faintly reddish annulate near apex; III, narrowly yellowish at base; hind lobe of pronotum becoming dark green basally; propleural stripe ending abruptly on medial pronotal constriction. Scutellum greenish at apex, both otherwise pale yellow. Rostrum just surpassing middle of mesosternum.

NYMPHS. *1st instar*. Unknown. *2nd instar*. Body: 1.50–1.70 long; 0.37–0.45 wide. Head: 0.52–0.55 wide. Antennal lengths: I, 0.35; II, 0.81; III, 0.80–0.88; IV, 1.08. Rostrum: 0.70 long. Pronotum: 0.23–0.32 long; 0.37–0.45 wide. Remainder as described for 2nd-instar nymph of *N. pilosus*. *3rd instar*. Body: 1.45–2.13 long; 0.41–0.62 wide. Head: 0.52–0.55 wide. Antennal lengths: I, 0.35–0.68; II, 0.70–2.23; III, 0.70–2.00; IV, 1.16–1.95. Rostrum: 0.70 long. Pronotum: 0.35–0.39 long ; 0.41–0.55 wide. Remainder as described for 2nd-instar nymph. *4th instar*. Body: 2.20–2.23 long; 0.66–0.95 wide. Head: 0.70–0.85 wide. Antennal lengths: I, 0.54–0.85; II, 1.67–2.24; III, 1.60–2.03; IV, 2.30–2.41. Rostrum: 1.25–1.68 long. Pronotum: 0.35–0.41 long; 0.66–0.95 wide. As described for 4th-instar nymph of *N. pilosus*, except setae shorter and sparser. *5th instar*. Body: 3.15–3.75 long; 0.85–0.95 wide. Head: 0.90–0.92 wide. Antennal lengths: I, 0.92–1.00; II, 2.15–2.98; III, 2.11–3.08; IV, 2.15–2.25. Rostrum: 1.65–1.70 long. Pronotum: 0.35–0.75 long; 0.75–0.90 wide. As described for 5th-instar nymph of *N. pilosus*, except setae shorter and sparser.

Distribution. E Maui.

Host. *Tetraplasandra meiandra* and its var. *ovalis*.

Etymology. Named for the long, curved right clasper of the male.

Type data. Holotype ♂ (BPBM 14,827), allotype ♀, paratypes 11♂, 10♀, 7N, **E MAUI**: Hāna Forest Reserve, 1370 m, 17.III.1974, on *Tetraplasandra meiandra*, W.C. Gagné. Additional paratypes with data differing from holotype, only as follows: 2♂, Waikamoi Stream, 14.I.1926, on

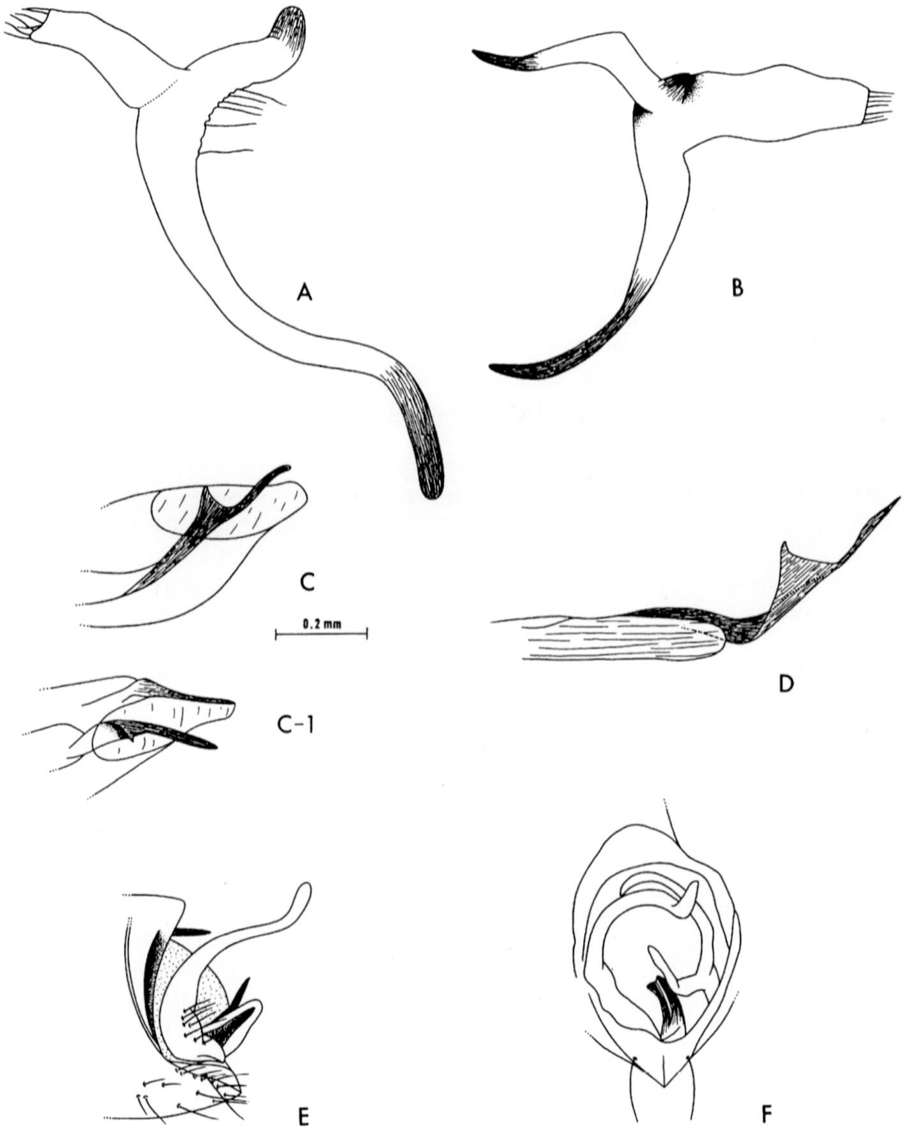

Fig. 35. *Nesiomiris curvatus,* n. sp., male genitalia. **a**, left clasper, internal; **b**, right clasper, internal; **c**, aedeagus and spiculum, left lateral; **c-1**, same, dorsal; **d**, spiculum; **e**, pygofer, left lateral; **f**, pygofer, terminal.

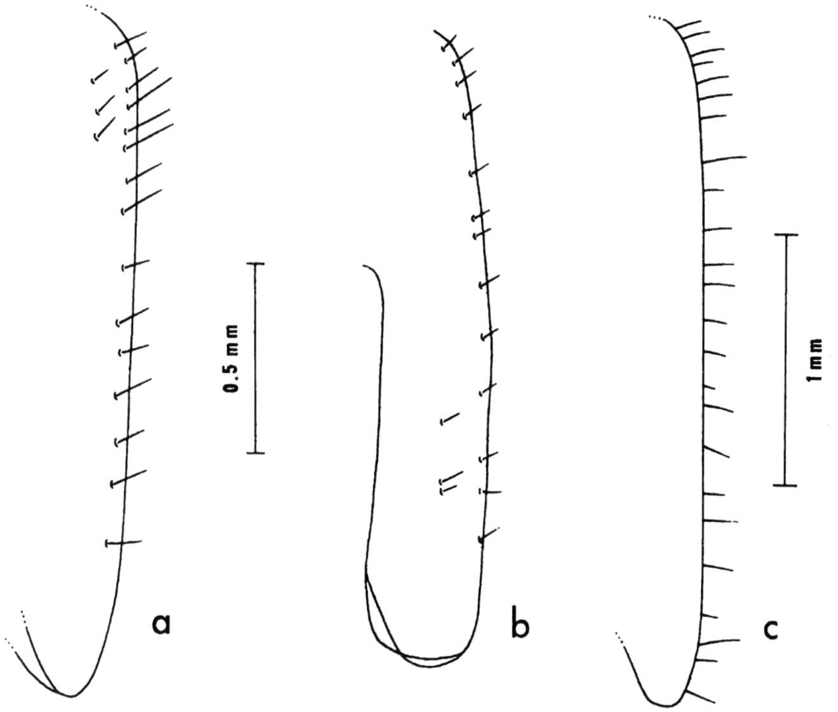

Fig. 36. *Nesiomiris* n. spp., 5th instar nymphs, right mesonotal wing bud, dorsal view. **a**, *swezeyi*; **b**, *parasinuatus*; **c**, *curvatus*.

Tetraplasandra sp., O.H. Swezey; 10♂, 6♀, 131 m, 6.VIII.1968, on *T. meiandra* var. *ovalis*; 1♂, 1♀, 131 m, 24.XII.1968, on *T. meiandra* var. *ovalis*;and many nymphs; 24.XII.1968, 2♂, 1♀, 9N (BPBM).

Variation. A male collected in fluid was dissected and found to have a sclerotized, dark line along the edge of the scutellum, making a V-shaped structure when viewed dorsally; also the inner apical margin of the hind wing is broadly infumate.

Discussion. There is a single male from W Maui with the following data: Kaulalewelewe-Pu'u Kukui Trail, 915 m, 28.V.1972, on *Tetraplasandra meiandra*, W.C. Gagné, which I provisionally consider as belonging here, although it is not included in the paratype series. This species seems to be strictly associated with *Tetraplasandra meiandra*. There are more potentially competing species of *Nesiomiris* on *Cheirodendron trigynum* on Maui, which may account for the apparent specificity of *curvatus* to *T. meiandra*; ecological displacement or preemption is implied.

3. *Nesiomiris pilosus* Gagné, **new species** Figs. 25b, 30c, 33b, 37

Nesiomiris sp. Timberlake, 1921, *Proc. Hawaii. Entomol. Soc.* 4: 472 (note, host).

The male genitalia, particularly, indicate a close relationship to *curvatus*. Both have the narrow, elongate form, and elongate outer arm of the right clasper. The phallus and spiculum are also similar. This species pair forms a sister group with the *sinuatus* complex, and both complexes together form a sister group with the *trifurcatus* subgroup.

Description. **MALE** (n = 7). Body: 6.80 (6.70–7.21) long; 1.32 (1.32–1.45) wide. Head 1.02 (0.99–1.05) wide; apex color, vestiture and median impression as described for *curvatus*; callosities dull; more distinctly carinate across base, this with a diffuse row of setae. Antennal lengths: I, 1.01 (1.01–1.45), yellowish and longer than head width; II, 2.44 (2.44–4.22), yellowish on basal 1/2, becoming blackish on apical 1/2; III, 1.43 (1.43–2.95), blackish; IV, 2.01 (1.15–2.40). Rostrum: 1.54 (1.50–1.60) long, just surpassing middle of mesosternum. Pronotum: 1.16 (0.95–1.16) long; 1.32 (1.32–1.45) wide; shape as illustrated (Fig. 33b); hind lobe basally greenish; remainder yellow; vestiture of scattered brown hairs dorsally, and pale ones ventrally. Scutellum entirely yellowish. Vestiture and morphology of remainder of dorsum as described for *N. curvatus*, excepting that the vestiture is denser and longer. Legs: femora and tibia pilose, about 3/4 the greatest thickness of hind femur, this reddish brown at apex, tarsi pygofer (Fig. 37e) and claspers as illustrated for *N. curvatus* (Fig. 37a, b).

FEMALE (n = 7). Body: 7.22 (7.13–7.45) long ; 1.95 (1.35–1.95) wide. Head: 0.90 (0.90–1.05) wide. Antennal lengths: I, 1.25 (1.25–1.40); II, 3.40 (3.40–4.32); III, 2.85 (2.62–3.15); IV, 2.38 (1.70–2.50). Rostrum: 1.55 (1.46–1.70) long. Pronotum: 2.02 (1.10–2.02) long; 1.95 (1.35–1.95) wide. Differing from male as follows: vestiture on and near apex of head comparatively dense. Ratio of antennal segment I to head width as 4: 3.25. antennal segment I becoming faintly greenish toward apex; antennal segment II yellowish on basal 3/4, blackish on apical 1/2. Hind lobe more broadly dark greenish. Scutellum faintly greenish as apex. Hind femur entirely yellow.

NYMPHS. *1st instar*. Unknown. *2nd instar* (n = 1). Body: 1.70 long; 0.38 wide. Antennal lengths: I, 0.29; II, 0.90; III, 0.77; IV, 1.10. Antennae I-IV with chaetotaxy and coloration as described for 3rd-instar nymph of *N. trifurcatus*. Rostrum: 0.70 long. Pronotum: 0.18 long; 0.38 wide; with a conspicuous row of 4 long, prominent bristles along the lateral edge continuing along the sides of the meso- and metanota; this line of bristle is darker than other bristles and the insertions of the setae appear darker; other setae scattered on the nota, but these are paler and are not darkened at their bases. Shape as described in 3rd-instar nymph of *N. trifurcatus*, except posterior margin of metanotum evenly concave. Legs: chaetotaxy as described for 3rd-instar nymph of *N. trifurcatus* except that there are now 2 more clearly discernable rows on the tibia. Abdomen: broadly immaculate along dorsal midline, but posterior margin of segments I-IV with an interrupted black line and a spot mesolaterally; chaetotaxy generally as for *N. trifurcatus*. *3rd instar*. Unknown. *4th instar* (n = 1). Body: 2.20 long; 0.51 wide. Head: 0.53 wide. Antennal lengths: I, 0.49; II, 1.30; III, 1.25; IV, 1.60. Rostrum: 0.94 long. Pronotum: 0.36 long; 0.51 wide. Not otherwise differing from 2nd instar. *5th instar* (n = 6). Body: 3.00–4.41 long; all 0.92 wide. Head: 0.90–0.92 wide; shape and chaetotaxy as for *N. trifurcatus*; frons striate, calli sclerotized. Antennal lengths: I, 0.92–1.07, with erect pilosity, not a mixture of short and long hairs; remainder as described for *N. trifurcatus* 5th-instar nymph; II, 2.77–3.20; III, 1.95–2.65; IV, 1.25–2.65. Rostrum: 1.50–1.70, reaching mesocoxae. Pronotum: 0.50–0.65 long; all 0.92 wide; disc broadly sclerotized on either side of broad ecdysial line. Pronotum subequal to length of meso- and metanota together. Chaetotaxy of femur and tibia comparatively densely pilose, as described for 2nd instar, but less contrastingly marked.

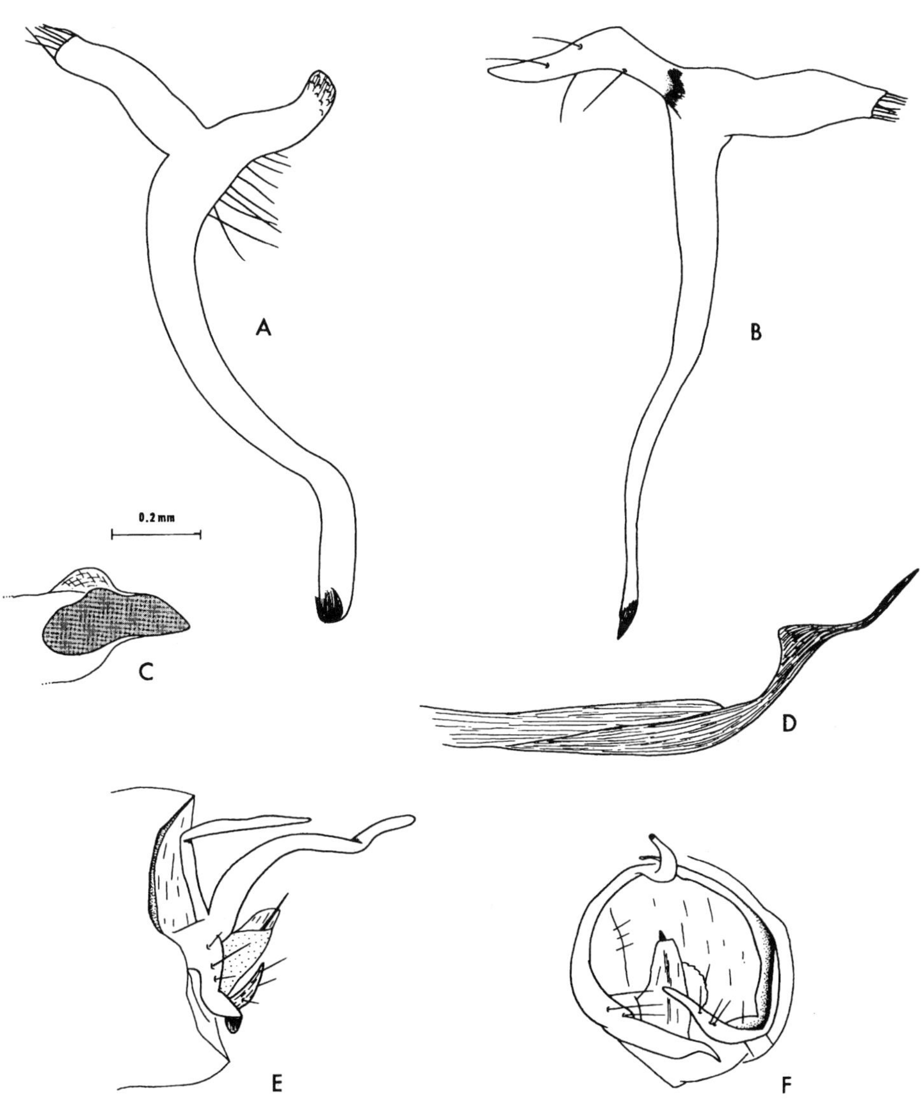

Fig. 37. *Nesiomiris pilosus,* n. sp., male genitalia. **a**, left clasper, internal; **b**, right clasper, internal; **c**, tip of aedeagus, spiculum removed; **d**, spiculum; **e**, pygofer, left lateral; **f**, pygofer, terminal.

Distribution. Hawai'i I.

Host. *Cheirodendron trigynum* (= *Gaudichaudii*), *Tetraplasandra meiandra* var. *tenuistyloides*, and *T. kavaiensis* var. *grandis*.

Etymology. Named for the densely pilose dorsum.

Type data. Holotype ♂ (BPBM 14,828) allotype ♀, paratypes 4♂, 9♀, 4N, **HAWAI'I**: Kūlani Prison Rd, 2500 ft. [763 m], 25.VII.1970, on *Tetraplasandra meiandra* var. *tenuistyloides*, W.C. Gagné. Additional paratypes with data differing only as follows: 2♂, 2♀, 2N, 1500 ft. [457 m]; 3♂, 1N, 3500 ft. [1066 m]; 21♂, 9♀, 2N, Kona, Kealakekua Bay, 1068 m, 16.VIII.1919, on *Cheirodendron Gaudichaudii*, P.H. Timberlake; 4♂, 2♀, 458 m; 3♂, 1068 m; 2♂, 6♀ 2N, Mt Hualālai, 1030 m, 22.I.1971, on *C. trigynum*; 2♂, 3♀, S Kona Dist, above Pāpā, 810 m, 7.II.1974, on *Tetraplasandra kavaiensis* var. *grandis*, 4♂, 1♀, 11N, Volcano, 1188 m, 18.V.1975, on *T. meiandra* (BPBM).

The *sinuatus* Complex

Description. **ADULTS**. Apex of head weakly incised to rounded; lower and inner arms of the right clasper reduced in length, its upper arm hooked; upper arm of left clasper long and S-shaped to evenly hooked; left apex of aedeagus thickened, bluntly flanged where thickened.

Distribution. Lāna'i, Maui, and Hawai'i Is.

Host. *Cheirodendron trigynum*.

Included species. Nos. 4–7.

4. *Nesiomiris sinuatus* Gagné, **new species** — Figs. 25c, 30d, 31c, 33c, 38

Description. **MALE** (n = 7). Body: 6.50 (6.10–6.50) long; 1.31 (1.22–1.31) wide. Head: 0.88 (0.86–0.90) wide, apex straight (Fig. 31c); yellowish brown dorsally; callosities dull; vestiture very short and sparse. Antennal lengths: I, 1.29 (1.09–1.29) long (Fig. 33c), yellowish longer than head width (ratio 9.5: 7), vestiture brown, recumbent setae comparatively short (less than 1/4 the width of segment I); II, 3.15 (3.15–3.55); III, 2.77 (2.03–2.77); IV, 2.96 (1.78–2.96), vestiture and color as described for *N. curvatus*. Rostrum: 1.52 (1.17–1.52) long; reaching middle of mesosternum. Pronotum: 1.10 (1.00–1.10) long; 1.31 (1.22–1.31) wide; shape as illustrated (Fig. 33c); glabrous. Mesoscutellum and scutellum paler than hemelytra, yellow, glabrous. Hemelytral vestiture comparatively short and sparse; areoles partly coriaceous as illustrated (Fig. 31c). Legs: yellowish with greenish tinge, vestiture comparatively short, recumbent, excepting a few longer hairs on ventral surfaces; vestiture is mostly less than 1/4 the width of the femur. Abdomen: atrium, aedeagus, spiculum, and claspers as illustrated (Fig. 38).

FEMALE (n = 7). Body: 7.21 (6.30–7.21) long; 1.95 (1.24–1.95) wide. Head: 1.00 (0.86–1.00) wide. Antennal lengths: I, 1.27 (1.09–1.27); II, 3.95 (3.20–3.95); III, 2.86 (2.24–2.86); IV, 2.25 (1.78–2.25). Rostrum: 1.51 (1.24–1.51) long. Pronotum: 1.16 (1.00–1.16) long; 1.95 (1.24–1.95) wide. Differing from male as follows: hind femur weakly reddish near apex. Internal genitalia as illustrate.

NYMPHS. *1st to 4th instars*. Unknown. *5th instar* (n = 8). Body: 2.76–3.24 long; 0.70–0.81 wide. Head: 0.70–0.81 wide. Apex rounded; chaetotaxy as for *N. trifurcatus*. Antennal lengths: I, 0.57–0.95; II, 2.31–2.76; III, 1.80–2.70; IV, 1.60–2.60; color as for *N. trifurcatus*; ratio of lengths

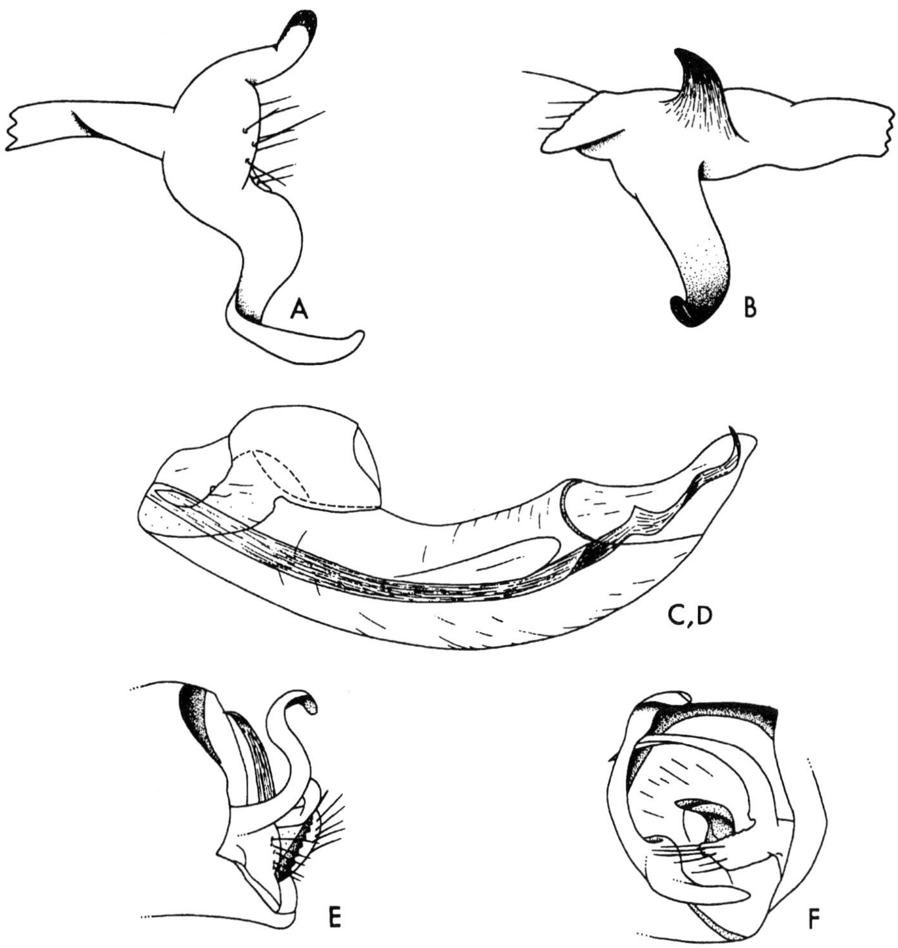

Fig. 38. *Nesiomiris sinuatus,* n. sp., male genitalia. **a**, left clasper, internal; **b**, right clasper, internal; **c, d**, aedeagus and spiculum, lateral; **e**, pygofer, left lateral; **f**, pygofer, terminal.

of segments I-IV as 2.5: 8: 7.5: 7. Rostrum: 1.17–1.25 long, reaching base of hind coxae. Pronotum: 0.35–0.66 long; 0.70–0.81 wide; shape as for *N. trifurcatus*; bristles that are denser at the humeral margins, with a row of 6 setae along the incipient, impressed medial constriction. Shape and chaetotaxy of mesonotal wing buds as illustrated (Fig. 36a), reaching 6th abdominal segment. Legs: setae present on all aspects, but longest erect ones are not exceeding 1/2 the thickness of femur; tarsi as described for *N. trifurcatus*. Abdomen: lacking infuscations; shape and chaetotaxy as described for *N. trifurcatus*, but latter is denser on all aspects of the last 2 segments.

***Distribution*.** W Maui.

Host. *Cheirodendron trigynum* var. *confertiflorum*.

Etymology. Named for the sinuate, S-shaped left clasper of the male.

Type data. Holotype ♂ (BPBM 14,829) allotype ♀ paratypes 7♀, **W MAUI**: ridge above Kaulalewelewe, Pu'u Kukui Trail, 3000–4000 ft. [914–1210 m], 28.V.1972, on *Cheirodendron trigynum* var. *confertiflorum*, W.C. Gagné. Additional paratypes with data differing only as follows: 1♀, 4.VIII.1964, D.E. Hardy; 2♂, [no elevation], 26.X.1966, P.D. Ashlock; 2♀, 3000–3400 ft. [914–1036 m], 24–26.X.1966, C.M. Yoshimoto; 10♂, 2♀, 16N, 3000 ft. [914 m], 8.VIII.1968; 15♂, 12♀, 915–1037 m, 8.VIII.1968; 2♂, 3♀, 915–1037 m, 17.III.1970; 1N, 700–1400 m, 17–19.V.1970, on *C. trigynum*; 9♂, 2♀, Kahoma Stream, 730 m, 16.IV.1971, on *C. trigynum* (BPBM).

Variation. Some males have the apices of the hind femur reddish, and the scutellum is at times greenish. Some specimens have an entirely pale pronotum. In the last-instar nymph the propleural stripe is lacking. However, specimens that are ready to moult to the adult have a propleural stripe visible through the overlying nymphal exuvium.

I have been unable to differentiate the females of *N. reductus* and *N. sinuatus*, so the paratype series of *N. sinuatus* may have some females referable to *N. reductus*.

5. ***Nesiomiris parasinuatus*** Gagné, **new species** Figs. 25d, 30e, 36b, 39

Description. **MALE** (n = 6). Body: 6.27 (5.67–6.27) long; 1.26 (1.21–1.28) wide. Head: 0.90 (0.86–0.95) wide; apex slightly bifid; color, callosities and vestiture as in *N. sinuatus*. Antennal lengths: I, 1.15 (1.05–1.15), greenish yellow, longer than head width (ratio 9: 7), vestiture as in *N. sinuatus*; II, 3.70 (3.30–3.70); III, 3.70 (2.14–3.70); IV, broken (1.68–2.13); vestiture and color of II-IV as in *N. curvatus*. Rostrum: 1.32 (1.28–1.32) long, reaching middle of mesosternum. Pronotum: 1.01 (0.90–1.03) long; 1.26 (1.21–1.28) wide, shape as in *N. sinuatus*, collar green, hind lobe greenish marmorate, remainder yellowish; a few scattered, inconspicuous brown setae. Meso- and metasternum yellowish, glabrous; scutellum greenish. Clavus greenish between veins; veins yellow, remainder of hemelytra greenish. Large areole and dorsal vestiture as illustrated for *N. sinuatus* (Fig. 31c); but small areole coriaceous. Legs: vestiture and color as described in *N. sinuatus*. Abdomen: atrium (Fig. 39e, f); claspers as illustrated (Fig. 39a, b).

FEMALE (n = 6). Body: 6.20 (6.18–6.40) long; 1.03 (1.05–1.28) wide. Head: 0.92 (0.75–0.92) wide. Antennal lengths: I, 1.10 (0.95–1.10); II, 3.20 (3.20–3.75); III, 2.45 (2.14–2.55); IV, 1.58 (1.35–2.37). Rostrum: 1.41 (1.25–1.41) long. Pronotum: 1.28 (1.03–1.10) long; 1.05 (1.05–1.31) wide. Differing from male as follows: apex of head straight; ratio of lengths of antennal segment I to head width as 8.28: 7.25; hind lobe of pronotum more narrowly greenish marmorate; propleural stripe not surpassing medial constriction. Internal genitalia as illustrated (Figs. 25d, 30e).

NYMPHS. *1st to 3rd instars*. Unknown. *4th instar* (n = 4). Body: 2.45–2.51 long; 0.45–0.65 wide. Head: 0.60–0.67 wide. Antennal lengths: I, 0.60–0.68; II, 1.57–1.95; III, 1.54–1.70; IV, 1.45–1.67. Rostrum: 0.94–1.03 long. Pronotum: 0.30–0.38 long; 0.45–0.65 wide. *5th instar.* Unknown.

Distribution. Lāna'i I.

Host. *Cheirodendron trigynum* vars. *Rockii* and *Hillebrandii*.

Etymology. Named for its close morphological affinities with *N. sinuatus*.

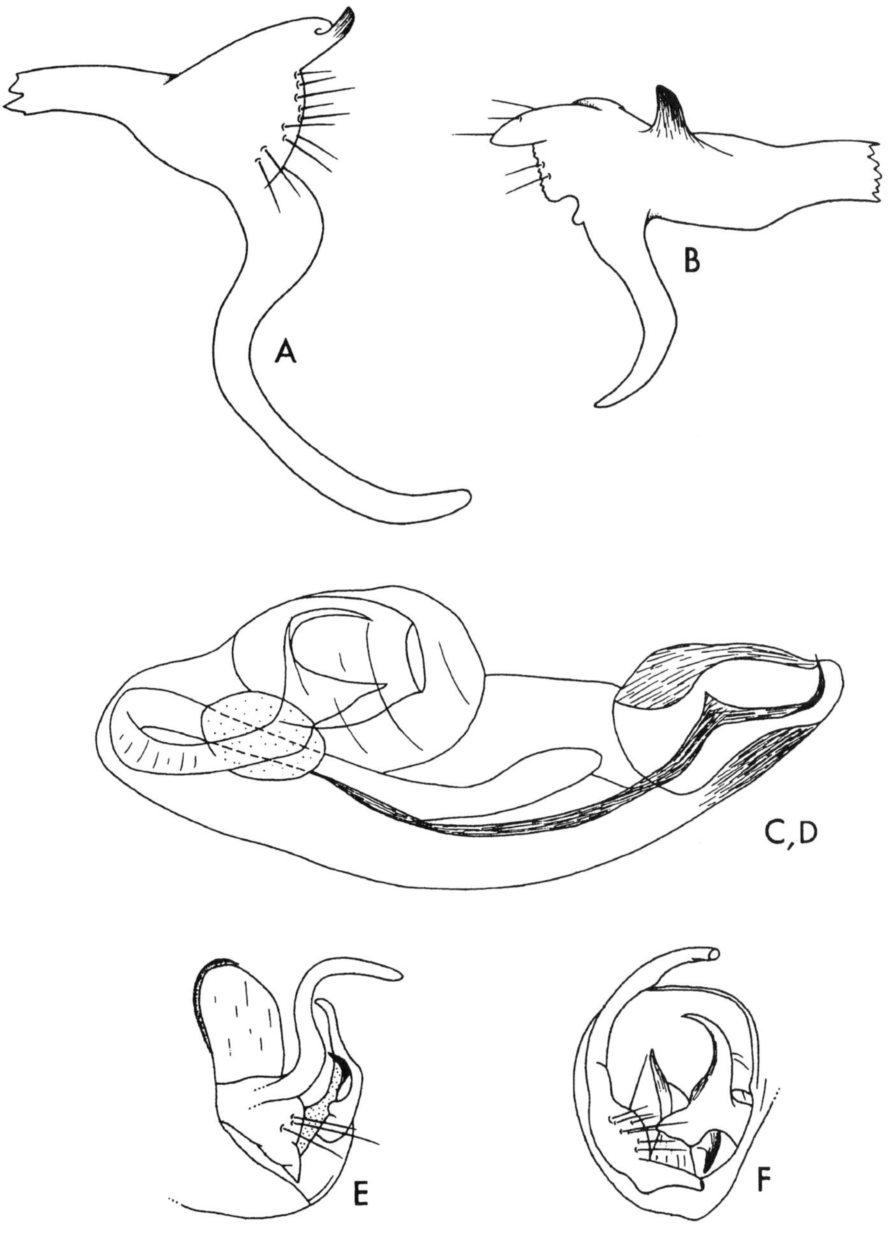

Fig. 39. *Nesiomiris parasinuatus,* n. sp., male genitalia. **a**, left clasper, internal; **b**, right clasper, internal; **c**, **d**, aedeagus and spiculum, lateral; **e**, pygofer, left lateral; **f**, pygofer, terminal.

Type data. Holotype ♂ (BPBM 14,830) allotype ♀, paratypes 25♂, 14♀, 15N, **LĀNA'I**: Lāna'ihale, 3300 ft. [1065 m], 12.VII.1968, on *Cheirodendron trigynum* var. *Rockii*, W.C. Gagné. Additional paratypes with data differing only as follows: 1♂, 1♀, 3000 ft. [914 m], 6.VI.1970, on *C. trigynum* var. *Hillebrandii* (BPBM).

Discussion. I was initially hesitant to separate, as species, the W Maui *sinuatus* from the Lāna'i populations *parasinuatus*, because of the close conformity of the male genitalia and the general similarity in habitus. But the differences in the genitalia appear to be constant, particularly the shape of the right claspers; in *N. sinuatus* the upper arm is long, surpassing the dorsum of the pygofer (Fig. 38e, f), while in *N. parasinuatus* this arm does not reach the dorsal wall of the atrium (Fig. 39f). Viewed internally, the differences in the left clasper are not as great as they would appear from the illustrations (Fig. 38a); since they are highly sinuate, it was very difficult to orient them in exactly the same plane for drawing. The aedeagus and spiculum (Figs. 38c, d; 39c, d) are also close. Differences in the female genitalia are also apparent (Fig. 30d, e), but I have not dissected a series of both populations (Lāna'i and W Maui) to study the variations.

6. ***Nesiomiris swezeyi*** Gagné, **new species** Figs. 25e, 30f, 31d, 33e, 36a, 40

Nesiomiris hawaiiensis: Swezey & Williams, 1932, *Proc. Hawaii. Entomol. Soc.* 7: 498 (host; misidentification); 1954, *B.P. Bishop Mus. Spec. Publ.* 44: 50 (host, misidentification, included also, other *Nesiomiris* species on *Cheirodendron* on Maui).

The male genitalia, size, habitus and other characters place *N. swezeyi* in the *sinuatus* complex. The upper arm of the left clasper is evenly hooked (Fig. 40a, a–1).

Description. **MALE** (n = 8). Body: 6.18 (5.67–6.18) long; 1.32 (1.24–1.32) wide. Head: 0.90 (0.81–0.90) wide; apex rounded; shape as illustrated (Fig. 33e); color as described for *curvatus*; callosities dull; basal carina extending to inner aspect of eye; vestiture as described for *N. curvatus*. Antennal lengths: I, (Fig. 33e), 1.07 (0.96–1.09), greenish yellow, longer than head width (ratio 7.75: 7), vestiture as described for *N. curvatus*; II, 3.25 (2.99–3.37), yellow on basal 1/2, becoming browner on apical 1/2; III, 2.15 (1.93–2.15); IV, broken (1.60–2.03); III and IV blackish. Rostrum: 1.28 (1.13–1.35) long; just surpassing middle of mesosternum. Pronotum: 1.03 (0.90–1.03) long; 1.32 (1.24–1.32) wide; shape as illustrated (Fig. 33e); hind lobe and collar dark green, remainder yellowish; propleural stripe extending from coxal cleft to medial constriction; propleuron not infuscate below propleural stripe anterior to coxal cleft; subglabrous and inconspicuous seta at each humeral angle. Mesoscutellum yellowish; scutellum pale greenish, latter with a few inconspicuous recumbent setae on disc. Hemelytra: uniformly green excepting base of clavus that is yellowish along claval vein; areoles partially coriaceous as illustrated (Fig. 31d); membrane infumate; vestiture as for *N. trifurcatus*. Legs: femora greenish, tibiae yellowish green; 1st and 2nd tarsal segments brown, 3rd segment black; vestiture as described for *N. sinuatus*. Abdomen: yellow; atrium, claspers, and aedeagus as illustrated (Fig. 40).

FEMALE (n = 7). Body: 6.50 (5.96–9.55) long; 1.38 (1.34–1.38) wide. Head: 0.92 (0.86–0.92) wide. Antennal lengths: I, 1.10 (0.95–1.10); II, 3.25 (2.84–3.26); III, 3.25 (2.03–3.25); IV, broken (1.93–2.06). Rostrum: 1.34 (1.17–1.35) long. Pronotum: 1.00 (0.92–1.09) long; 1.38 (1.34–1.40) wide. Differing from male as follows: callosities of head weakly shining, weakly pulverulent

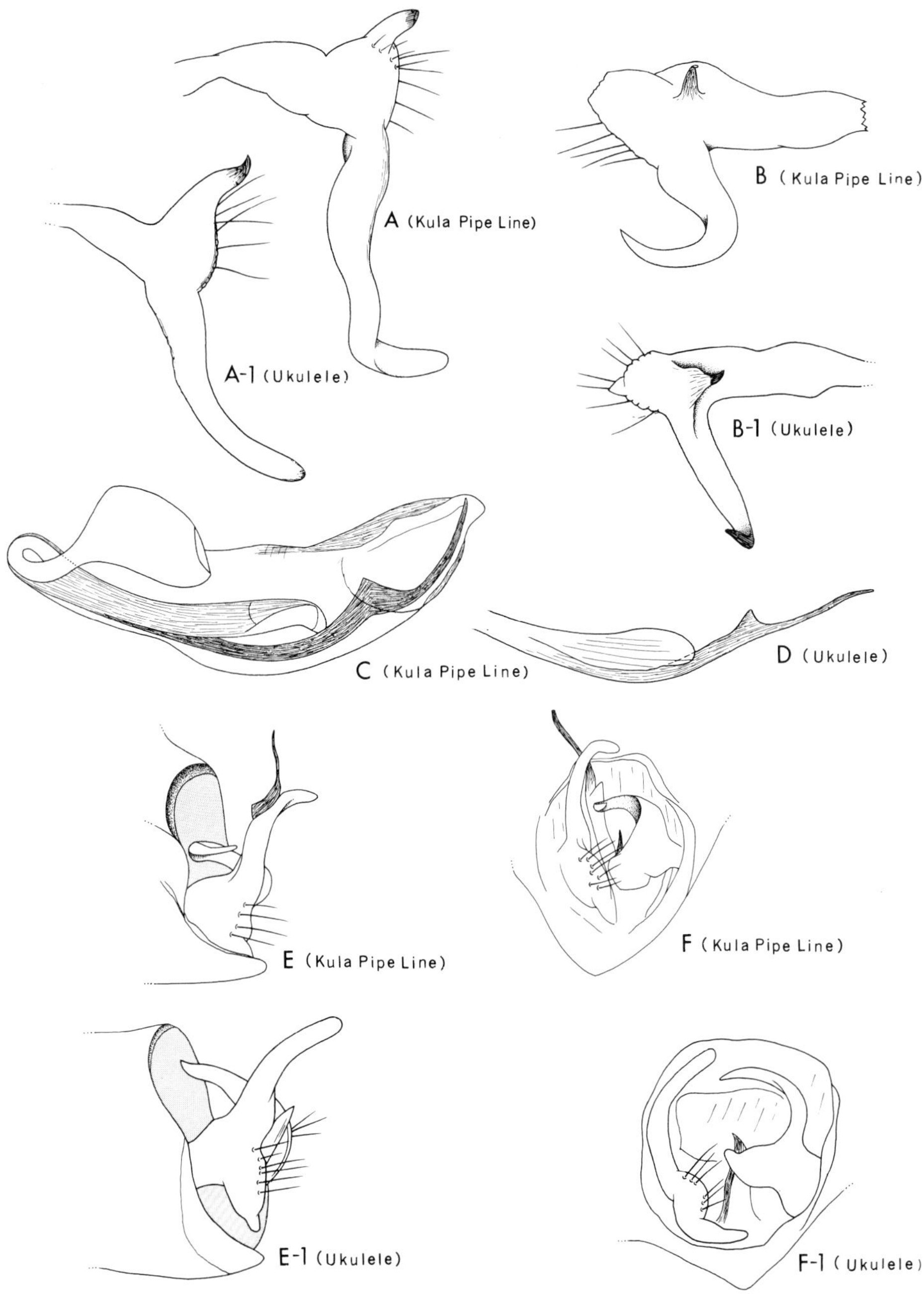

Fig. 40. *Nesiomiris swezeyi,* n. sp., male genitalia, showing population variation of Kula Pipe Line and 'Ukulele, both E Maui. **a**, **a-1**, left clasper, internal; **b**, **b-1**, right clasper, internal; **c, d**, aedeagus and spiculum, lateral; **e**, **e-1**, pygofer, left lateral; **f**, **f-1**, pygofer, terminal.

anterior to these; a row of 6 short setae on basal carina of head; ratio of antennal segment I to head width as 8: 7.5. Scutellum basally yellowish, apical 1/2 pale green; femora yellowish green; tibiae yellowish.

NYMPHS. Unknown.

Distribution. E Maui.

Host. *Cheirodendron trigynum* var. *confertiflorum*.

Etymology. Named in honor of Dr. Otto H. Swezey (1869–1959), one of Hawai'i's most distinguished and deeply respected scientists and entomologists. Dr. Swezey collected this and several other *Nesiomiris* species, keeping careful record of their host associations, thereby making my task, in many ways, easier.

Type data. Holotype ♂ (BPBM 14,831), allotype ♀, paratypes, 1♂, 2♀ **E MAUI**: Ko'olau Forest Reserve, 1585 m, 8.VIII.1973, on *Cheirodendron trigynum*, W.C. Gagné. Additional paratypes: 6♂, 1♀, Ko'olau Forest Reserve, 1680–1980 m, 8.VIII.1973, on *Cheirodendron trigynum*, W.C. Gagné; 3♂, 5♀, Kula Pipeline, 14.I.1926, on *Cheirodendron* sp., O.H. Swezey; 1♂, 'Ukulele Pipeline, Haleakalā Mt, 5000 ft. [1525 m], 14.VII.1919, on *Astelia veratroides*, P.H. Timberlake; 1♂, 1♀, Kula Pipeline, 12.VI.1927, on *Cheirodendron* sp., O.H. Swezey; 3♀, Kula Pipeline, 13.VI.1927, on *Cheirodendron* sp., O.H. Swezey; 1♂, Olinda, [no elevation], 13.V.1926, on *Cheirodendron* sp., O.H. Swezey; 4♂, 3♀, Olinda, 27.II.1926, on *Cheirodendron* sp., O.H. Swezey; 1♂, NW Haleakalā, 6000–6500 ft. [1830–1938 m], 18.VIII.1937, beating, [no host], E.C. Zimmerman; 2♂, 1♀, 2N, Waikamoi, 15.X.1965, D.E. Hardy; 2♂, 1♀, Waikamoi Stream, 4300 ft. [1313 m], 6.VIII.1968, on *C. trigynum* var. *confertiflorum*; 9♂, Waikamoi Stream, 1313 m, 6.VIII.1968, on *Tetraplasandra meiandra* var. *polystigmata*; 1♂, 1♀, 'Ukelele, 5200 ft. [1586 m], 22.XII.1968, on *C. trigynum* var. *confertiflorum*; 2♂, Makawao Forest Reserve, Waiohiwi Gulch, W branch, 4205 ft. [1281 m], 15.IV.1971; 1♂, Hāna Forest Reserve, 2100 m, 4.VIII.1973; 6♂, 3♀, 1N, Hāna Forest Reserve, 5600 ft. [1710 m], 6.VIII.1973, on *C. trigynum*; 3♂, 5♀, Hāna Forest Reserve, 4.VIII.1974, 6600–6850 ft. [2010–2090 m], T. Parman and S. Conant; 3♂, Kīpahulu Valley near Palikea, 670 m, 2.VII.1975; 2♂, 18N, Waikamoi Stream, 2035 m, 21.VII.1975 (BPBM).

Variation. Some of the observed variation in the conformation of the male genitalia, especially the right clasper, the lower arm of which varies from blunt to acute, is shown in Figure 40a, a–1, e, e–1. I interpret this variation to be incipient speciation on the comparatively large landmass of E Maui. The rain forest habitat of the host plant forms, or at least presumably once formed, a continuum around the windward periphery of the mountain between 300 m and 2000 m elevation. The variation observed would indicate that there is not extensive gene flow between populations on the W and NW sides of the mountain with those on the NE side of the mountain.

Discussion. I consider the occurrence of a single male on *Tetraplasandra* and *Astelia* to be merely perching records, since no evidence was found of breeding on these hosts.

7. *Nesiomiris williamsi* Gagné, **new species** Figs. 25f, 29a, 30g, 34b, 41

Nesiomiris sp. nr. *hawaiiensis*: Swezey & Williams, 1932, *Proc. Hawaii. Entomol. Soc.* 8: 179 (note, host).

Nesiomiris hawaiiensis: Zimmerman, 1948b, *Insects of Hawaii* 3: 217, fig. 96 (in part). Swezey, 1954, *B.P. Bishop Mus. Spec. Publ.* 44: 50 (host; misidentification).

This species occurs in the same habitat as *N. hawaiiensis*, but it feeds on *Cheirodendron trigynum* rather than *Ilex anomala*, the host of *N. hawaiiensis*. Since specimens of both species were collected by Fullaway at that time without indication of a host, excepting "*Cheirodendron*" for 1 of the true *N. hawaiiensis* specimens, the confusion is understandable.

This species is most closely related to *N. swezeyi* of E Maui I. It may be differentiated from *N. swezeyi* by details of the male genitalia.

Description. MALE (n = 7). Body: 5.82 (5.41–5.87) long; 1.22 (1.10–1.28) wide. Head: 0.85 (0.81–0.85) wide; apex rounded, basal carina extending only to a line drawn through inner aspect of antennal fossa perpendicular to carina; callosities and vestiture as described for *N. curvatus*. Antennal segment lengths: I, 1.01 (0.99–1.06), color and vestiture as described for *N. curvatus*, longer than head width; II, 3.20 (2.99–3.30), light brown on basal 1/2 becoming black toward apex; III, 2.15 (1.85–2.31); IV, broken (0.90–1.95); III and IV blackish. Rostrum: 1.27 (1.19–1.35) long; reaching apex of mesosternum. Pronotum: 0.90 (0.78–0.94) long; 1.22 (1.10–1.28) wide; shape as illustrated for *N. swezeyi* (Fig. 33e); hind lobe pale green, remainder yellow; propleuron as described for *N. swezeyi*, vestiture consisting of a few recumbent brown setae at each humeral angle. Mesoscutellum and scutellum yellow; vestiture as described in *N. swezeyi*. Hemelytra: predominantly yellowish green on basal 1/2, greener on apical 1/2; areoles, membrane, and vestiture as described in *N. swezeyi*. Hind wing infuscate along 1A and anal fold. Legs: yellowish, tarsi and vestiture as described for *N. swezeyi*. Atrium, claspers, and aedeagus as illustrated (Fig. 41).

FEMALE (n = 7). Body: 6.16 (5.31–6.18) long; 1.29 (1.17–1.35) wide. Head: 0.85 (0.84–0.90) wide. Antennal lengths: I, 1.07 (0.90–1.07); II, 3.05 (2.90–3.20); III, 1.80 (1.13–2.28); IV, broken (0.90–1.07). Rostrum: 1.28 (1.23–1.35) long. Pronotum: 0.94 (0.90–1.00) long; 1.29 (1.17–1.35) wide. Differing from male as follows: rostrum reaching apex of mesosternum; propleural stripe indistinct, but discernable at medial constriction; hemelytra generally darker green.

NYMPHS. 1st and 2nd instars. Unknown. 3rd instar. (n = 1). Body: 2.10 long; 0.50 wide. Head: 0.53 wide; apex rounded; chaetotaxy: setae generally distributed around periphery of dorsum. Antennal segment lengths: I, 0.50, setae about equal in length to thickness of the segment; II, 1.41; III, 1.25; IV, 1.45; color and chaetotaxy as distributed around periphery of dorsum. Antennal lengths: I, 0.50, setae about equal in length to thickness of the segment; II, 1.41; III, 1.25; IV, 1.45; color and chaetotaxy as described for *N. trifurcatus* 3rd instar. Rostrum: 0.78 long; reaching base of abdomen. Pronotum: 0.30 long; 0.50 wide; anterior margin convex, posterior margin straight, lateral margin convex; lacking medial constriction; chaetotaxy consisting of a bristle at each humeral angle, but disc glabrous. Meso- and metanota with convex lateral and anterior margins, lacking anterolateral constriction, wing bud evident; chaetotaxy: humeral and lateral bristle, and 4 on disc forming a quadrangle; ratio of lengths of pronotum; mesonotum: metanotum about 1: 1: 0.5. Legs: as described for *N. trifurcatus* 3rd instar. Abdomen: a lateral and medial pair of bristle on all but last segment, last segment with a strong black bristle and several other shorter, paler bristles; shape and ventral chaetotaxy as described for *N. trifurcatus* 3rd instar. 4th instar (n = 6). Body: 2.31–2.99 long; 0.45–0.70 wide. Head: 0.60–0.70 wide. Antennal lengths: I, 0.50–0.71; II, 1.85–2.02; III, 1.55–1.60; IV, 1.35–1.95. Rostrum: 0.85–1.15 long; reaching mesocoxae. Pronotum: 0.25–0.35 long; 0.45–0.70 wide. Differing from 3rd instar as follows: setae shorter, sparser and less conspicuous, especially on abdomen; mesonotal wing bud now with a row of setae along the exterior edge, and a metanotal wing bud with a few setae near apex; abdomen lacking infuscate pattern. 5th instar (n = 8). Body: 3.00–3.90 long; 0.66–0.85 wide. Antennal lengths: I, 0.80–0.92; II, 2.40–2.60; III, 2.05–2.41; IV, 1.60–2.15. Rostrum: 1.20–1.41 long; reaching middle of mesosternum. Pro-

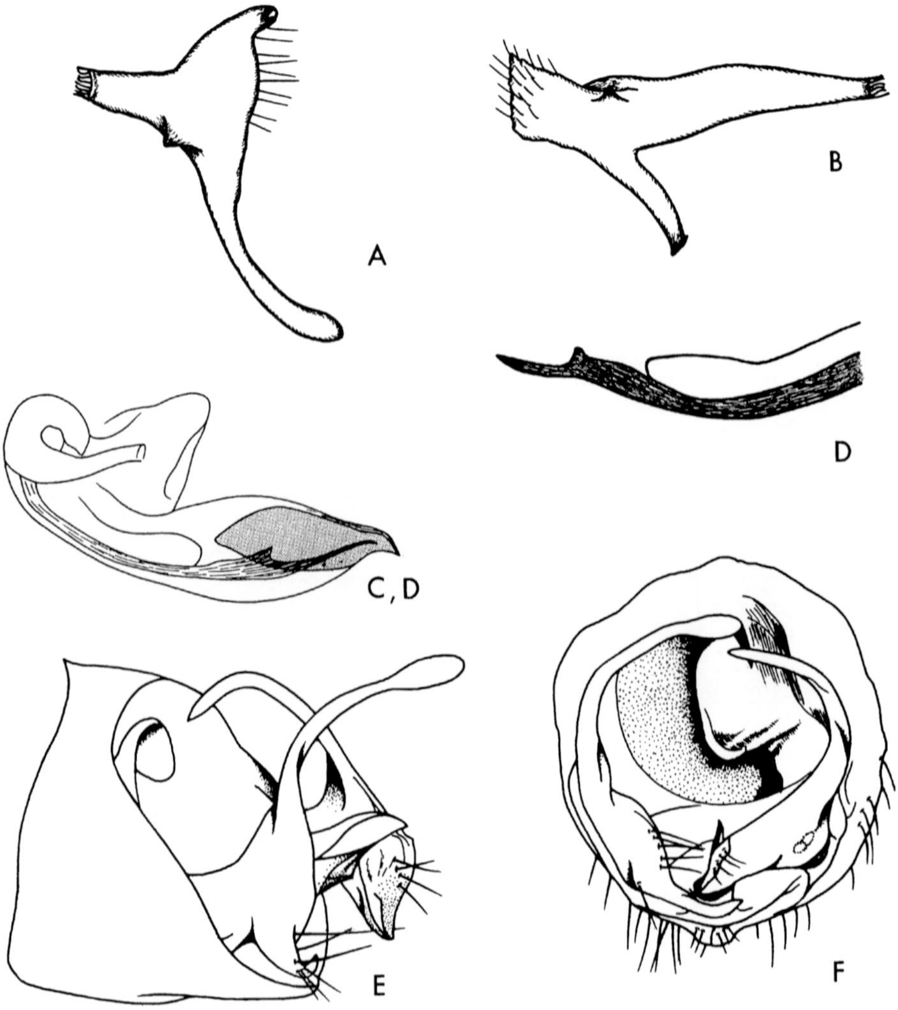

Fig. 41. *Nesiomiris williamsi,* n. sp., male genitalia. **a**, left clasper, internal; **b**, right clasper, internal; **c, d**, aedeagus and spiculum, lateral; **e**, pygofer, left lateral; **f**, pygofer, terminal.

notum: 0.41–0.55 long; 0.66–0.85 wide. Differing from 4th instar as follows: apex of head straight; shorter and sparser, approximately that of the adult; wing buds reaching at least 4th abdominal segment; in more mature specimens the incipient wing membrane is noticeably infuscate; pronotum generally shaped as in adult, but the medial constriction is only weakly developed; last tarsal segment dark.

Distribution: Hawai'i I.

Host. *Cheirodendron trigynum* vars. *trigynum* and *acuminatum*.

Etymology. Named for Francis X. Williams (1882–1967), a leading entomologist whose work on the systematics and biology of higher Hymenoptera and the biology of aquatic and semiaquatic Hawaiian insects, among others, are outstanding contributions. He and O.H. Swezey were among the 1st to collect this species and to recognize it as distinct from *N. hawaiiensis*.

Type data. Holotype ♂ (BPBM 14,832), allotype ♀, paratypes 2♂, 11♀, **HAWAI'I**: Hilo Forest Reserve, Laupāhoehoe, 4200 ft. [1280 m], 25.II.1971, on *Cheirodendron trigynum*, W.C. Gagné. Additional paratypes with data differing only as follows: 1♂, 1♀, Kīlauea, VII.1906, R.C.L. Perkins; 1♂, 2♀, 29 mi on 'Ōla'a Highway, 2300 ft. [702 m], 12.IX.1919, D.T. Fullaway, Plesiotype *N. hawaiiensis*; 2♂, 1♀, Nauhi Gulch, 5000–8500 ft. [1525–2593 m], 29.IX.1931, on *Cheirodendron*, O.H. Swezey & F.X. Williams; 1♂, 3♀, Hawai'i Volcanoes Nat. Park, Kīpuka Puaulu, 4000 ft. [1220 m], 24.XI.1968, on *C. trigynum* var. *trigynum*; 1♂, 5♀, 'Ōla'a Forest Reserve, Wright Rd, 28.XI.1968, on *C. trigynum* var. *trigynum*; 5♂, 4♀, Kīpuka #7, Saddle Rd, 18.4 mi W of Hilo, 30.XI.1968, on *C. trigynum* var. *trigynum*; 8♂, 8♀, Kipuka #8, Saddle Rd, 19 mi W of Hilo, [no elevation], 30.XI.1968, on *C. trigynum* var. *trigynum*; 1♂, 1♀, Kipuka #9, Saddle Rd, 19 mi W of Hilo, [no elevation], 30.XI.1968, on *C. trigynum* var. *trigynum*; 3♂, 3♀, Kipuka [no #], Saddle Rd, 19 mi W of Hilo, [no elevation], 30.XI.1968, on *C. trigynum* var. *trigynum*; 1♀, Kipuka #6, Saddle Rd, [no elevation], 1.XII.1968, on *C. trigynum* var. *trigynum*; 1♂, 2♀, Kipuka #7, Saddle Rd, 18.5 mi W of Hilo, 4900 ft. [1493 m], 1.XII.1968, on *C. trigynum* var. *trigynum*; 3♂, 5N, Kipuka #8, Saddle Rd, 19 mi W of Hilo, 5020 ft. [1530 m], 1.XII.1968, on *C. trigynum* var. *trigynum*; 1♂, Kipuka [no #], Saddle Rd, 15.5 mi W of Hilo, [no elevation], 21.VII.1970, on *C. trigynum* var. *trigynum*; 1♀, Waikaloa Stream above Waimea, 3200 ft. [975 m], 2.XII.1968, on *C. trigynum* var. *trigynum*; 2♂, 5♀, 4N, Kīlauea Forest Reserve, 5400 ft. [1547 m], 23.VII.1970, on *C. trigynum* var. *trigynum*; 2♂, 14♀, 8N, Kīlauea Forest Reserve, 5400 ft. [1547 m], 15.IX.1970, *C. trigynum* var. *trigynum*; 2♂, Kīpuka Puaulu, [no elevation], 7.I.971, on *C. trigynum* var. *trigynum*; 2♂, 3♀, Kohala Mts, Upper Hāmākua Ditch Trail, 24.I.1971; 4♀, Kīlauea Forest Reserve, 1650 m, 26.X.1971; 1♂, I.B.P. Study Site, Kīlauea Forest Reserve, 1586 m, 15.XII.1970, Malaise trap #2, F. Bianchi & J.L. Gressitt; 2♂, I.B.P. Study Site, Kīlauea Forest Reserve, 1586 m, 21–28.XII.1970, Malaise trap #2, J.L. Gressitt & W.C. Gagné; 1♂, Kīlauea Forest Reserve, I.B.P. Study Site, 1586 m, 14–21.VI.1971, Malaise trap #2, L. Goff & W.A. Steffan; 2♂, 1♀, Kīlauea Forest Reserve, I.B.P. Study Site, 1586 m, 21–28.VI.1971, Malaise trap #2, L. Goff & W.A. Steffan; 1♀, Kīlauea Forest Reserve, I.B.P. Study Site, 1586 m, 19.VII.1971 Malaise trap; 2♂, 1N, Kīlauea Forest Reserve, I.B.P. Study Site, 1586 m, 7–14.VI.1971, Malaise trap #2, L. Goff & W.A. Steffan; 1♂, 1♀, Kīlauea Forest Reserve, I.B.P. Study Site, 1586 m, 10–17.VI.1972, Malaise trap #6, W.A. Steffan; 1♂, Kīlauea Forest Reserve, I.B.P. Study Site, 1586 m, 30.VIII–6.IX.1971, Malaise trap; 1♀, Kīlauea Forest Reserve, I.B.P. Study Site, 1586 m, 18–25.IX.1972, Malaise trap (BPBM).

Variation. There is considerable variation in the upper, outer arm of the right clasper of the male. This variation is not correlated with distribution as dissected in series, for example, from the *kīpuka* (vegetated "islands" surrounded by more recent lava flows) along the Saddle Road. But for the other relatively constant morphology of this clasper, the Hawai'i I population might be more logically treated as Hawai'i I populations of *N.*

swezeyi. The upper arm of the right clasper forms an obtuse, rather than a right angle with the base of the clasper.

The pronotum varies from green on the posterior lobe to entirely yellow. The mesoscutellum and scutellum are yellow. The apex of the head is sometimes weakly striate dorsally.

Discussion. I have examined the specimen designated as "Plesiotype, *N. hawaiiensis*" (in Dr. R.L. Usinger's handwriting) for Zimmerman's (1948b) *Insects of Hawaii* series, illustrated on page 217 (fig. 96). I find that it is a new species rather than *N. hawaiiensis* Kirkaldy. This specimen has the following data: "29 miles/Olaa Haw./2300 ft/D.T. Fullaway/12.IX.19 Plesiotype/*N. hawaiiensis*" (latter penciled on blue label).

Behavioral observations. Several males and females were observed on the underside of leaves of *Cheirodendron trigynum* in the Kīlauea Forest Reserve from (0900h–1000h) on 21 July 1973, in clear, calm conditions. The tree, about 3 m in height, was in the shaded understory of an *Acacia koa* forest. The males were observed to "chase" one another on a single leaf and then stop and "thump" the abdomen rapidly against the substrate for less than a second. Sometimes another male apparently would respond by similarly "thumping" the abdomen. At other times there would no contact or "chasing" between males, but the thumping would be imitated. The females on the same leaf did not exhibit this behavior. Resting nymphs in several instars also appeared oblivious, as did several feeding adults and nymphs. Others were cleaning their antennae.

The *reductus* Complex

Description. **ADULTS**. Apex of head rounded; right clasper with all arms reduced; left clasper evenly hooked, not sinuate or S-shaped; aedeagus with a flange or thickened process on the left apex.

Distribution. W Maui.

Host. *Cheirodendron trigynum.*

Included species. No. 8

8. *Nesiomiris reductus* Gagné, **new species** Fig. 42

This species is very close to *sinuatus* in all morphological respects except the male genitalia. The upper, outer arm of the right clasper (Fig. 42b) is much reduced, the left clasper (Fig. 42a) is evenly hooked rather than sinuate, and the apex of the aedeagus (Fig. 42c, c–1) has a broad flange, or keel-like process. Since several ♂ specimens have been found to bear this identical combination of genital characters, plus others outlined below, I conclude that I am dealing with a species, and not merely a mutant or morph of the sympatric *N. sinuatus*.

Until a pair is collected *in copula*, I hesitate to associate the sexes, therefore, ♀ paratypes have not been designated.

***Description*.** **MALE** (n = 4). Body: 6.13 (6.06–6.37) long; 1.33 (1.28–1.33) wide. Head: 0.85 (0.84–0.89) wide; apex rounded; yellowish; chaetotaxy as for *N. swezeyi*; callosities weakly shining, minutely rugose. Antennal lengths: I, 0.94 (0.94–1.00), yellow, longer than width of head, vestiture as in *N. sinuatus*; II, 3.24 (2.86–3.25), yellowish on basal 3/4 becoming black on apical 1/2; III, 2.22 (2.22–2.31); IV, 2.10 (1.07–2.10); III and IV blackish. Rostrum: 1.21 (1.18–1.28) long, reaching middle of mesosternum. Pronotum: 1.00 (1.00–1.09) long; 1.33 (1.28–1.33) wide; shape close to *N. sinuatus*, yellowish; propleural stripe, propleuron and chaetotaxy as in *N. swezeyi*. Both mesoscutellum and scutellum greenish, disc of hemelytron greenish yellow, vestiture as for *N. sinuatus*. Areoles coriaceous and as illustrated for *N. swezeyi*; membrane weakly infuscate. Legs: as in *N. sinuatus*. Abdomen: genital atrium, claspers and aedeagus as illustrated (Fig. 42).

FEMALE. Unknown.

NYMPHS. Unknown.

***Distribution*.** W Maui.

***Host*.** *Cheirodendron trigynum* var. *confertiflorum*.

***Etymology*.** Named for the reduced state of the right clasper of the male.

***Type data*.** Holotype ♂ (BPBM 14,833), **W MAUI**: Kaulalewelewe-Pu‘u Kukui Trail, 1434 m, 28.V.1972. on *Cheirodendron trigynum*, W.C. Gagné. Paratypes: 3♂, same data as holotype, except as follows: 915–1037 m, on *C. trigynum* var. *confertiflorum* (BPBM).

***Other material examined*.** A single 5th-instar nymph, with a prominent propleural stripe, collected with a series of *N. sinuatus* adults and nymphs is provisionally placed here. Data differs from the holotype of *N. reductus* as follows: 914 m, 8.VIII.1968. It is not considered a paratype.

***Discussion*.** The similar aedeagi, at least 1 unreduced arm of the right clasper, and the flange or process on the apex of the aedeagus resemble a frequently seen morphology in the *elongatus* subgroup where *N. reductus* might be placed, but for its greater morphological affinities with the *sinuatus* complex.

The *elongatus* Subgroup

***Description*.** **ADULTS**. Apex of head bifid to rounded; vestiture on dorsum reclined; antennal segment I and apices of hind femora sometimes rufescent, usually yellowish brown; in green species, hind lobe of pronotum and sometimes collar, green in fresh specimens, and propleural stripe variously expressed (see complexes); pygofer unarmed; upper arm of left clasper long and apically hooked; arms of right clasper much reduced, at most with 2 blunt arms; upper arm of left clasper with or without prominent teeth; female with outer lobe of *K* structure usually longer than inner lobe (Fig. 25g-n); meso-dorsal evagination of *K* structure usually minutely spiculate.

NYMPHS. Propleural stripe present or absent; apices of hind femora usually pallid.

***Distribution*.** Kaua‘i I to Hawai‘i I.

***Host*.** *Cheirodendron trigynum*, *Munroidendron racemosum*, and various species of *Reynoldsia* and *Tetraplasandra*.

***Included species*.** Nos. 9–18.

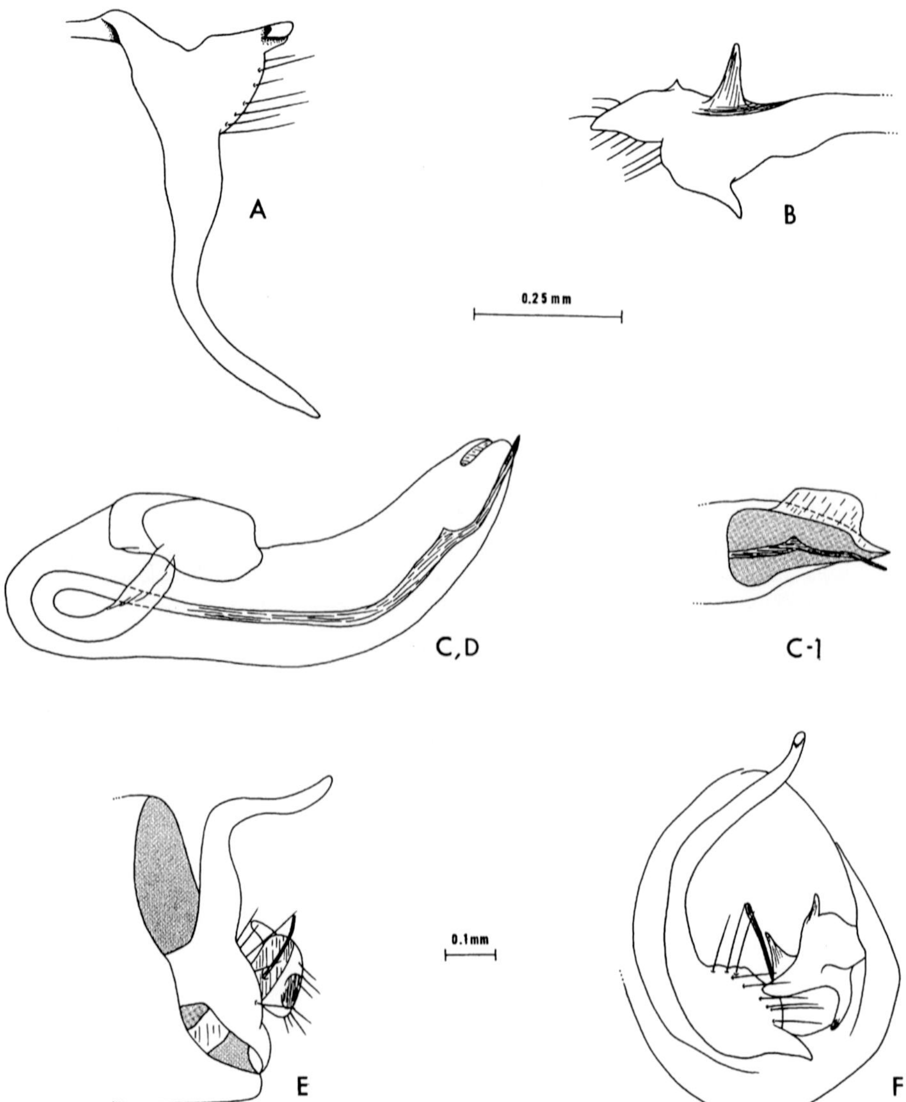

Fig. 42. *Nesiomiris reductus,* n. sp., male genitalia. **a**, left clasper, internal; **b**, right clasper, internal; **c, d**, aedeagus and spiculum, lateral; **c-1**, apex of aedeagus and spiculum; **e**, pygofer, left lateral; **f**, pygofer, terminal.

The *elongatus* Complex

Description. **ADULTS**. Greenish species; apex of head bifid; propleural stripe present, extending from collar to mesal constriction; membrane of forewing infumate; hemelytra and posteri-

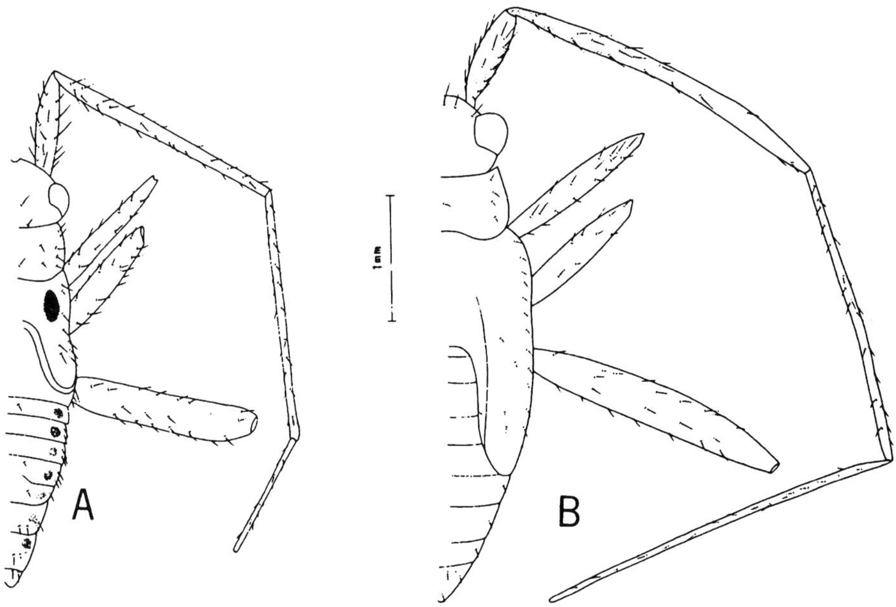

Fig. 43. *Nesiomiris elongatus,* n. sp., nymphs, dorsal view. **a**, 4th instar; **b**, 5th instar. Maculate areas on abdomen are green.

or lobe of pronotum greenish; hind femora of adults weakly incrassate; vestiture of legs and antennal segment I not long and dense; apex and atrium of aedeagus not twisted to face left claspers; apex of aedeagus not keeled; right clasper 2 armed, consisting of an outwardly facing apical arm and an inwardly facing fingerlike inner arm.

NYMPHS. Apices of antennal segment I, femora and bases of tibiae sometimes reddish (Fig. 43); propleural stripe usually present.

Distribution. Kaua'i I.
Host. *Tetraplasandra* spp. (section *Pterotropia*), *Munroidendron racemosum.*
Included species. No. 9.

9. ***Nesiomiris elongatus*** Gagné, **new species** Figs. 25i, 30j, 33g, 43, 44

Swezey (1954: 177) recorded a "large green mirid . . . on *Pterotropia* at Halemanu, Kaua'i". This species possesses many morphological characteristics that I consider plesiomorphic: the large, elongate narrow body, the often reddish apices of the hind femora of the adults and nymphs, the toothed apex of the aedeagus, the long flabellum on the spiculum, the bifid apex of the head, the partially coriaceous areoles, and the long, sinuous, unreduced left clasper.

The general conformation of the male genitalia link *N. elongatus* with *N. fasciatus* and *N. maculatus*, green spotted species of the *lihuensis* group, which also attack *Tetraplasandra kavaiensis*.

The hind legs are weakly incrassate. This feature is particularly striking in the *curvipes* complex, and since the general conformation of the male genitalia is similar, the *curvipes* complex appears to have evolved from an *elongatus*-like ancestor and is here placed as a sister group of the *elongatus* complex (see Fig. 96).

***Description*.** **MALE** (n = 7). Body: 6.86 (6.81–7.21) long; 1.25 (1.25–1.45) wide. Head: (Fig. 33g) 1.05 (1.01–1.07) wide; apex bifid; shining; basal carina as illustrated (Fig. 33g); yellowish green. Antennal segment lengths: I, 1.05 (1.05–1.10); II, 3.40 (3.27–3.55); III, 2.66 (2.49–2.66); IV, 1.78 (1.78–1.95); setae and color as described for *N. ohe*; I, subequal to head width. Rostrum: 1.45 (1.45–1.54) long; reaching middle of mesosternum. Pronotum: 1.19 (1.09–1.19) long; 1.25 (1.25–1.45) wide; shining; subglabrous, the few setae pale, inconspicuous; propleural stripe extending from collar to mesal constriction but not beyond; apex of propleuron anterior and posterior to coxal cleft, infuscate; most of pronotum, mesoscutellum and scutellum faded to yellowish; coriaceous parts of hemelytra greenish, the veins more so, fading to yellow anterior to apex of scutellum; vestiture, areoles, and membrane as described for *N. ohe*; hind legs weakly incrassate; vestiture on femora, longer and denser than *N. ohe*, those on posterior aspects of hind femora erect, its apex and base tibia weakly reddish annulate. Abdomen, atrium of pygofer unarmed; clasper and aedeagus as illustrated (Fig. 44).

FEMALE (n = 7). Body: 6.91 (6.90–7.35) long; 1.42 (1.33–1.55) wide. Head 1.05 (1.01–1.05) wide. Antennal lengths: I, 1.05 (1.05–1.10); II, 3.15 (3.00–3.27); III, 2.35 (2.31–2.85); IV, 1.78 (1.78–2.12). Rostrum: 1.45 (1.35–1.54) long. Pronotum: 1.18 (1.09–1.20) long; 1.42 (1.33–1.55) wide. Differing from male as follows: more greenish, i.e., is less faded on head, antennal segment I and basal 1/2 of segment II, hind lobe of pronotum, and in spots along middle of scutellum; antennal segment I about equal in length to head width; internal genitalia as illustrated (Figs. 25i, 30j).

NYMPHS. *1st and 2nd instars*. Unknown. *3rd instar* (n = 6). Body: 2.05–2.45 long; 0.55–0.70 wide. Head: 0.45–0.66 wide. Antennal lengths: I, 0.35–0.65; II, 1.06–1.45; III, 1.00–1.25; IV, 1.00–1.75. Rostrum: 0.90–1.05 long. Pronotum: 0.25–0.35 long; 0.35–0.56 wide. Differing from same instar of *N. ohe* as follows: apex of head indented; setae on body distinct, long, erect, posteriorly directed on terminal abdominal segments, brown, mostly set on small, brown papillae, especially those along lateral margins, distributed as in *N. curvipes*; those on appendages pale, more recumbent; propleural stripe usually present; lateral margin of mesonotum straight. *4th instar* (n = 10). (Fig. 43a). Body 2.35–3.20 long; 0.67–0.77 wide. Head: 0.65–0.91 wide. Antennal lengths: I, 0.61–0.75; II, 1.95–2.17; III, 1.68–2.00; IV, 1.68–2.00. Rostrum: 1.00–1.25 long. Pronotum: 0.30–0.43 long; 0.50–0.71 wide. Differing from 3rd instar as follows: head subglabrous; pronotum, and those on mesonotal wing bud only, on papillae; apices of pro- and metafemur and base of tibia reddish brown; metafemur also reddish on apical 1/2, posteriorly. *5th instar* (n = 10). (Fig. 43b). Body: 3.35–4.79 long; 0.91–1.05 wide. Head: 0.95–1.06 wide. Antennal lengths: I, 0.88–1.10; II, 2.60–3.22; III, 2.35–2.85; IV, 1.25–2.15. Rostrum: 1.41–1.41 long. Pronotum: 0.60–0.79 long; 0.91–1.05 wide. Differing from 4th instar as follows: propleural stripe not reaching posterior border; pronotum infuscate on posterior border between humeral angle and mesal cavity, a row of setae on anterior and lateral border only; mesonotal wing bud infuscate just posterior to humeral angle; stout setae present mostly along lateral border with few on disc, forming a line through apex of

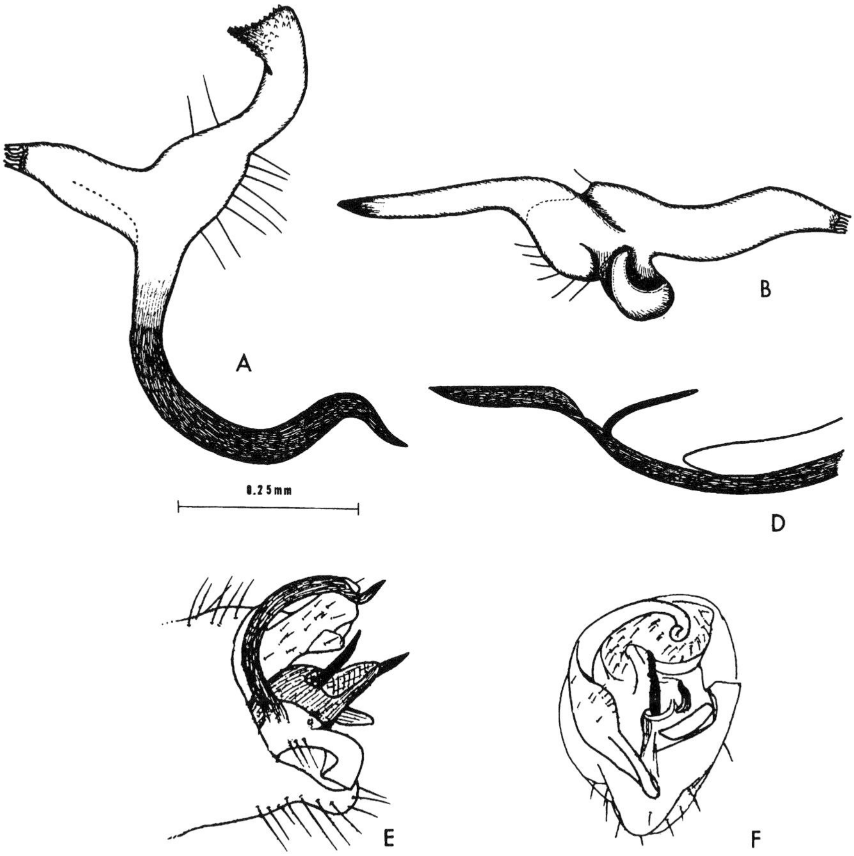

Fig. 44. *Nesiomiris elongatus,* n. sp., male genitalia. **a**, left clasper, internal; **b**, right clasper, internal; **d**, spiculum, lateral; **e**, pygofer, left lateral; **f**, pygofer, terminal.

metanotum; metanotal wing bud similarly infuscate; fore- and mid-legs only weakly infuscate; rostrum reaching at most to mesocoxae, but usually only to middle of mesosternum.

Distribution. Kaua'i I.

Host. *Munroidendron racemosum*, *Tetraplasandra kavaiensis* (= *Pterotropia* sp.).

Etymology. Named for the comparatively elongate, narrow body.

Type data. Holotype ♂ (BPBM 14,834), allotype ♀, paratypes 11♂, 15♀, 33N, **KAUA'I**: Kōke'e, 15–20.XI.1968, on *Tetraplasandra kavaiensis*, W.C. Gagné. Additional paratypes with data differing only as follows: 1♂, 2♀, 4N, 25.VIII.1970; 3♂, Halemanu, 29.VI.1932, on *Pterotropia*, O.H. Swezey; 1♀, 22–25.VII.1968, at light; 13♂, 22♀, 8N, Nā Pali Coast, Kalalau Trail, 18.XII.1968, on *Munroidendron racemosum*; 7♂, 9♀, 10N, Pole Line Trail near Kualapa, 2100 ft. [641 m], 4.IX.1970; 2N, Wai'ale'ale Trail, 3500 ft. [1103 m], 25.VIII.1970 (BPBM).

Variation. There was a varying degree of expression of red at the apex and base of the hind femora and tibia, respectively. It varied from deep red to absent, supplanted then with the normal greenish yellow.

The specimens from low elevations (about 200 m) on *Munroidendron racemosum* along the Nā Pali coast, particularly the males, are distinctly shorter than those at higher elevations on *Tetraplasandra kavaiensis*, but the 2 populations appear to be identical in all other morphological details.

Discussion. This species, like *N. fasciatus* and *N. maculatus* of Kaua'i and *N. ohe* of O'ahu, feeds on *Tetraplasandra*, predominantly of section *Pterotropia* (*sensu* Sherff 1955). *Tetraplasandra kavaiensis* and *T. gymnocarpa* have a stellate-tomentose lower leaf surface, a substrate to which these bugs show apparent affinity and specificity. Another host, *Munroidendron racemosum* (formerly included in *Tetraplasandra*) has the lower leaf surface densely white-tomentose. Unfortunately, there has been no generic study along evolutionary lines of *Tetraplasandra* and its relatives to determine what is primitive or advanced; therefore, a more comprehensive analysis of possible coevolution of bugs and hosts cannot be attempted.

Although *N. fasciatus* and *N. maculatus* have been found on *Tetraplasandra kavaiensis*, they have not been taken with *N. elongatus*, so allopatry may be involved here. *Nesiomiris elongatus* is the only species that has been found to attack *Munroidendron racemosum*. This ecology parallels that of species attacking summer-deciduous *Reynoldsia* on other islands.

The *curvipes* Complex

Description. **ADULTS**. Pale yellow species with faint greenish areas on dorsum; apex of head rounded, propleural stripe, when present, represented by a spot at mesal constriction; membrane of forewing not infumate; hind femora of adults and late instar nymphs incrassate; vestiture long, dense on legs and antennal segment I, subequal to greatest diameter of associated segments; right clasper and aedeagus as described for *elongatus* complex.

NYMPHS. Legs and apices of antennal segment I pallid yellowish; propleural stripe absent.

Distribution. Maui, Moloka'i, and Hawai'i Is.

Host. *Tetraplasandra hawaiiensis* and its varieties (i.e., Section *Eutetraplasandra*).

Included species. Nos. 10–11.

Discussion. The hind legs of *N. elongatus* of Kaua'i are weakly incrassate. The general conformation of the claspers is quite similar.

10. *Nesiomiris curvipes* Gagné, **new species** Figs. 25n, 45c, 46b, 47

Nesiomiris hawaiiensis: Swezey & Bryan, 1929, *Proc. Hawaii. Entomol. Soc.* 7(2): 303, 311 (host; misidentification, also included *N. secularis*). Swezey, 1954, *B.P. Bishop Mus. Spec. Publ.* 44: 317 (host, misidentification).

The incrassate hind femora in both adults and late instar nymphs, and the pallid body coloration with only faint traces of green in the living specimens at once distinguish this obligate species on *Tetraplasandra hawaiiensis*. It is also closest to *N. pallasatus*, of Hawai'i I, that is on the same host and shares the strongly incrassate femora as a character and from which it may be distinguished by the lack of infuscations at the bases of the hemelytral setae. In the late instar nymphs, each dorsal seta is not set on a raised papilla. It may be distinguished from the superficially similar *N. pallidus* as noted under that species.

Description. **MALE** (n = 7). Body: 6.80 (6.50–7.03) long; 1.56 (1.50–1.57) wide; mostly pallid with faint greenish areas on pronotum, scutellum, and hemelytra; pubescence on dorsum short, brown, and reclined. Head: (Fig. 45c) 0.99 (0.95–1.00) wide; apex rounded, striolate; subglabrous; tylus shining, yellowish; basal carina extending in a concave arc almost from eye to eye; pallid, shallowly impressed along middle. Antennal lengths: I, 1.27 (0.90–1.27); II, 3.21 (2.61–3.21); III, 1.70 (1.31–1.71); IV, 0.78 (0.78–1.04); I-III, pallid, excepting reddish brown apex of II and III; IV, brownish; pubescence on I and basal 1/2 of II pale, inclined, remainder as described for *N. pallidus*; I, much longer than head width (ratio 9.5: 7.75). Rostrum: 1.54 (1.54–1.66) long; reaching middle of mesosternum. Pronotum: 1.14 (1.05–1.14) long; 1.56 (1.50–1.57) wide; callosities weakly shining, remainder dull; collar partly faint greenish; hind lobe partially faint greenish, shallowly rugose-punctate; propleural stripe evident only as a spot at the mesal constriction; a few sparse, short, pale setae present. Legs: pallid; pubescence pale, erect, hirsute on femora, about 1/4 the thickness of femur, but more inclined on tibiae; femora incrassate (Fig. 46b); infuscate at inner apex. Mesonotum and disc of scutellum yellowish, apex of latter greenish. Hemelytra: pubescence of short, recumbent brown setae mostly along veins, those on cuneus pale; faint greenish areas on basal 1/2 between veins; membrane infumate only on opposite anal ridge, remainder creamy; areolar veins pallid; areoles narrowly coriaceous along inner border. Abdomen: pallid; pygofer unarmed; claspers and aedeagus as illustrated (Fig. 47).

FEMALE (n = 7). Body: 6.60 (6.45–7.38) long; 1.45 (1.45–1.55) wide. Head: 0.95 (0.95–1.01) wide. Antennal lengths, I, 1.10 (1.10–1.21); II, 2.87 (2.69–3.10); III, 1.60 (1.25–1.60); IV, 0.89 (0.81–0.91). Rostrum: 1.60 (1.35–1.64) long. Pronotum: 1.07 (1.07–1.11) long; 1.45 (1.45–1.55) wide. Differing from male as follows: dorsum entirely pallid; antennal segment I subequal in length to head; rostrum reaching apex of mesosternum. Bursa copulatrix as illustrated (Fig. 25n).

NYMPHS. *1st and 2nd instars*. Unknown. *3rd instar* (n = 1). Body: 1.67 long; 0.75 wide. Head: 0.61 wide; base and frons with erect, stout, brown setae, remainder glabrous. Antennal segment lengths: I, 0.56; II, 0.38; III, 0.45; IV, 0.45; I and II with setae as described for dorsum, interspersed with inclined, pale, finer setae on II; III with shorter, more inclined and denser setae, those on IV still shorter and recumbent; apices of II-IV reddish, I totally pallid. Rostrum: 1.06 long, reaching middle of abdomen. Pronotum: 0.35 long; 0.65 wide; 6 ill-defined longitudinal rows of stout setae, the row of setae along lateral edge comprised of 4 setae, 2 inner rows on disc of 3 setae, 1st seta set back from anterior margin, and the other 2 rows astride the ecdysial line set back from both anterior and posterior margins; propleural stripe absent. Mesonotum with 2 ill-defined horizontal rows of stout setae and a row along its lateral border. Metanotum and abdominal tergites each with 1 horizontal row of 6 stout setae that are nearest the posterior border, those on apical segments much shorter; 1st abdominal tergite without a lateral set, only 4 central setae. Legs: pallid; femora and tibiae with stout setae, those of tibiae in about 4 ill-defined rows on outer aspects; hind femora not incrassate but slightly thickest setae on inner aspects shorter, paler and more reclined, resembling those on tarsi. *4th instar* (n = 8). Body: 2.25–3.26 long; 0.75–1.06 wide. Head: 0.65–0.80 wide,

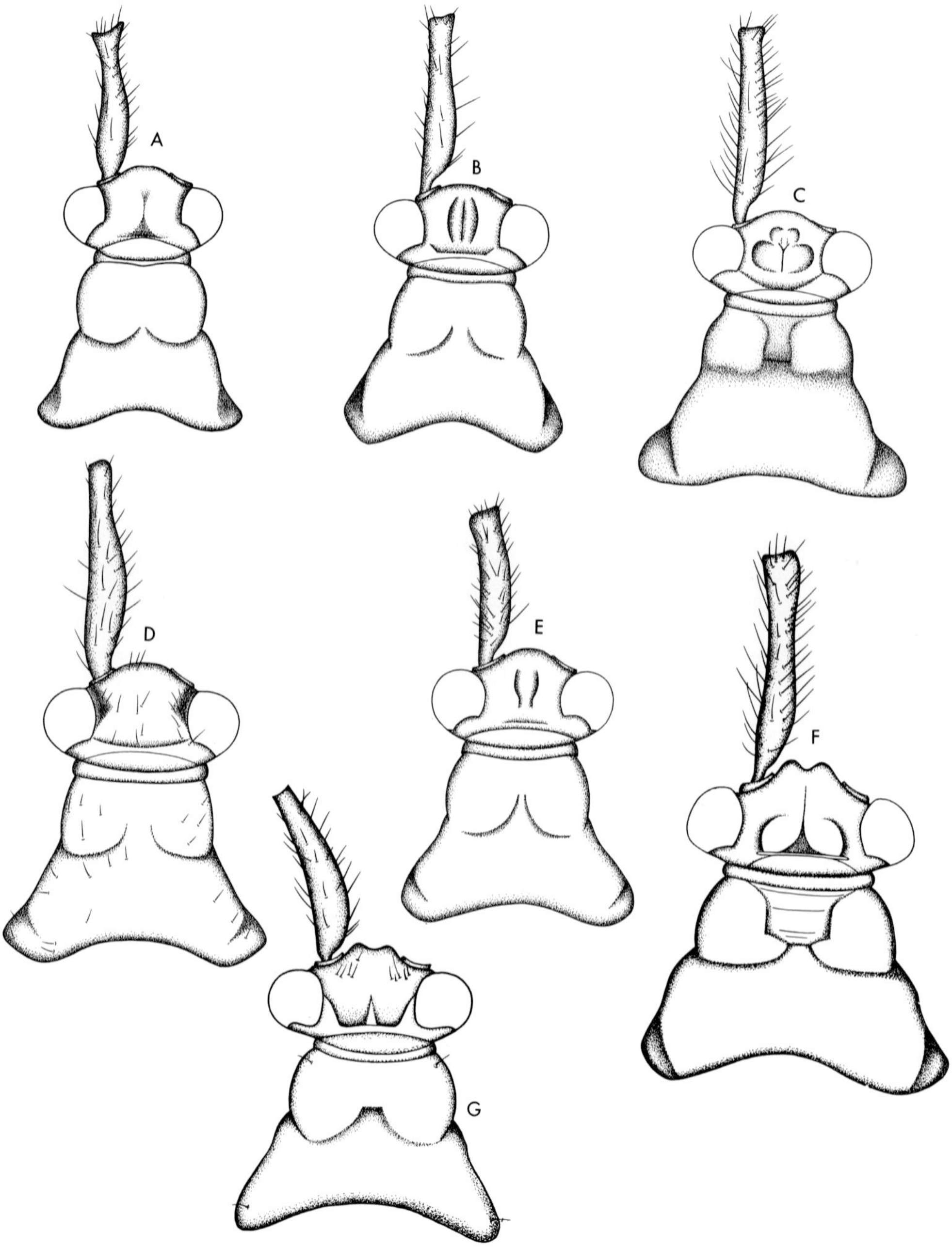

Fig. 45. *Nesiomiris* n. spp., antennal segment I, head, and pronotum, dorsal view. **a**, *.timberlakei* (O'ahu population); **b**, *timberlakei* (Hawai'i I population); **c**, *curvipes;* **d**, *trinesiotes;* **e**, *pallidus;* **f**, *planatus;* **g**, *tetraplasandra.* Drawn to same scale.

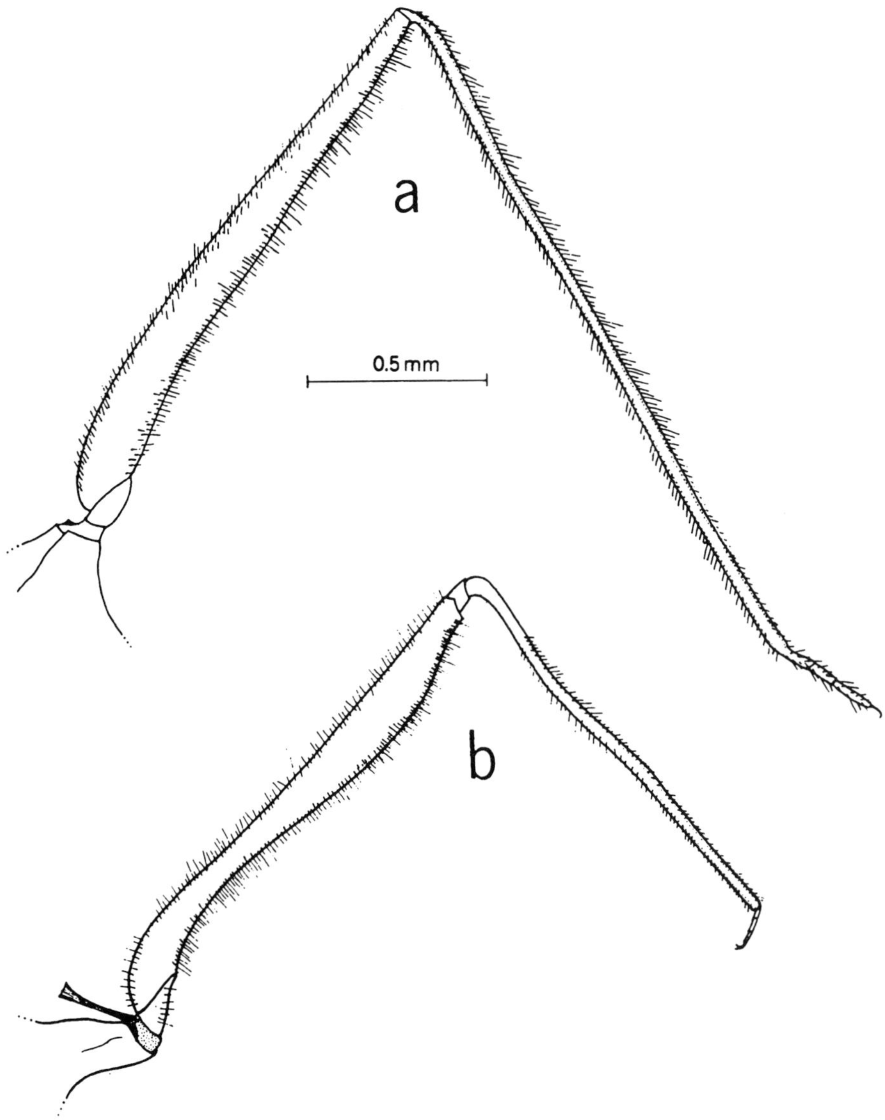

Fig. 46. *Nesiomiris* hind legs, dorsal. **a**, *hawaiiensis* Kirkaldy; **b**, *curvipes*, n. sp.

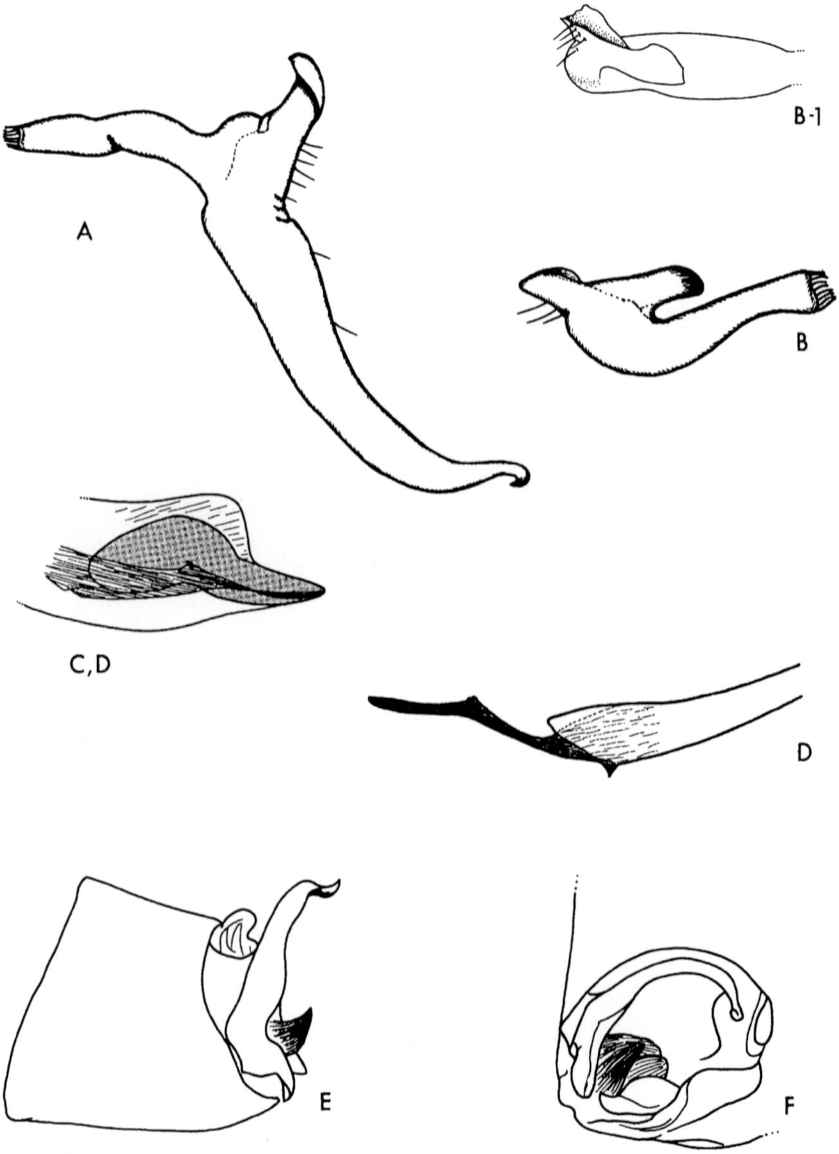

Fig. 47. *Nesiomiris curvipes,* n. sp., male genitalia. **a,** left clasper, internal; **b,** right clasper, dorsal; **b-1**, right clasper, internal; **c, d**, aedeagus and spiculum, lateral; **d**, spiculum, lateral; **e**, pygofer, left lateral; **f**, pygofer, terminal.

apex striolate. Antennal lengths: I, 0.45–0.60; II, 0.90–1.25; III, 0.60–0.91; IV, 0.60–0.71. Rostrum: 1.21–1.39 long, just surpassing hind coxae. Pronotum: 0.35–0.50 long; 0.67–0.81 wide; dorsal setae not on raised papillae. Differing from 3rd instar as follows: mesonotal wing buds with a row of stout, lateral setae and others generally distributed over disc; hind femora distinctly thicker than other femora; stout setae on apical 1/2 of hind femora. *5th instar* (n = 8). Body: 3.37–4.65 long; 1.25–1.41 wide. Head: 0.85–1.06 wide. Antennal lengths: I, 0.80–0.91; II, 1.68–2.38; III, 1.08–1.42; IV, 0.70–0.95. Rostrum: 1.42–1.50 long. Pronotum: 0.45–0.67 long; 1.00–1.07 wide. Differing from 4th instar as follows: pubescence strikingly shorter, inconspicuous, paler and inclined on dorsum and appendages; pronotum with only 1 row of stouter setae across base, plus a few paler ones laterally; hind legs incrassate; setae in rows on abdominal tergites becoming stouter and darker apically.

Distribution. Moloka‘i and W Maui Is.

Host. *Tetraplasandra hawaiiensis* and its var. *gracilis*, *T*. sp. (prob. *hawaiiensis*).

Etymology. Named for the curvate or incrassate hind legs.

Type data. Holotype ♂ (BPBM 14,835), allotype ♀, paratypes 4♂, 2♀, 9N, **W MAUI**: Makamaka‘ole Stream, 420 m, 11.IV.1971, on *Tetraplasandra hawaiiensis* var. *gracilis*, W.C. Gagnè. Additional paratypes with data differing only as follows: 1♂, 2♀, 5N, 600 m, 1.IV.1970, on *T. hawaiiensis* var. *gracilis*; 4♂, 1♀, 420 m, 2.II.1970, on *T. hawaiiensis* var. *gracilis*; 2♂, 6♀, 31.III.1970; 13♂, 17♀, Kaulalewelewe, 915 m, 27.V.1972 on *T. hawaiiensis*; **MOLOKA‘I I**: 20♂, 17♀, 2N, Kainalu Gulch, 1600 ft. [488 m], 23.VII.1927, on *Tetraplasandra*, E.H. Bryan, Jr.; 6♂, 12♀, 4N, Waihānau Diversion Dam, 2200 ft. [672 m], 20.I.1973, on *T. hawaiiensis*, S.L. Montgomery; 9♂, 2♀, Kainalu Gulch, 400 m, 28.I.1973, on *T. hawaiiensis*; 2♂, 2♀, Pia Gulch, 580 m, 29.I.1973 (BPBM).

Variation. All adults of the Moloka‘i population examined lack the propleural spot, whereas most adults of the Maui population have the propleural spot at the mesal constriction. Adult specimens in alcohol have the apices of antennal segments II and III reddish; the reddish appears to have faded to brownish in the dry mounted specimens.

Discussion. Swezey and Bryan (1929) recorded this species as *Nesiomiris hawaiiensis* on "*Tetraplasandra*" on Moloka‘i. I have examined their specimens and find them to be *N. curvipes*. It is puzzling that this species has not been found on E Maui. In some ways it parallels the distribution of *N. trinesiotes*, which occurs on Moloka‘i, Lāna‘i, and W Maui; however, the latter is not host-specific to *Tetraplasandra hawaiiensis*, which may be due to recent extinction, since the population of host plants seems to have been drastically reduced by human activities within the last 200 years.

11. ***Nesiomiris pallasatus*** Gagné, **new species** Figs. 31g, 48

Externally, the adults of this species are readily differentiated from the closely related *N. curvipes* by the presence of spots at the bases of the setae on the hemelytra, and in the later-instar nymphs the dorsal abdominal setae are each on a raised papilla. The males differ in particulars of the genitalia; *N. pallasatus*, lacks any indication of a flabellum near the apex of the spiculum (Fig. 48c, d), and the spiculum is comparatively straight and thin. In most other respects, the 2 species appear to be very similar.

Description. **MALE** (n = 7). Body: 5.78 (5.57–7.11) long; 1.30 (1.22–1.55) wide. Head: 0.94 (0.89–0.94) wide. Antennal segment lengths: I, 1.05 (0.90–1.19); II, 2.76 (2.49–2.79); III, 1.55 (1.43–1.78); IV, 1.02 (1.02–1.25). Rostrum: 1.43 (1.43–1.55) long. Pronotum: 0.93 (0.85–1.12) long; 1.30 (1.22–1.30) wide. Differing from *N. curvipes* as follows: more slender, elongate; green color evident only in areolar veins, otherwise all pallid yellow; a small spot in the center of the mesal constriction of the pronotum; irregularly distributed recumbent, brown setae on clavus and corium internal to radial vein, with brown spots about their bases, those on corium external to radial vein and on cuneus, pale about their bases, the spots present about setae extending roughly from apex of scutellum to a line drawn parallel to base of cuneus; setae on hind legs longer, about 1/2 the greatest thickness of the hind femora; ratio of length of antennal segment I to head width as 8.5: 7.5; rostrum reaching apices of mesocoxae; membrane only weakly infumate opposite anal ridge; cuneus yellowish brown at outer apex. Pygofer, claspers and aedeagus as illustrated (Fig. 48).

FEMALE (n = 6). Body: 5.65 (5.65–7.11) long; 1.23 (1.23–1.55) wide. Head 0.93 (0.93–1.12) wide. Antennal segment lengths: I, 1.06 (1.06–1.15); II, 2.42 (2.25–2.70); III, 1.42 (1.42–1.65); IV, 0.99 (0.99–1.28). Rostrum: 1.33 (1.09–1.62) long. Pronotum: 0.93 (0.93–1.10) long; 1.23 (1.23–1.78) wide. Not otherwise differing from male.

NYMPHS. *1st and 2nd instars*. Not examined. *3rd instar* (n = 1). Body: 2.01 long; 0.85 wide; pallid. Head: 0.65 wide. Antennal segment lengths: I, 0.37; II, 1.69; III, 0.51; IV, 0.55. Rostrum: 1.07 long. Pronotum: 0.30 long; 0.52 wide. Dorsum, lateral margins and basal appendages beset with erect, short, apically blunt, brown setae, each on elevated base, otherwise different from 3rd-instar nymph of *N. curvipes* as follows: apex of head not striolate; stout setae on antennae mostly pale; stout setae on pronotum not set back from anterior margin; 1st abdominal tergite with only a pair of setae astride the ecdysial line, abdominal tergites 2–4 with 2 lateral and a pair of mesal setae making a row of 6 setae along posterior border of each tergite, meso-lateral setae on remaining apical tergites. Rostrum reaching apex of hind coxae; setae on legs stout, sparser. *4th instar* (n = 5). Body: 2.14–2.62 long; 1.00–1.15 wide. Head: 0.73–0.80 wide. Antennal segment lengths: I, 0.45–0.64; II, 1.00–1.22; III, 0.64–0.78; IV, 0.73–0.80 Rostrum: 1.13–1.20 long. Pronotum: 0.37– 0.43 long; 0.65–0.80 wide. Differing from 3rd instar as follows: setae on dorsum of body especially, shorter, each on a slightly raised, brown papilla; pronotal callosities glabrous, setae only around periphery; apical area of mesonotal wing bud with shorter setae, these about 1/2 the length of setae on lateral border, not on raised papillae; rostrum reaching middle of mesocoxae. *5th instar* (n = 8). Body: 3.35–4.45 long; 1.32–1.45 wide. Head: 0.59–0.95 wide. Antennal segment lengths: I, 0.82–0.96; II, 1.53–1.85; III, 1.00–1.32; IV, 0.55–1.07. Rostrum: 1.25–1.45 long. Pronotum: 0.50–0.67 long; 0.86–0.97 wide. Differing from 4th instar as follows: pubescence strikingly shorter and comparatively paler except on lateral margins of pronotum, wing buds, and apical abdominal tergites; those on tergites 7–9 on raised papillae; other abdominal setae shorter and more reclined; hind femora distinctly incrassate.

Distribution. Hawai'i I.

Host. *Tetraplasandra hawaiiensis* vars. *hawaiiensis* and *microcarpa*.

Etymology. The name refers to the pattern of spots on the pale dorsum, which give a "salt and pepper" or pallasate appearance.

Type data. Holotype ♂ (BPBM 14,836), allotype ♀ paratypes 20♂, 12♀, 10N, **HAWAI'I**: Ka'ū District, Manukā Pk, 1700 ft. [518 m], 29.VII.1970, on *Tetraplasandra hawaiiensis*, W.C. Gagné. Additional paratypes with data differing only as follows: 13♂, 10♀, 9N, Kohala Mts, Kupehau Ridge, 800 ft. [244 m], 5.XII.1968, on *T. hawaiiensis* var. *microcarpa*; 16♂, 24♀, 14N, Kohala Mts, Honopue Valley, 2000 ft. [605 m], 9.VI.1970; 18♂, 22♀, 15 N, Hwy. 13 at 1955 lava

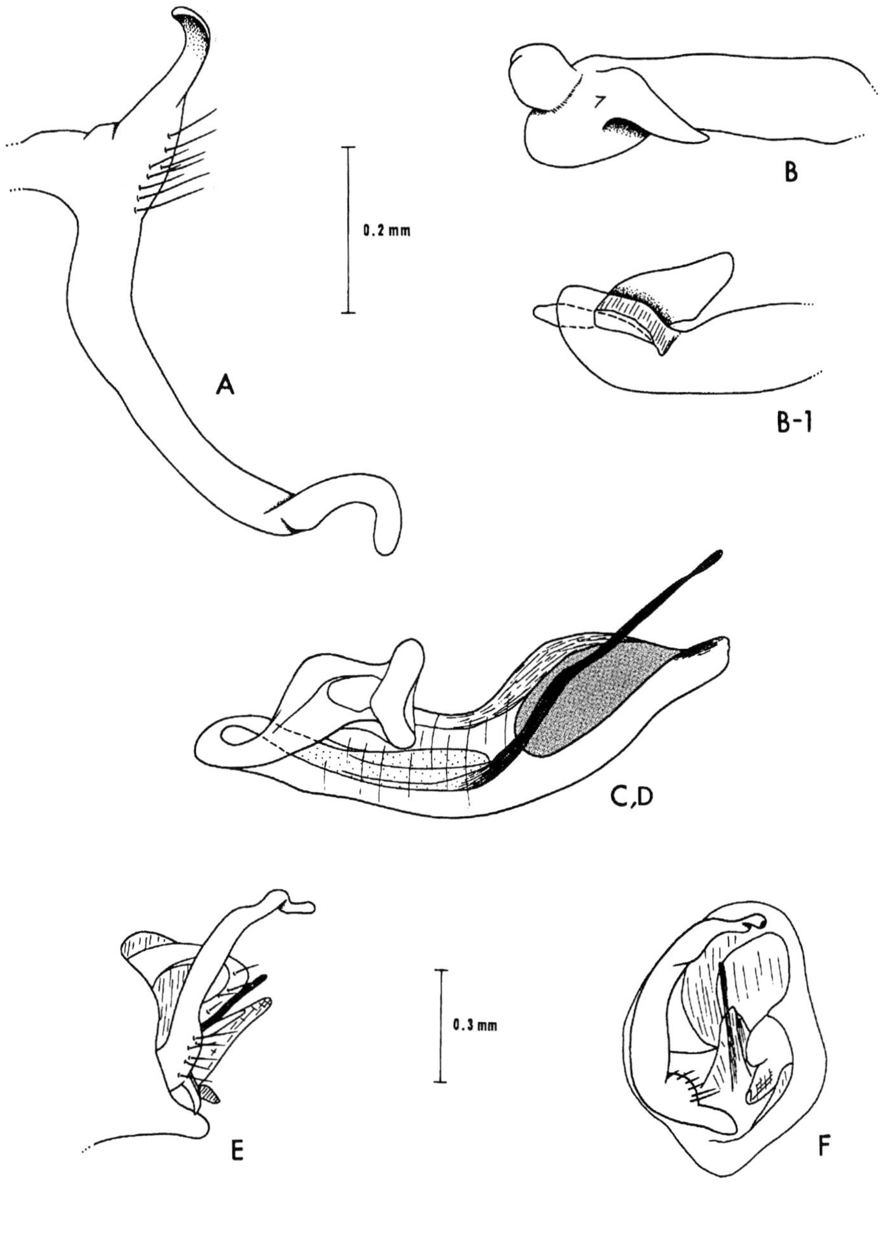

Fig. 48. *Nesiomiris pallasatus,* n. sp., male genitalia. **a**, left clasper, internal; **b**, right clasper, dorsal; **b-1**, right clasper, internal; **c, d**, aedeagus and spiculum, lateral; **e**, pygofer, left lateral; **f**, pygofer, terminal.

flow, (between Kaimū and Pāhoa), 800 ft. [244 m], 15.IX.1970; 11♂, 20♀, 8N, Kohala Mts, Kupehau Ridge, E side, 800 ft. [244 m], 13.VI.1970; 2♂, Hīlea Gulch, 605 m, 25.VII.1970; 1♂, 1200 ft. [366 m], 6.I.1971; 4♂, 1760 ft. [536 m], 30.III.1971; 22♂, 15♀, 6N, Hīlea Gulch, 2000 ft. [605 m], 8.I.1971; 7♂, 8♀, Hīlea Gulch, 732 m, 17.XI.1973, S.L. Montgomery; 2♂, Kulani Prison Rd., 800 ft. [243 m], 25.VII.1970 (BPBM).

Discussion. The "salt and pepper" pattern on the hemelytra renders these bugs cryptic on the host plant. They blend into the background of fecal spotting on the white undersurface of the leaf. *Nesiomiris fasciatus* and *N. maculatus* on Kaua'i are similarly cryptically patterned, but they have green dorsal spotting on the integument; *Tetraplasandra kavaiensis* also has pale undersides of the leaves. All these pale species tend to fly or move quickly to the underside of another leaf when disturbed and then "freeze."

The pallasate pattern appears to be retention of a juvenile character of adaptive value into the adult stadium; in this case the raised, brownish papillae with elevated setae is partially retained on the wings of adults.

The distribution of *N. pallasatus* parallels that of the host plant on Hawai'i I. There probably was considerable gene flow between populations at least until recently; when habitat alteration by humans created many disjunct populations of host plants around the island.

The *cheliferus* Complex

Description. **ADULTS**. Greenish species; hemelytra and posterior lobe of pronotum greenish; apex of head rounded to weakly indentate; aedeagus thickened or produced; hind legs straight, not incrassate; vestiture of legs and antennal segment I not long, dense; propleural stripe present; membrane of forewing infuscate; right clasper and aedeagus as described for *elongatus* complex; inner arm thumblike (*N. spatulatus*), bifid (*N. cheliferus*), or bladelike (*N. spatulatus*).

NYMPHS. Propleural stripe present or absent, apices of antennal segment I and legs pallid.

Distribution. Moloka'i I to Hawai'i I.

Host. *Cheirodendron trigynum* and various species of *Reynoldsia* and *Tetraplasandra*.

Included species. Nos. 12–14.

12. ***Nesiomiris cheliferus*** Gagné, **new species** Figs. 25h, 30i, 31f, 33d, 49a, 50

Nesiomiris hawaiiensis: Swezey, 1954, *B.P. Bishop Mus. Spec. Publ.* 44: 50 (host; misidentification, included other *Nesiomiris* spp. on *Cheirodendron* "on Hawai'i, Maui, and Kaua'i").

Nesiomiris prob. *Nesiomiris* sp.: Bianchi, 1967, *Proc. Hawaii. Entomol. Soc.* 19(2): 131 (host; in part, also included *N. secularis*, n. sp.)

The chelate process on the apex of the aedeagus relates *N. cheliferus* to *N. spatulatus* of Moloka'i that has the aedeagus similarly developed. Both appear to be apomorphic with respect to this character on comparison with *N. secularis*. This suggests that the 2 form a sister group with *N. secularis*. Nymphs of *N. cheliferus* are quite different from others, as discussed below.

The apex of the chelate process can be observed externally, and if not confused with the spiculum (which was not exserted in the comparatively long series of this species available), it and the reduced right clasper are quite diagnostic. It is sympatric with *N. rufonotus* of the *usingeri* complex.

Description. **MALE** (n = 6). Body: 7.01 (5.67–7.22) long; 1.48 (1.36–1.56) wide. Head: 1.05 (1.00–1.06); apex rounded, weakly brown, striolate; callosities weakly shining; shape as illustrated (Fig. 33d); basal carina indistinct; setae generally distributed between and in front of eyes anterior to glabrous callosities, on mesal impression and across base; head brown. Antennal segment lengths: I, 0.98 (0.75–1.07), subequal to the width of the head (ratio 0.9: 1.0); II, 3.31 (2.60–3.31); III, 1.95 (1.42–2.85); IV, 1.60 (1.60–1.88); color and vestitures as for *N. curvatus*. Rostrum: 1.27 (1.27–1.45) long; reaching middle of mesosternum. Pronotum: 1.12 (1.06–1.25) long; 1.48 (1.36–1.56) wide; shape as illustrated (Fig. 33d); color and propleuron as in *N. spatulatus*, except basal lobe paler across apex. Mesonotum yellow with a couple of scattered setae. Scutellum greenish; vestiture as in *N. reductus*. Hemelytra: generally as described for *N. swezeyi*; areoles coriaceous as illustrated (Figs. 31f, 49a); membrane infuscate except adjacent to apex of cuneus. Legs: femora and outer apices of protibiae greenish, remainder of legs yellowish, excepting dark brown terminal tarsal segments; vestiture as for *N. sinuatus*. Abdomen: yellowish, shining; atrium, claspers and aedeagus as illustrated (Fig. 50); apex of phallotheca extending over the right clasper as a sclerotized, chelate process.

FEMALE (n = 6). Body: 7.32 (6.40–7.45) long; 1.64 (1.40–1.64) wide. Head: 1.06 (1.00–1.06). Antennal segment lengths: I, 1.15 (0.77–1.15); II, 3.38 (2.50–3.38); III, 1.68 (1.60–2.15); IV, 1.66 (1.45–1.66). Rostrum: 1.45 (1.31–1.55) long. Pronotum: 1.21 (1.10–1.21) long; 1.64 (1.40–1.64) wide. Differing from male as follows: antennal segment I longer than width of head (ratio 9: 8.5); apex of head weakly bifid; antennal segment I greenish yellow; segment II brownish; mesonotum glabrous; scutellum pale on basal 1/2; protibia generally greenish yellow. Bursa copulatrix as illustrated (Figs. 25h, 30i).

NYMPHS. *1st and 2nd instars.* Unknown. *3rd instar* (n = 4). Body: 1.77–2.06 long; 0.50–0.53 wide. Head: 0.50–0.53; apex rounded, a row of setae encircling an area homologous to the callosities of the adult, and a few more at the apex. Antennal segment lengths: I, 0.35–0.39; chaetotaxy of this and subsequent segments very sparse, short, pale, and inconspicuous, much shorter than 1/2 the width of antennal segment I; segments I-III pallid, segment IV brownish black; II, 0.76–0.89; III, 0.77–0.87; IV, 0.33–0.70. Rostrum: 0.70–0.77, reaching mesocoxae. Pronotum: 0.23–0.28 long; 0.50–0.53 wide, subquadrate, lateral margin slightly sinuate; narrowing slightly anteriorly; a row of 4 evenly spaced setae on the anterior margin, and a row of 6 evenly spaced setae along the posterior margin; those at the posterior humeral angle are darkened about the base; propleural stripe present or medially convex posteriorly, wing bud overlapping anterior 1/2 of metanotum; a concave arc of 5 setae across the posterior area, and 1 on the margin behind the anterior humeral angle. Metanotum: shape generally as for mesonotum, excepting the concave posterior margin, about 1/2 as large as mesonotum; a seta near each humeral angle and a pair on each side of the midline. Legs: pale, with short inconspicuous setae as described for the antennae; last tarsal segment distally dark. Abdomen: pallid, chaetotaxy a medial pair of setae and 1 on meso-lateral margins, together making 4 vertical rows dorsally on all but 2 glabrous terminal segments. *4th instar* (n = 8). Body: 2.27–2.85 long; 0.56–0.71 wide. Head: 0.64–0.71 wide. Antennal segment lengths: I, 0.50–0.55; II, 1.15–1.43; III, 1.06–1.29; IV, 0.55–0.56. Rostrum: 0.85–0.89 long; reaching middle of mesosternum. Pronotum: 0.29–0.36 long; 0.56–0.71 wide. Differing from 3rd instar as follows: apex of head slightly bifid, glabrous; antennal segment I with 1 stronger seta near inner apex, setae inconspicuous; propleural stripe present or absent; mesonotal wing bud broadly but incompletely overlapping metanotal wing bud, the latter reaching hind border of 2nd abdominal segment; only apex of abdomen reddish. *5th instar* (n = 10). Body: 2.95–3.91 long; 0.81–1.05 wide. Head: 0.76–0.91 wide. Antennal segment lengths: I, 0.67–0.78; II, 1.60–2.16; III, 1.46–1.80; IV, all 1.42. Rostrum: 0.95–1.25 long; reaching mesocoxae. Pronotum: 0.42–0.63 long; 0.81–1.05 wide. Differing from 4th instar as follows: antenna segment I with a few stronger setae on inner aspects; dorsum apparently glabrous; metanotal wing bud reaching 4th abdominal segment; propleural stripe present or absent.

Distribution. E Maui.

Host. *Cheirodendron trigynum* and possibly *Reynoldsia sandwicensis*.

Etymology. Named for the chelate process at the apex of the aedeagus.

Type data. Holotype ♂ (BPBM 14,837), allotype ♀, paratypes 17♂, 16♀, 18N, **E MAUI**: SW Rift Zone, Pu‘u Mākua, 1543 m, 14.IV.1971, on *Cheirodendron trigynum*, W.C. Gagné. Additional paratypes with data differing only as follows: 6♂, 7♀, 12N, S of ‘Ulupalakua Ranch, ± 600 m, 13.IV.1965, on *Reynoldsia sandwicensis*, F.A. Bianchi; 30♂, 28♀, 14N, ‘Ulupalakua, 25.VIII.1965, on *Cheirodendron* sp., R.L. Usinger; 1♂, 1♀, Kula Pipe Line, 12.VI.1972, on *Cheirodendron* sp., O.H. Swezey; 2♂, 2♀, Auwahi, 1200 m, 20.VII.1965, C.M. Yoshimoto; 1♂, Auwahi, 3700 ft. [1128 m], 20.VII.1965, C.M. Yoshimoto [no host]. 5♂, 4♀, Auwahi, 8500 ft. [2590 m], 4.IV.1970, on *C. trigynum* (BPBM).

Variation. In some males and females the rostrum just surpasses the apices of the procoxae, and the profemora have faded to yellow in most of the specimens.

Discussion. A single nymph, a penultimate female of another *Nesiomiris* species, was collected with *N. cheliferus* nymphs. I consider that this nymph is *N. rufonotus* since it is obviously a last-instar nymph and yet is about 1/2 the size of the penultimate nymphs of *N. cheliferus*. Adults of *N. rufonotus* that were collected with the type series of *N. cheliferus* are very small in comparison with *N. cheliferus*; they are also about 1/2 as long as *N. cheliferus* in the comparable instar.

Although the general conformation of the male genitalia, as noted above, might indicate a close association with the *reductus* complex, the fact that the 3 nymphal instars examined do not have reddish apices of antennal segments I and II, and are practically glabrous in the 4th and 5th instars, suggests that this species better belongs in the *elongatus* subgroup.

The material obtained by Bianchi, and apparently that reported by him (Bianchi 1967) as having been collected on *Reynoldsia sandwicensis*, contains both *N. cheliferus* and *N. secularis*. Recollecting in this area I did not recover any *cheliferus* from *Reynoldsia*, but only from *Cheirodendron*. He may have confused the hosts. I have not collected *N. secularis* on anything but *Reynoldsia* in this region so I conclude that *N. cheliferus* is restricted to *Cheirodendron trigynum*.

13. ***Nesiomiris spatulatus*** Gagné, **new species** — Figs. 25g, 30h, 31e, 51

This species appears to be most closely related to *N. cheliferus* of E Maui. Its most diagnostic feature is the spatulate process on the apex of the aedeagus, a feature that is similarly developed in *N. cheliferus*. These 2 species, then, appear to form a natural group

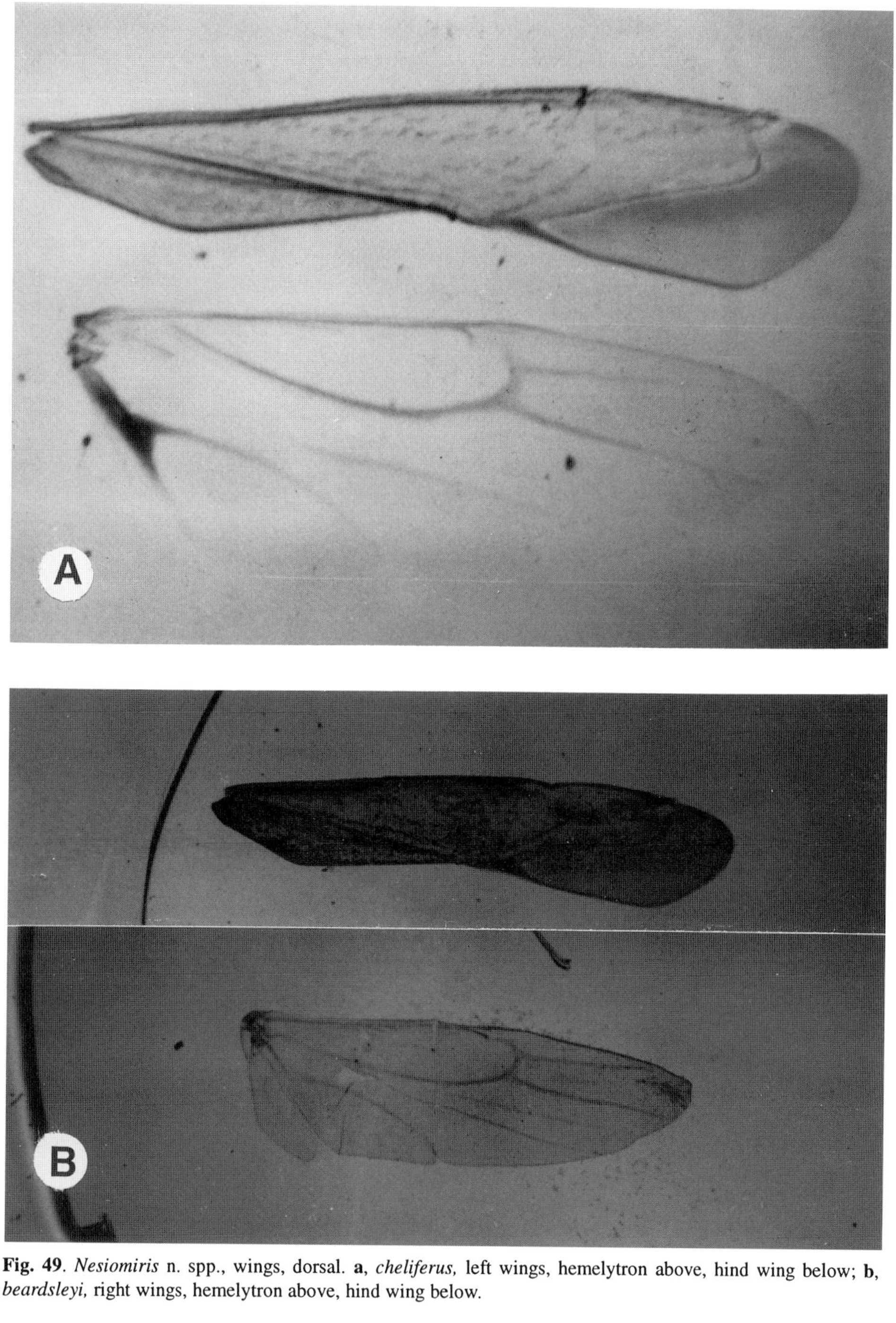

Fig. 49. *Nesiomiris* n. spp., wings, dorsal. **a**, *cheliferus,* left wings, hemelytron above, hind wing below; **b**, *beardsleyi,* right wings, hemelytron above, hind wing below.

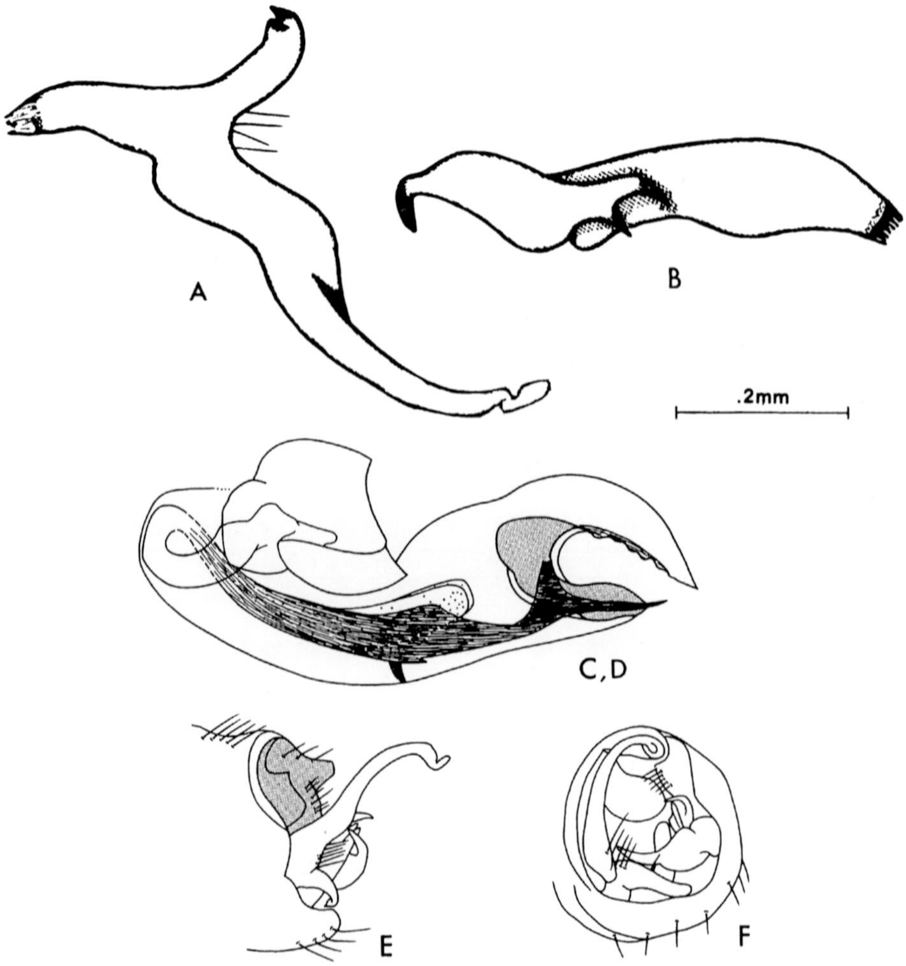

Fig. 50. *Nesiomiris cheliferus*, n. sp., male genitalia. **a**, left clasper, internal; **b**, right clasper, internal; **c, d**, aedeagus and spiculum, lateral; **e**, pygofer, left lateral; **f**, pygofer, terminal.

that is a sister group of *N. secularis*.

Description. **MALE** (n = 5). Body: 6.35 (5.96–6.35) long; 1.28 (1.28–1.35) wide. Head: 0.92 (0.89–0.92) wide; apex slightly bifurcate; other aspects as for *N. reductus*. Antennal segment lengths: I, 1.01 (0.92–1.09), longer than head width (ratio 8: 7); II, 3.36 (3.23–3.36); III, 2.20 (2.14–2.49); IV, 1.95 (1.93–2.14); color and vestiture as for *N. curvatus*. Rostrum: 1.29 (1.28–1.31) long; reaching middle of mesosternum. Pronotum: 1.28 (0.92–1.28) long; 1.28 (1.28–1.35) wide, shape as in *N. sinuatus*, posterior lobe and collar pale greenish, remainder yellowish; propleuron and propleural stripe as is *N. parasinuatus*. Mesoscutellum and base of scutellum greenish; vestiture as is

N. reductus. Hemelytra: generally as described for *N. swezeyi*; areoles coriaceous as illustrated (Fig. 31e); membrane infuscate; vestiture as for *N. sinuatus*. Legs: as in *N. sinuatus*. Abdomen: atrium, claspers, and aedeagus as illustrated (Fig. 51); spatulate process of aedeagus extending exteriorly and posteriorly beyond the claspers.

FEMALE (n = 4). Body: 6.70 (6.00–6.70) long; 1.32 (1.32–1.44) wide. Head: 0.92 (0.90–0.95) wide. Antennal segment lengths: I, 1.05 (1.00–1.09); II, 3.30 (3.10–3.30); III, 2.37 (2.14–2.42); IV, 1.84 (1.84–2.03). Rostrum: 1.23 (1.23–1.35) long. Pronotum: 1.05 (1.05–1.09) long; 1.28 (1.28–1.44) wide. Differing from male as follows: apex of head straight; head greenish yellow dorsally and on frons. Antennal segment I yellow on basal 1/2, greenish yellow on apical 1/2, ratio of length of antennal segment I to head width as 8: 7.25; antennal segment II yellowish, brown at apex. Area of propleuron anterior to coxal cleft and below propleural stripe not infuscate. Apex of scutellum greenish, remainder of scutellum and mesoscutellum yellow; hemelytra as for male *N. swezeyi*; legs yellowish. Bursa copulatrix as illustrated (Fig. 25g).

NYMPHS. Unknown.

Distribution. Moloka'i I.

Host. *Cheirodendron trigynum* var. *molokaiensis*.

Etymology. Named for the spatulate process on the apex of the aedeagus.

Type data. Holotype ♂ (BPBM 14,838), allotype ♀, paratypes 2♂, 1♀, **MOLOKA'I**: Pu'u Kolekole Cabin, 3900 ft. [1190 m], 10.VII.1968, on *Cheirodendron trigynum* var. *molokaiensis*, W.C. Gagné & P.D. Ashlock. Additional paratypes with data differing only as follows: 2♂, 1♀, 3950 ft. [1204 m], 10.VII.1968, P.D. Ashlock; 4♂, 1♀, Kawela Gulch, 1068–1144 m, 8–10.VII.1968, W.C. Gagné; 1♀, Kawela Gulch, 1068–1144 m, 8–10.VII.1968, D.M. Tsuda; 1♀, [no specific locality], 6000 ft. [1220 m], VI.1896, [no host], R.C.L. Perkins; 3♀, 1080 m, 14.VIII.1971, on *C. trigynum*, W.C. Gagné (BPBM).

14. ***Nesiomiris secularis*** Gagné, **new species** Figs. 24a; 25j, k; 52a; 53

Nesiomiris hawaiiensis: Swezey & Bryan, 1929, *Proc. Hawaii. Entomol. Soc.* 7(2): 303, 311 (host; misidentification, also included *N. curvipes*, n. sp.). Swezey, 1954, *B.P. Bishop Mus. Spec. Publ.* 44: 217 (host; misidentification, also included *N. curvipes* , n. sp.).

Nesiomiris prob. *Nesiomiris* sp.: Bianchi, 1967, *Proc. Hawaii. Entomol. Soc.* 19: 132 (note; host; in part, also included *N. cheliferus*, n. sp.).

The apex of the aedeagus is comparatively simple, the reduced right clasper has an inner excavate, bladelike process, and the nymphs have usually retained the propleural stripe, especially in the earlier instars. Coupled with its comparatively extensive distribution, these features readily distinguish this species.

Description. **MALE** (n = 7). Body: 5.95 (5.36–6.40) long; 1.39 (1.28–1.44) wide. Head: 1.00 (0.96–1.02); apex straight, weakly carinate across base, remainder of head as described for *N. cheliferus*. Antennal segment lengths: I, 0.89 (0.81–1.00), shorter than head width, numerous erect setae on inner aspect mixed with brown, reclined setae, yellow; II, 2.66 (2.55–3.05), yellow on basal 1/3 becoming brown on apical 2/3; III, 1.79 (1.60–1.85), color of vestiture as for *N. sinuatus*; IV, 1.45 (1.25–1.45). Rostrum: 1.25 (1.14–1.28) long; reaching apex of mesosternum. Pronotum: 1.07 (0.92–1.12) long; 1.39 (1.28–1.44) wide; greenish yellow; propleuron as described for *N. spatulatus*. Mesonotum and scutellum yellow, brown setae on both. Hemelytra: as described for *N. swezeyi*; setae distributed as illustrated (Fig. 52a); areoles and membrane as illustrated for *N. cheliferus* (Fig.

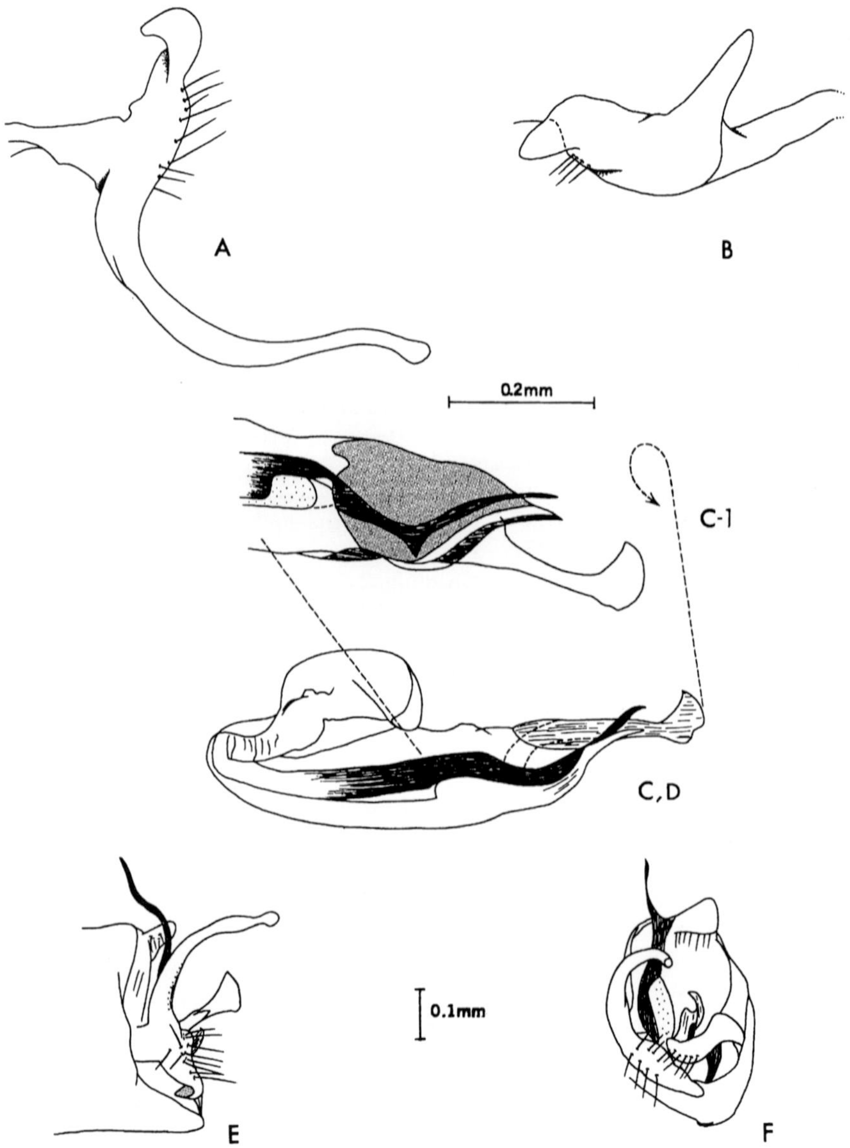

Fig. 51. *Nesiomiris spatulatus,* n. sp., male genitalia. **a**, left clasper, internal; **b**, right clasper, internal; **c, d**, aedeagus and spiculum, lateral; **c-1**, apex of spiculum, lateral; **e**, pygofer, left lateral; **f**, pygofer, terminal.

31f). Legs: presumably all faded to yellow, excepting dark brown terminal tarsal segments; vestiture as for *N. sinuatus*. Abdomen: pygofer, claspers and aedeagus as illustrated (Fig. 53).

FEMALE (n = 7). Body: 6.07 (5.87–6.80) long; 1.39 (1.35–1.50) wide. Head: 0.98 (0.96–1.02) wide. Antennal lengths: I, 0.83 (0.81–1.03); II, 2.45 (2.45–3.01); III, 1.81 (1.60–1.85); IV, 1.25 (1.06–1.45). Rostrum: 1.31 (1.29–1.32) long. Pronotum: 1.00 (1.00–1.17) long; 1.39 (1.35–1.50) wide. Differing from male as follows: apex of head slightly bifid; antennal segment I longer than head width; antennal segment II, yellowish with brownish apex; III, yellow on basal 1/4 becoming darker apically; rostrum reaching middle of mesosternum; internal genitalia as illustrated (Figs. 24a, 25j).

NYMPHS. *1st and 2nd instars*. Unknown. *3rd instar* (n = 2). Body 2.30–2.37 long; 0.64–0.65 wide. Head: 0.56–0.58 wide; apex rounded; lacking a basal carina; setae forming an arc from the inner posterior margin of the eye and around the apex, absent across base and on callosities. Antennal segment lengths: I, 0.35–0.37, setae sparse, pale, about equal in length to thickness of this segment; II, 0.81–0.89, setae subrecumbent, shorter and denser than on I, some scattered, erect; III, 0.70–0.78, setae shorter; IV, all 0.96, subinfumate; setae still shorter and denser. Rostrum: all 0.67 long; reaching mesocoxae. Pronotum: 0.31–0.37 long; 0.54–0.55 wide; shape as described for *N. cheliferus*; propleural stripe present or absent, when present extends from anterior margin of propleuron to a point just anterior to setae on posterior humeral angle, which is darkened about the base. Mesonotum: convex on lateral and anterior margins, straight between incipient wing buds on posterior margins, wings just overlapping margins of metanotum, ratio, along median line, of mesonotum to metanotum, about 1.5: 0.5. Metanotum: shape as for mesonotum; chaetotaxy consisting of a pair of setae astride midline, 1 near posterior humeral angle, 1 at posterior humeral angle, and 1 mesolateral. Legs: setae as described for *N. trifurcatus*, but much sparser. Abdomen: tergal setae consisting of a pair on either side of midline of each segment, mesally, and 1 mesally on margin of all but posterior tubular segments. *4th instar* (n = 1). Body: 2.52 long; 0.86 wide. Head: 0.75. Antennal segment lengths: I, 0.48; II, 1.17; III, 1.02; IV, broken. Rostrum: 0.81 long. Pronotum: 0.37 long; 0.65 wide. Differing from 3rd instar as follows: 6 setae present across base of head; setae of antennal segment II with a line of erect, longer setae on basal 1/2 of inner aspect, with reclined setae; rostrum reaching posterior coxae; pronotal margin concave at middle, becoming convex at humeral angles, setae generally distributed on disc except for mesal area on each side of midline; propleural stripe present or absent; when present extends full length of pronotal margin, broader; setae on an indistinct line along lateral margin of mesonotum and darkened about their bases, irregularly distributed but denser on lateral margin; mesonotal wing bud reaching apex of 3rd abdominal segment, a pair of setae at outer aspect of wing bud and 4 more at regular intervals across the base between the metanotal wing bud; a pair of setae on each abdominal margin, dorsally. *5th instar* (n = 4). Body: 3.40–4.02 long; 1.10–1.24 wide. Head: 0.86–0.90 wide. Antennal segment lengths: I, 0.60–0.65; II, 1.72–1.92; III, 1.31–1.59; IV, 1.22–1.45. Rostrum: 1.00–1.12 long. Pronotum: 0.51–0.56 long; 0.86–0.92 wide. Differing from 4th instar as follows: apex of head slightly bifid; 4 middle setae on posterior margin of head noticeably shorter than others on head; antennal setae darker on I and II; rostrum reaching middle coxae; propleural stripe usually absent, when present, found only at humeral angle and intermittently to middle of lateral margin; setae generally distributed on mesonotal wing buds, and not noticeably concentrated along lateral margins, much sparser on mesonotum between the wing buds, these reaching to about 6th or 7th abdominal segments.

Distribution. Moloka‘i, E Maui, and Hawai‘i Is.

Host. *Reynoldsia mauiensis*, *R. Hillebrandii*, *R. huehuensis*, *R. sandwicensis*,

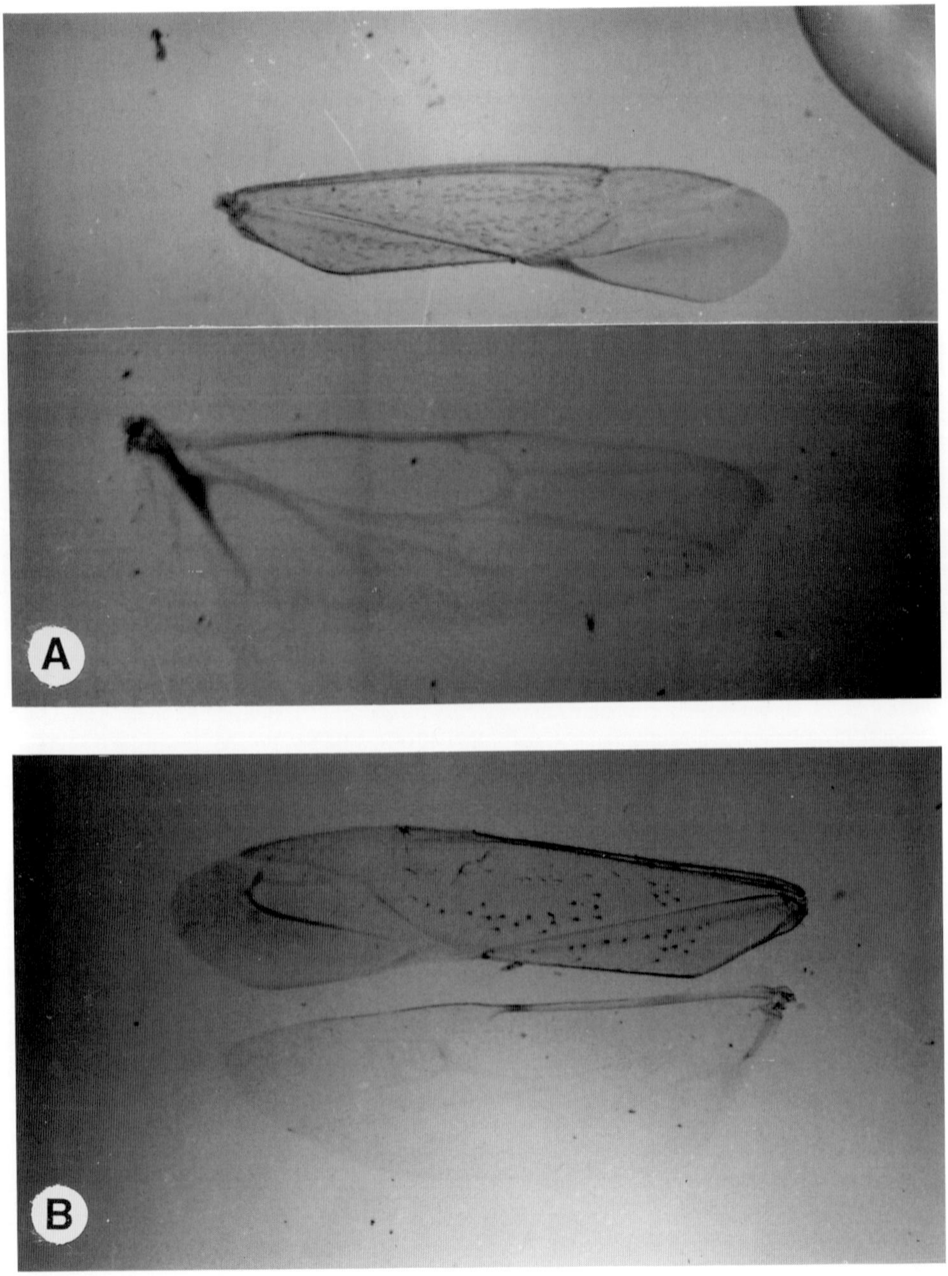

Fig. 52. *Nesiomiris* n. spp., wings, dorsal. **a**, *secularis,* right wings, hemelytron above, hind wing below; **b**, *pallasatus,* right wings, hemelytron above, hind wing below.

Tetraplasandra meiandra var. *mauiensis*, and *T.* sp. (prob. *hawaiiensis*).

Etymology. The specific epithet means "worldly," referring to the comparatively broad distribution of this species.

Type data. Holotype ♂ (BPBM 14,839), allotype ♀, paratypes 72♂, 54♀, 55N, **E MAUI**: 'Ulupalakua, 26.V.1965, on *Reynoldsia sandwicensis*, J.W. Beardsley. Additional paratypes with data differing only as follows: 2♀, 13N, ca. 1 mi S of 'Ulupalakua on lava flows, [no elevation], 7.VIII.1968, on *R. sandwicensis,* W.C. Gagné; 1♂, Haleakalā, 9400 ft. [2867 m], 23.III.1932, on *Cyathodes* sp., O. Bryant; 1♂, 2♀, 549 m, 7.VIII.1968, on *R. mauiensis*, W.C. Gagné; 1♂, 1♀, 2N, 1 mi S of 'Ulupalakua, [elevation illegible], 21.XII.1968, on *R. mauiensis*, W.C. Gagné; 8♂, 3♀, 2 mi E of 'Ulupalakua, 520 m, 21.II.1970, on *R. mauiensis*, W.C. Gagné; 1♀, 11N, 2 mi E of 'Ulupalakua, 500 m, 21.II.1970, on *R. mauiensis*, W.C. Gagné; 5♂, 1♀, 7N, 5 mi E of 'Ulupalakua, 1600 ft. [488 m, 5.IV.1970, on *R. mauiensis*, W.C. Gagné; 20♂, 25♀, 9N, 5 mi E of 'Ulupalakua, 500 m], 21.II.1970, on *R. mauiensis*, W.C. Gagné; 5♂, 10♀5 mi E of 'Ulupalakua, 500 m, 4.IV.1970, on *R. mauiensis*, W.C. Gagné; 8♂, 13♀, Auwahi, 3500 ft. [1066 m], 4.IV.1970, on *R. meiandra* var. *mauiensis*, W.C. Gagné; 4♂, 9♀, 5N, Auwahi, 2200 ft. [670 m], 4.IV.1970, on *R. mauiensis*, W.C. Gagné; 1♂, 1♀, SW Rift Zone, Pu'u Māhoe Arboretum, 730 m, 17.IV.1971, on *T. hawaiiensis*, W.C. Gagné; 4♂, 2♀, Waiho'i Valley, 610 m, 18.VII.1972, on *T. hawaiiensis*, W.C. Gagné; 3♂, 1♀, S of 'Ulupalakua Ranch, ± 600 m, 13.IV.1965, F.A. Bianchi. **HAWAI'I**: 2♂, 1♀, N Kona District, Pu'u Wa'awa'a, 3500 ft. [1068 m], 20.I.1972, on *T. meiandra,* S.L. Montgomery; 2♂, 3♀, 7N, 600 m, 6.XII.1968, on *R. huehuensis*, W.C. Gagné; 16♂, 11♀, [no elevation], 6.XII.1968, on *R. huehuensis*, W.C. Gagné; 2♂, near Wauha'ula Heiau (Royal Gardens Subdivision) [no elevation], 26.VII.1970, on *R. Hillebrandii*, W.C. Gagné; 9♂, 11♀, 4N, near Wauha'ula Heiau (Royal Gardens Subdivision) near sea level, 7.XII.1968, on *R. Hillebrandii*, W.C. Gagné; 2♂, 2♀ 3N, Royal Gardens Subdivision, near Kupa'ahu, 60 m, 26.VII.1970, on *R. Hillebrandii*, W.C. Gagné; 6♂, 12♀, 14N, Pu'u Wa'awa'a, 23.I.1971, on *R. huehuensis*, W.C. Gagné; 1♂, 5N, Ka'ū District, Ocean View Estates, 10.VIII.1973, on *R. sandwicensis*, S.L. Montgomery and G. Schattauer. **MOLOKA'I**: 1♂, Kainalu, 1600 ft. [488 m], 24–27.VII.1927, E.H. Bryan, Jr., on *Tetraplasandra* (presumably *T. hawaiiensis*) (BPBM).

Discussion. The specimen collected by Bryan on Moloka'i was taken with his series of *N. curvipes*. It may seem incongruous that no specimens have been taken on W Maui. Suitable hosts are there, notably *Reynoldsia* and *Tetraplasandra hawaiiensis*, but the dry forest on the leeward portions has been so decimated and the host plants are now so few and scattered that possible populations on them have probably become extinct. I have sampled both hosts and failed to find any evidence of the bugs. *T. hawaiiensis* on the wet windward side of W Maui is inhabited by *N. pallidus*, *N. curvipes*, and *N. trinesiotes*, but *N. secularis* has not been found there.

I have selected the holotype from Beardsley's collection on *Reynoldsia sandwicensis* because this series is the most extensive, even though the type locality "'Ulupalakua" is not precise, although it probably lies a few miles east of 'Ulupalakua where I subsequently made collections of this species. I consider Bryant's aberrant host record "on *Cyathodes*" (= *Styphelia tameiameiae*) to be in error, or merely a perching record. I have never found *Nesiomiris* species breeding on or even perching on this shrub.

Although adults of this species have been collected on *Tetraplasandra*, until nymphs are also taken, I shall regard this as involving only adult feeding and not indicative of a complete life cycle on this host.

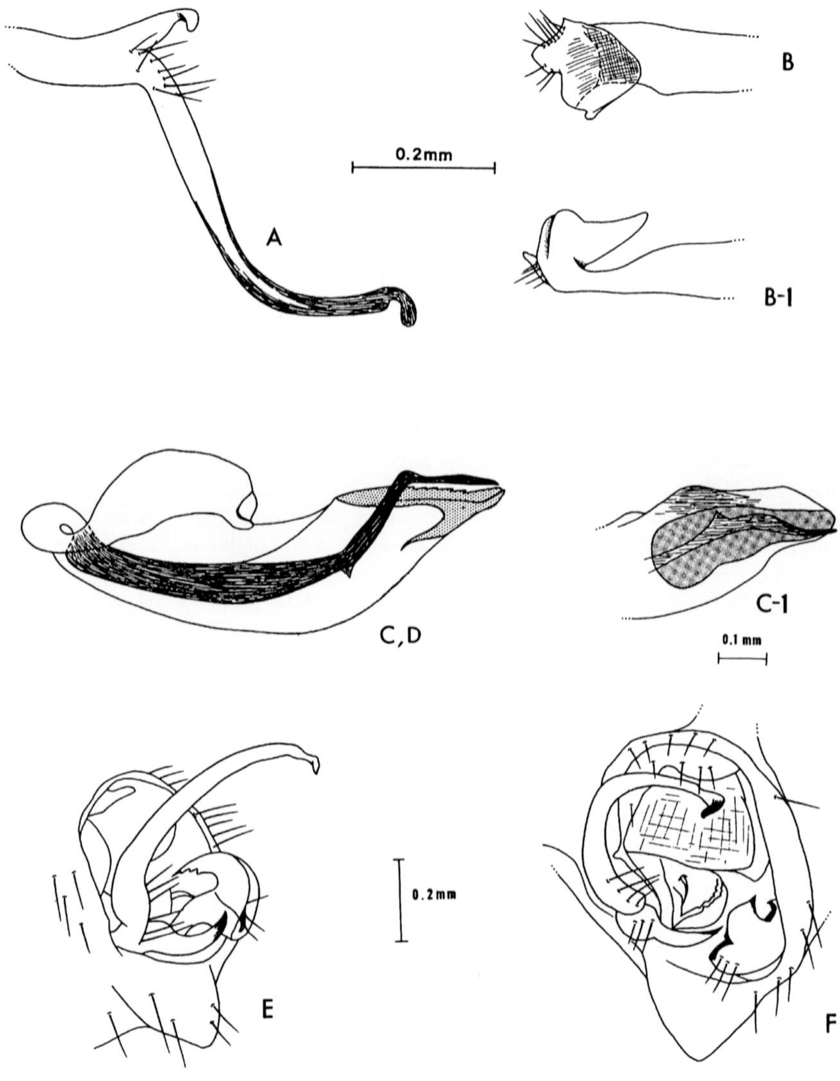

Fig. 53. *Nesiomiris secularis,* n. sp., male genitalia. **a**, left clasper, internal; **b**, right clasper, dorsal; **b-1**, right clasper, internal; **c, d**, aedeagus and spiculum, lateral; **c-1**, apex of spiculum, lateral; **e**, pygofer, left lateral; **f**, pygofer, terminal.

This species is important phylogenetically in that the male genitalia are somewhat intermediate to *N. elongatus* and the *curvipes* complex. With respect to host relationships, the *secularis* complex represents a break from the comparatively strict monophagy of other species complexes on *Cheirodendron* or *Tetraplasandra*, to more stenophagous species, which additionally may feed on *Reynoldsia*, and have "forsaken" *Cheirodendron* in several cases.

There appears to be some overlap in the distributions of *N. secularis* and *N. cheliferus* on E Maui where the latter occurs sparingly on *Tetraplasandra meiandra* and possibly also *Reynoldsia*. Since there is some possibility of females and nymphs of these 2 species being mixed, I have designated females as paratypes only when associated males were all of 1 species.

The *ohe* Complex

***Description*.** **ADULTS**. Predominantly greenish species (except *N. pallidus*), with posterior lobe of pronotum and hemelytra greenish; apex of head weakly indentate to rounded; propleural stripe present, extending from collar to mesal constriction (interrupted at middle in *pallidus*); vestiture of legs and antenna segment I not long and dense; hind femora of adults and 5th-instar nymphs straight, not incrassate; apex and atrium of aedeagus twisted to face left clasper; apex of aedeagus keeled; right clasper lobate, lacking an inwardly projecting arm.

NYMPHS. Propleural stripe present or absent; antenna segment I and apices of II and III and hind femora sometimes rufescent.

***Distribution*.** O'ahu, Moloka'i, Lāna'i, Maui, and Hawai'i Is.

***Host*.** *Cheirodendron trigynum* and various species of *Reynoldsia* and *Tetraplasandra*.

***Included species*.** Nos. 15–18.

15. ***Nesiomiris ohe*** Gagné, **new species** Figs. 29b, 33f, 54

This species is unique in that the pygofer has a tooth above the right clasper, rather than above the left clasper or on the dorsum, as is the case for all other species possessing a toothed pygofer. This enigmatic species appears to be related to *N. elongatus*, based on the general conformation of the male genitalia and the comparatively large size of both species. Species having similarly reduced right claspers and a keeled, fingerlike projection on the left apex of the aedeagus are placed in the *secularis* complex. *Nesiomiris ohe* differs additionally from *N. elongatus* in having a very short to obsolescent flabellum on the spiculum.

***Description*.** **MALE** (n = 6). Body: 6.30 (6.30–6.50) long; 1.44 (1.39–1.45) wide; apex bifid, brownish striolate; brownish on base of frons; distinctly carinate across base, shining especially on callosities; vestiture brownish, densest between eyes and antennal bases, subglabrous elsewhere dorsally. Antennal lengths: I, 1.13 (1.09–1.17), greenish yellow; II, 3.51 (3.20–3.55), yellow on basal 1/2, becoming darker brown apically; III, 2.65 (2.49–3.37), brown at base, remainder black;

IV, 1.95 (1.93–2.49), black. Rostrum: 1.78 (1.61–1.78) long; just surpassing apex of mesosternum. Pronotum: (Fig. 33f) 1.19 (1.09–1.19) long; 1.44 (1.39–1.45) wide; hind lobe pale greenish, remainder yellowish; propleural stripe beginning behind collar and distinctly extending to mesal constriction then continuing on posterior lobe as a weakly impressed, shining line; apex of propleuron below propleural stripe immediately anterior to coxal cleft infuscate as in *N. sinuatus*; propleuron wrinkled on posterior lobe; a seta at each humeral angle, otherwise glabrous. Mesonotum and scutellum sparsely setaceous; yellowish. Hemelytra: uniformly greenish; large areole coriaceous on inner 1/2; small areole coriaceous on basal 1/2; vestiture dark, reclined; membrane weakly infumed in large areole and more strongly so on remainder, except area immediately adjacent to apex of cuneus, which is clear. Femora and tibiae greenish yellow; tarsi dark brown; vestiture brownish, reclined. Abdomen: atrium of pygofer with a blunt tooth above right clasper; claspers, atrium, and aedeagus as illustrated (Fig. 54).

FEMALE (n = 6). Body: 6.70 (6.40–6.70) long; 1.44 (1.44–1.46) wide. Head: 1.06 (1.03–1.06) wide. Antennal lengths: I, 1.11 (1.09–1.17); II, 3.39 (3.20–4.05); III, 2.82 (2.65–2.95); IV, 1.78 (1.78–2.49). Rostrum: 1.78 (1.55–1.78) long. Pronotum: 1.15 (1.00–1.19) long; 1.44 (1.44–1.46) wide. Differing from male as follows: antennal segment II greenish yellow at base, becoming brown mesally, then black at apex; pronotum additionally with a few short, erect, pale setae dorsally; edge of propleuron additionally infuscate just posterior to coxal cleft.

NYMPHS. *1st and 2nd instars*. Unknown. *3rd instar* (n = 1). Body: 2.14 long; 0.69 wide. Head: 0.60 wide, apex broadly rounded; lacking basal carina; setae indistinct, only 1 seta apparent between eye and antennal socket. Antennal lengths: I, 0.46; II, 1.28; III, 1.22; IV, 1.62; I rufescent; II and III reddish at apices; setae of I brown, others pale; setae as described for *N. timberlakei*. Rostrum: 1.00 long; reaching middle of hind coxae. Pronotum 0.64 long; 0.51 wide; narrower across apex than base, a few setae along lateral borders only; propleural stripe absent. Lateral borders of meso- and metanota convex. Legs: pale throughout; setae sparse, pale. Abdomen: pale throughout. *4th instar* (n = 1). Body: 2.70 long; 0.95 wide. Head: 0.64 wide. Antennal segment lengths: I, 0.80; II, 1.28; III, 1.22; IV, 1.62. Rostrum: 1.00 long. Pronotum: 0.64 long; 0.54 wide. Differing from 3rd instar as follows: a couple of setae at apex of head; antennal segment II weakly rufescent, reddish at apex; setae on dorsal aspects of femora darker. *5th instar* (n = 5). Body: 3.26 (other specimens mutilated) long; 1.19–1.31 wide. Head: 0.81–0.92 wide. Antennal segment lengths: I, 0.81–1.00; II, 2.65–2.80; III, 1.17–2.65; IV, 1.85 (others broken). Rostrum: 1.19–1.39 long. Pronotum: 0.48–0.54 long; 0.85–0.95 wide. Differing from 4th instar as follows: apex of head concave, weakly striate; mesonotal wing bud with a row of setae along the lateral border; reaching base of abdominal segment VI; apex of femora at times weakly rufescent.

Distribution. O'ahu I (Ko'olau Mts).

Host. *Tetraplasandra gymnocarpa*.

Etymology. A Hawaiian name for some members of the genus *Tetraplasandra* is *ohe*, which is used as a noun in apposition.

Type data. Holotype ♂ (BPBM 14,840), allotype ♀, paratypes 7♂, 5♀ (including 1 pair *in copula*), 8N, **O'AHU**: Ko'olau Mts, Kōnāhuanui, Pu'u 'Ōhia Trail, 1500 ft. [757 m], 19.IX.1971, on *Tetraplasandra gymnocarpa*, W.C. Gagné. Additional paratypes: 7♂, 5♀, 7N, Castle Trail, 790 m, 30.V.1971, on *Tetraplasandra gymnocarpa*, W.C. Gagné (BPBM).

Variation. A projection on the left apical side of the phallotheca varied, in 3 males dissected, from incipient to clearly evident (see Fig. 54). At its strongest, this projection approached that of *N. elongatus*, although in no case was it as prominent, heavily sclerotized, or acuminate.

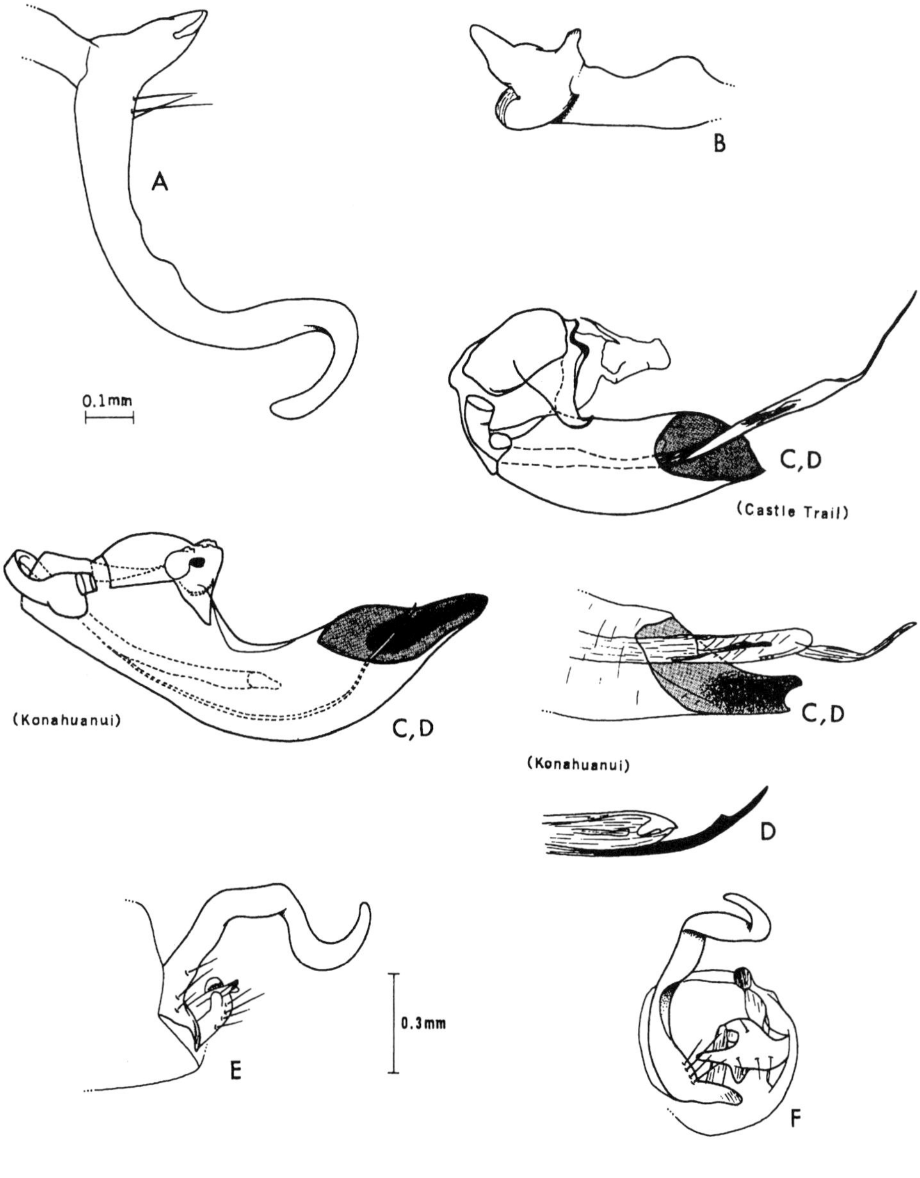

Fig. 54. *Nesiomiris ohe,* n. sp., male genitalia, showing variation of aedeagus and spiculi of Kōnāhuanui and Castle Trail populations. **a**, left clasper, internal; **b**, right clasper, internal; **c, d**, aedeagus and spiculum, lateral; **d**, apex of spiculum, lateral; **e**, pygofer, left lateral; **f**, pygofer, terminal.

Discussion. One ecological aspect used in hypothesizing the ancestral relative of *N. ohe* was the host association. Both of the *Tetraplasandra* host species of *N. ohe* and *N. elongatus*, *T. gymnocarpa* and *T. kavaiensis*, respectively, are densely stellate-tomentose on the lower leaf surfaces, young shoots, and inflorescences. They were formerly included, with several other *Tetraplasandra* species, in *Pterotropia*, which is now considered to be a section of *Tetraplasandra* (Sherff 1955). Other than on O'ahu and Kaua'i, these trees are now increasingly rare, especially due to the depredation of feral pigs. Only 1 tree was found on W Maui, about 6 trees on E Maui, and 5 on Hawai'i; none has been recorded from Moloka'i, and those once occurring on the now ravished island of Lāna'i have apparently been extirpated. Of these populations only the Hawai'i population harbored *Nesiomiris* (*pilosus*), which occurs also on other Araliaceae.

16. ***Nesiomiris trinesiotes*** Gagné, **new species** Figs. 251, 27d, 30k, 45d, 55

The general conformation of the claspers and aedeagus place this species in the *ohe* complex in that the right clasper is much reduced and simply lobate, and the apex of the aedeagus is twisted so that its atrium faces the left clasper.

Among the species having the above combination of characters, it is at once distinguished by the teeth and denticles on the thickened upper arm of the left clasper. Since this arm of the clasper of *N. timberlakei* is at times "lumpy," *N. trinesiotes* is considered closest to, and sharing a common ancestor with, *N. timberlakei*.

Description. **MALE** (n = 7). Body: 6.45 (6.10–6.48) long; 1.36 (1.32–1.39) wide. Head: 1.00 (0.99–1.02) wide. Antennal segment lengths: I, 1.08 (1.03–1.14); II, 3.37 (3.20–3.55); III, 2.81 (2.14–2.81); IV, 1.71 (1.71–1.93); I longer than head width. Rostrum: 1.29 (1.28–1.35) long. Pronotum: 1.05 (1.05–1.17) long; 1.36 (1.32–1.39) wide. Differing from *N. timberlakei* as follows: apex of head more deeply excavate; setae on head dense and darker, a distinct row of these along the basal carina; antennal segment II becoming blackish toward apex; pronotal setae darker, stouter, particularly more abundant on hind lobe; apex of propleuron anterior to coxal cleft infuscate; rostrum reaching middle of mesosternum; pygofer, claspers and aedeagus as illustrated (Fig. 55).

FEMALE (n = 7). Body: 6.40 (6.30–6.70) long; 1.39 (1.35–1.48) wide. Head: 0.97 (0.96–1.02) wide. Antennal segment lengths: I, 1.05 (1.05–1.15); II, 3.39 (3.01–3.52); III, 2.91 (2.20–2.91); IV, 1.79 (1.54–2.03); ratio of I to head width as 8.5: 8.0. Rostrum: 1.36 (1.35–1.45) long. Pronotum: 1.01 (1.01–1.22) long; 1.39 (1.35–1.45) wide. Not otherwise differing from male. Internal genitalia as illustrated (Figs. 251, 30k).

NYMPHS. *1st instar*. Unknown. *2nd instar* (n = 1). Body: 1.35 long; 0.51 wide. Head: 0.51 wide. Antennal segment lengths: I, 0.27; II, 0.61; III, 0.60; IV, 1.21. Rostrum: 0.73 long. Pronotum: 0.25 long; 0.46 wide. Not appreciably differing from same instar of *N. timberlakei* excepting that setae on body and appendages appear to be less prominent and femora proportionately longer, apices of hind femora exceeding apex of abdomen. *3rd instar* (n = 1). Body: 2.18 long; 0.66 wide. Head: 0.64 wide. Antennal segment lengths: I, 0.51; II, 1.28; III, 1.12; IV, 1.61. Rostrum: 0.89 long. Pronotum: 0.30 long; 0.56 wide. Differing from 2nd instar as follows: propleural stripe wider at anterior humeral angle, sometimes interrupted at middle; rostrum not exceeding apices of hind coxae. *4th instar* (n = 5). Body: 2.24–2.77 long; all 0.81 wide. Head: 0.69–0.75 wide. Antennal lengths: I, 0.64–0.66; II, 1.53–1.61; III, 1.40–1.45; IV, 1.55–1.61. Rostrum: 0.90–1.00 long. Pronotum: 0.39–

0.44 long; 0.61–0.64 wide. Differing from 3rd instar as follows: mesonotal wing buds with a row of strong setae along the external borders and rows of weaker setae along incipient wing veins; metanotal wing bud with a strong seta at apex; rostrum reaching apices of mesocoxae; dorsal abdominal setae more prevalent, each one longer than the length of the segment on which it occurs. *5th instar* (n = 3). (Fig. 27d). Body: 3.01–3.70 long; 1.07–1.21 wide. Head: 0.85–0.90 wide. Antennal segment lengths: I, 0.91–0.92; II, 2.14–2.55; III, 1.78–1.92; IV, 1.81–2.00. Rostrum: 1.17–1.31 long. Pronotum: 0.50–0.54 long; 0.75–0.90 wide. Differing from 4th instar as follows: chaetotaxy and shape as illustrated (Fig. 27d); apices of hind coxae weakly reddish annulate; propleural stripe usually interrupted at middle.

Distribution. Lāna'i, ,W Maui, and Moloka'i Is.

Host. *Tetraplasandra hawaiiensis* and its var. *gracilis*, *T. meiandra* and its var. *ramosior*, *Reynoldsia sandwicensis*.

Etymology. The name means "dweller on 3 islands," referring to the distribution of this species.

Type data. Holotype ♂ (BPBM 14,841), allotype ♀, paratypes 6♂, 3♀, **LĀNA'I**: Kaiholena Gulch, 750 m, 9.VI.1971, on *Tetraplasandra hawaiiensis*, W.C. Gagné. Additional paratypes with data differing only as follows: 2♂, 3♀, Lāna'ihale, near Pu'u 'A'ali'i, 2700 ft. [818 m], 6.VI.1971; 6♂, 4♀, Lāna'ihale, 3000 ft. [914 m], 6.VI.1971, on *T. meiandra*. **MOLOKA'I**: 5♂, 8♀, 3N, Kalaupapa, 60 m, 10.III.1974, on *Reynoldsia sandwicensis*; 6♂, 3♀, Hālawa Valley, 300 ft. [91 m], 5.III.1972, on *R. sandwicensis*; 4♂, Mākolelau Valley, 700 m, 4.III.1972, on *R. sandwicensis*; 10♂, 8♀, Pēpē'ōpae, 4000 ft. [1219 m], 13.VIII.1971, on *Tetraplasandra meiandra*; 13♂, 10♀, 8N, Kamiloloa Gulch, 700 m, 15.VII.1971, on *R. sandwicensis*. **W MAUI**: 1♂, Kaulalewelewe, 915 m, 27.V.1972; 1♂, Pōhākea Gulch, 1190 m, 13.IV.1971, on *Cheirodendron trigynum*; 1♀, Makamaka'ole Stream, 420 m, 31.III.1970, on *T. hawaiiensis* var. *gracilis*; 2♀, 4N, Kaulalewelewe, 915–1037 m, 23.XII.1968, on *T. meiandra* var. *ramosior*; 1♂, 1♀, 13N, Kaulalewelewe, 915–1037 m, 8.VIII.1968, on *T. meiandra* var. *ramosior*.

Variation. (Fig. 55a-d). The upper arm of the left clasper varies in thickness and in the number and placement of the teeth and denticle. Some males of the Lāna'i and Moloka'i populations have the fewer teeth and the apex of the arm is not sinuate as the W Maui populations. The shape of the right clasper, aedeagus, and spiculum exhibited less discernable variation. A right clasper of the Kamiloloa Gulch, Moloka'i, population is also illustrated (Fig. 55b) to show the slightly different shape on the inner surface and the placement of the setae. This variation suggests 3 populations that still have some gene flow between them, or have only been recently isolated.

Discussion. No specimens have been taken on *Tetraplasandra hawaiiensis* on Moloka'i. *Nesiomiris trinesiotes* has been taken sporadically on *T. hawaiiensis* on W Maui, where it coexists with *N. pallidus* and *N. curvipes*.

I consider the collection of a single male on *Cheirodendron trigynum* to be a perching record. Adults and nymphs have been found only on *Reynoldsia sandwicensis*, *Tetraplasandra hawaiiensis* and *T. meiandra*, and these I consider to be true host associations.

On *T. hawaiiensis*, this species seems to predominate on the greenish and glabrous leaf surfaces, water shoots or suckers.

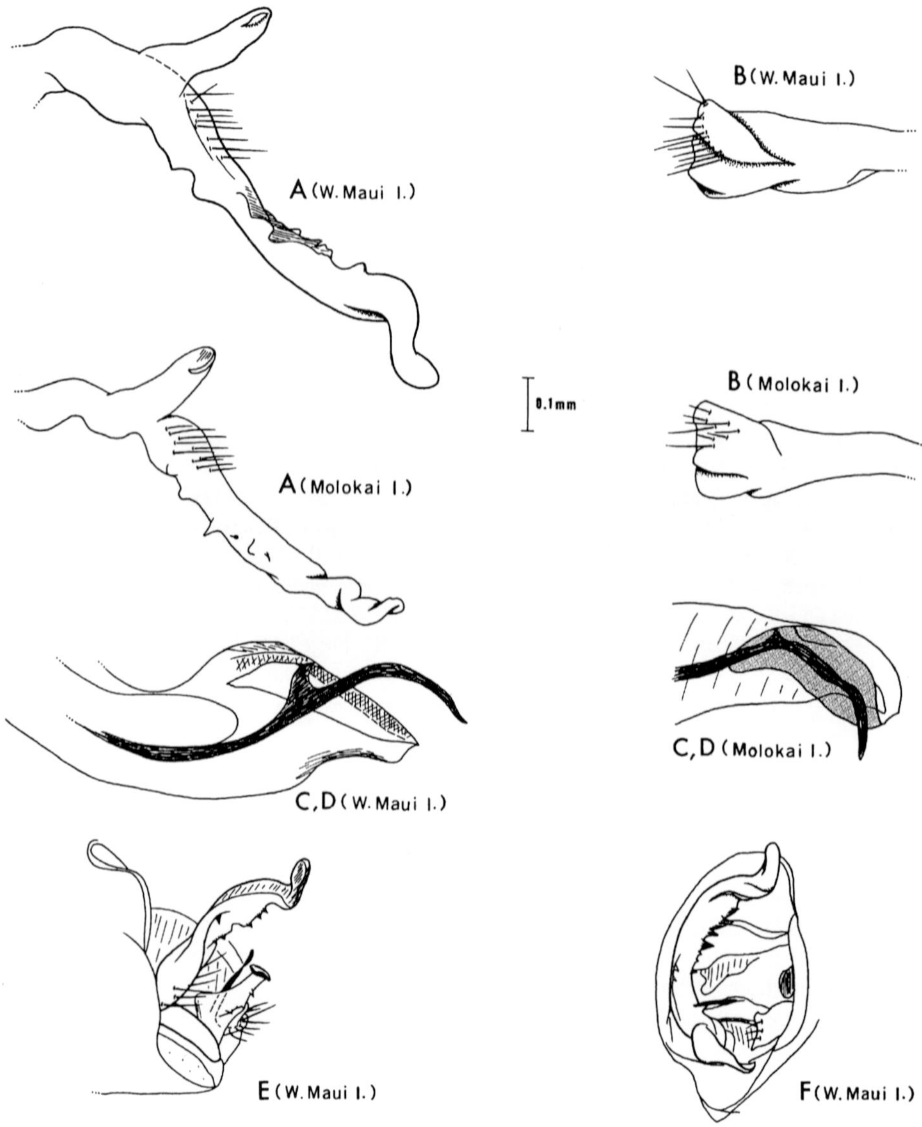

Fig. 55. *Nesiomiris trinesiotes,* n. sp., male genitalia, showing variation of clasper, aedeagus, and spiculi of Moloka'i and W Maui populations. **a**, left clasper, internal; **b**, right clasper, internal; **c, d**, aedeagus and spiculum, lateral; **e**, pygofer, left lateral; **f**, pygofer, terminal.

17. ***Nesiomiris timberlakei*** Gagné, **new species** Figs. 27c; 45a, b; 56

Nesiomiris sp.: Timberlake, 1921, *Proc. Hawaii. Entomol. Soc.* 4: 472 (note, host).

Timberlake (1921) noted that the populations on *Reynoldsia* on O'ahu and Hawai'i Islands were the same species. This species is about the same size as *N. secularis* and is separated from it with difficulty other than for the details of the right clasper. *Nesiomiris timberlakei* has a very reduced right clasper that lacks an internally directed blade (Fig. 56b). In this respect, *N. timberlakei* is closest to *N. pallidus* (Fig. 57b) and *N. trinesiotes* (Fig. 55b). The arms of the claspers are reduced to 3 small, sclerotized knobs at the periphery of the clasper. *Nesiomiris timberlakei* would appear to form a sister species with *N. pallidus*.

Description. **MALE** (n = 7). Body; 6.08 (5.05–6.08) long; 1.20 (1.15–1.28) wide. Head: 0.93 (0.86–0.93) wide; apex slightly indented (Fig. 45a, b); basal carina obsolescent. Antennal segment lengths: I, 0.90 (0.81–0.96); II, 2.97 (2.42–3.13); III, 1.95 (1.54–2.30); IV, 1.81 (1.43–2.06); I shorter than head width, ratio 7: 7.25. Rostrum: 1.21 (1.17–1.31) long. Pronotum: 0.93 (0.86–1.00) long; 1.20 (1.14–1.28) wide. Differing from *N. secularis* as follows: setae sparser and paler; hind lobe more finely rugose-punctate; less strikingly truncated posteriorly; mesosternum not differing in color from rest of venter; hind femora more slender-elongate, ratio length to width as 8.2: 0.75 (8.0: 0.80 in *N. secularis*), more pilose, suberect hairs more abundant. Abdomen: pygofer, clasper and aedeagus as illustrated (Fig. 56); all arms of right clasper reduced to small knobs; atrium and apex of aedeagus twisted to left (Fig. 56c).

FEMALE (n = 7). Body: 5.84 (4.44–5.84) long; 1.25 (1.19–1.31) wide. Head: 0.93 (0.90–0.93) wide. Antennal lengths: I, 0.93 (0.93–1.09); II, 2.85 (2.42–2.86); III, 2.07 (1.89–2.30); IV, 1.53 (1.53–2.14). Rostrum: 1.21 (1.10–1.31) long. Pronotum: 0.93 (0.93–1.02) long; 1.25 (1.09–1.31) wide. Differing from male as follows: antennal segment II paler, posterior pronotal lobes pale green, and more declivent anteriorly.

NYMPHS. *1st instar* (n = 1). Body: 1.77 long; 0.34 wide. Head: 0.37 wide; apex broadly rounded; lacking basal carina, setae pale, indistinct, a couple of stronger setae at apex only. Antennal segment lengths: I, 0.21; II, 0.30; III, 0.31; IV, broken; II and III weakly infuscate at apices, otherwise pale; I and II with a few erect setae on inner aspects mixed with reclined, paler setae that become shorter and denser on more apical segments. Rostrum: 0.59 long; reaching bases of hind coxae. Pronotum: 0.15 long; 0.34 wide; wider than meso- and metanota; not truncated and posterior borders straight; propleural stripe absent; glabrous, as well as remainder of body; pro-, meso-, and metanota subequal in size and shape. Legs: pale throughout; with a few suberect setae. *2nd instar* (n = 1). Body: 1.55 long; 0.47 wide. Head: 0.46 wide. Antennal segment lengths: I, 0.26; II, 0.55; III, 0.49; IV, 1.00. Rostrum: 0.65 long. Pronotum: 0.20 long; 0.40 wide. Differing from 1st instar as follows: head additionally with a seta at posterior-lateral angle, and another between eye and antennal base; rostrum just surpassing apices of mesocoxae; propleural stripe present, more distinct on anterior 1/2; anterior margin of pronotum concave; ratio of lengths of pronotum: mesonotum: metanotum along midline, as 1: 1: 0.5; hind margin of metanotum concave; abdominal dorsum sometimes with infumate pattern. *3rd instar* (n = 1). Body: 1.80 long; 0.56 wide. Head: 0.54 wide. Antennal segment lengths: I, 0.37; II, 0.92; III, 0.86; IV, 1.17. Rostrum: 0.75 long. Pronotum: 0.21 long; 0.48 wide. Differing from 2nd instar as follows: meso- and metanota together forming a continuous convex border, wider than pronotum and infuscate on anterior humeral border, with 1 stronger seta on infuscation on anterior humeral angle; infuscations on antennae weakly reddish. *4th instar* (n = 1). Body: 2.74 long; 0.73 wide. Head: 0.70 wide. Antennal segment lengths: I, 0.46; II,

1.59; III, 1.39; IV, 1.61. Rostrum: 1.00 long. Pronotum: 0.37 long; 0.61 wide. Differing from 3rd instar as follows: generally more pilose, setae generally pale; posterior margin of pronotum weakly concave, slightly constricted at humeral angle; mesonotal wing bud with a row of stronger, erect setae along the convex lateral edge, but not forming a discernable pattern. *5th instar* (n = 1). Body: 3.52 long; 1.07 wide. Head: 0.79 wide. Antennal segment lengths: I, 0.76; II, 2.35; III, 3.37; IV, 1.74. Rostrum: 1.00 long. Pronotum: 0.48 long; 0.75 wide. Differing from 4th instar as follows: apex of head weakly striolate; rostrum reaching middle of mesosternum; propleural stripe interrupted at middle; pronotum more deeply concave posteriorly; apex of head slightly concave; mesonotal wing bud with a row of stronger erect setae along the exterior edge.

Distribution. O'ahu and Hawai'i Is.

Host. *Cheirodendron trigynum* and its vars. *trigynum* and *halawanum*, *Reynoldsia sandwicensis*, *Tetraplasandra gymnocarpa*, *T. hawaiiensis* and its var. *microcarpa*, *T. kaalae* and its var. *multiplex*, *T. turbans*, *T. waianensis*.

Etymology. Named for the late Dr. P.H. Timberlake, renowned hymenopterist, based at the University of California, Riverside, for over half a century. He was among the first to collect this species, and the first to recognize the species as distinct from the generotype, *N. hawaiiensis*, and further, recognize the populations on O'ahu and Hawai'i Islands as being conspecific.

Type data. Holotype ♂, (BPBM 14,842) allotype ♀, paratypes 14♂, 14♀, **O'AHU**: Wai'anae Mts, gulch near Pu'u Pane, 2200 ft. [666 m], 17.V.1970, on *Tetraplasandra kaalae*, W.C. Gagné. Additional paratypes with data differing only as follows: 3♂, 1N, Wailupe, 21.IV.1918, on *Reynoldsia* sp., O.H. Bryant; 1♀, Wai'anae, VIII.1955, Ford; 1♂, Wailupe Valley, 24.V.1960, J.W. Beardsley; 1♂, Mt Ka'ala, 16.V.1967, C.M. Yoshimoto; 1♂, hills behind Dillingham A.F.B., X.1968, on *R. sandwicensis*; 1♂, hills behind Dillingham A.F.B., 14.XII.1968, on *R. sandwicensis*; 3♂, 4♀, hills behind Dillingham A.F.B., 21.IX.1969, on *R. sandwicensis*, E.F. Drake; 3♂, 2♀, 26N, hills behind Dillingham A.F.B., sea level, 28.II.1970, on *R. sandwicensis*; 16♂, 6♀, [no elevation], 28.II.1970, on *R. sandwicensis*; 5♂, 3♀, 1500 ft. [457 m], 1.III.1970, on *R. sandwicensis*; 2♂, 1500 ft. [457 m], 3.V.1970, on *R. sandwicensis*; 1♂, 1♀, Honouliuli Forest Reserve, W of S fork of Kalua'a Gulch, 800 m, 26.IV.1970; 1♂, Honouliuli Forest Reserve, Kalua'a-Waieli Ridge, 760 m, 26.IV.1970; 3♂, 2♀, 13N, Mokulē'ia Forest Reserve, gulch SSE of Pu'u Iki, 1500 ft. [457 m], 1.III.1970, on *R. sandwicensis*; 23N, Mokulē'ia Forest Reserve, gulch SSE of Pu'u Iki, 1500 ft. [457 m], 3.V.1970; 5♂, 9N, Pu'u Pane, 2000 ft. 609 m], 17.V.1970; 3♀, Dupont Trail, 1096 m, 21.V.1970, on *Cheirodendron trigynum* var. *halawanum*; 2♂, 1♀, Mt Ka'ala, Dupont Trail, 1200 m, 21.V.1970, on *C. trigynum* var. *halawanum*; 1♀, 1N, Kumaipo-Makaha Valley, 930 m, 28.V.1970, on C. trigynum; 54N, Mokulē'ia Forest Reserve, Makaleha Stream, E branch, 457m, 19.VII.1970; 18♂, 16♀, 2200 ft. [666 m], 29.V.1970; 12♂, 16♀, Mākaha Valley, 800 m, 28.V.1970; 1♂, 1♀, hills behind Dillingham A.F.B., [no elevation], 8.VI.1970, on *R. sandwicensis*;1♂, 3♀, 4N, hills behind Dillingham A.F.B., 200 ft. [61 m], 8.VI.1970, on *R. sandwicensis*; 1♂, 1♀, 'Ōpae'ula Gulch, 1000 ft. [305 m], 12.VI.1970, on *T. gymnocarpa*, S.L. Montgomery; 5♀, Mokulē'ia Forest Reserve, Kaimuhole Gulch, 1400 ft. [427 m], 8.VI.1970, on *R. sandwicensis*; 10♂, 5♀, Mokulē'ia Forest Reserve, Palikea Gulch, 300 m, 18.VI.1970, on *T. turbans*, S.L. Montgomery; 5♂, 14♀, 24N, Mokulē'ia Forest Reserve, Palikea Gulch, 425 m, 1.VII.1970, on *T. turbans*, S.L. Montgomery; 8♂, 4♀, 3N, Wai'anae Kai Valley, N end, 800 m, 5.VII.1970, on *T.*

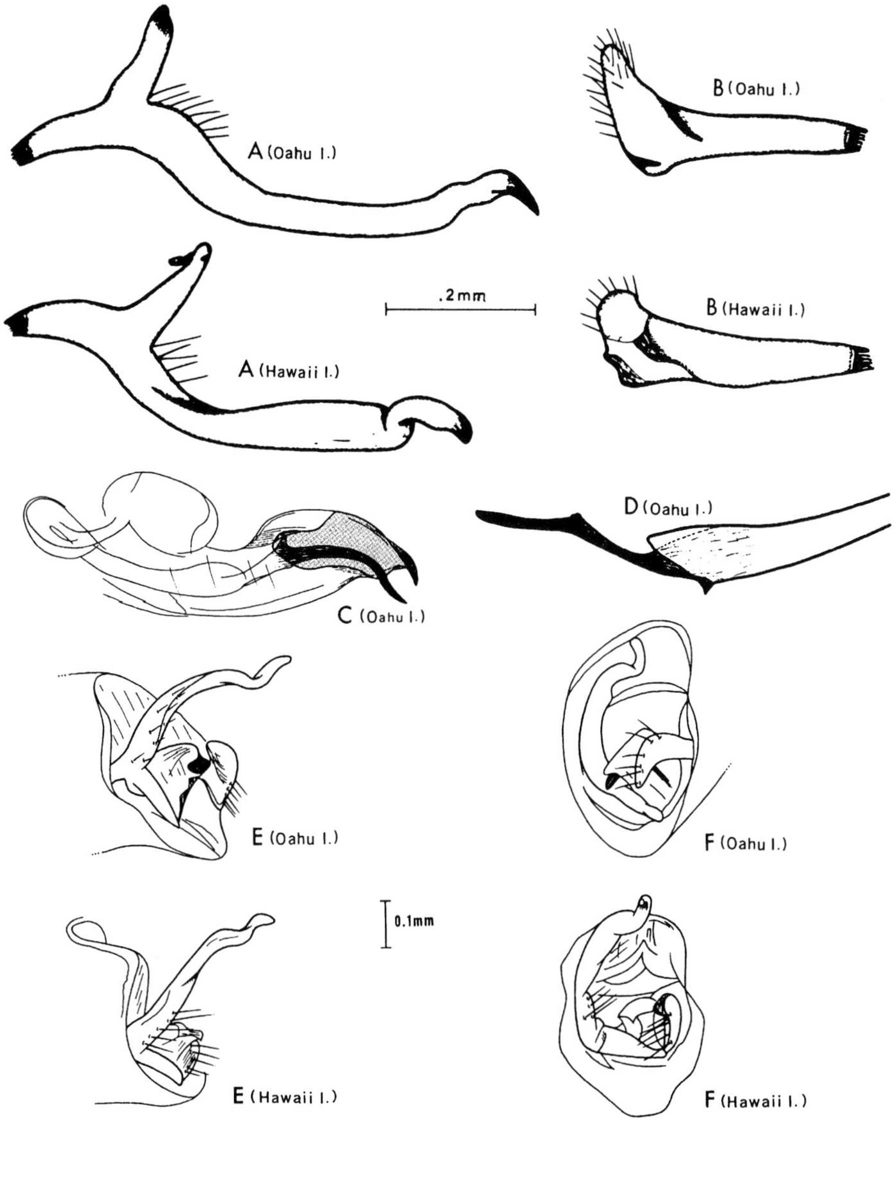

Fig. 56. *Nesiomiris timberlakei,* n. sp., male genitalia, showing variation of claspers of O'ahu and Hawai'i I populations. **a**, left clasper, internal; **b**, right clasper, internal; **c,** aedeagus and spiculum, lateral; **d**, spiculum, lateral; **e**, pygofer, left lateral; **f**, pygofer, terminal.

kaalae var. *multiplex*; 6♀, 5N, ʻŌpaeʻula Stream, 1000 ft. [305 m], 16.VII.1970, on *T. gymnocarpa*; 13♂, 6♀, Makaleha Stream, E branch, 19.VIII.1970; 3♂, 2♀, 50N, Mokulēʻia Forest Reserve, Makaleha Stream, E branch, 1500 ft. [457 m], 19.VIII.1970; 1♂, 1♀, Kaua Trail, Puʻu Kaua, 2200 ft. [666 m], 2.VIII.1970;1♂, 13N, Palikea Gulch, below Puʻu Pane, 425 m, 1.VII.1970, on *T. kaalae*; 2N, Palikea Gulch, below Puʻu Pane, 425 m, 24.X.1970, on *R. sandwicensis*; 1♂, 6N, Palikea Gulch, below Puʻu Pane, 425 m; 4N, hills behind Dillingham A.F.B., 100 ft. [31 m], 15.XI.1970, on *R. sandwicensis*; 1♀, Mokulēʻia Forest Reserve, Palikea Gulch, 425 m, 24.XI.1970; 2♂, 7♀, 7N, Koʻolau Hts., Kuluʻī Gulch, 225 m, 31.I.1971, on *R. sandwicensis*; 1♂, 1♀, Makaleha Stream, E branch, 300 m, 30.I.1971; 1♀, 9N, Makaleha Stream, central branch, 1000 ft. [305 m], 30.I.1971; 4♂, 8♀, 8N, Kuaokalā Forest Reserve, Keʻekeʻe, Gulch, 1600 ft. [488 m], 26.IX.1971; 17♂, 5♀, Mokulēʻia Forest Reserve, Manuwai Valley, 6.VIII.1972, on *T. waianensis*, J. Obata; 1♂, 1♀, Mākaha Valley, E side, 800 m, 19.VI.1972, on *T. micrantha*; 1♂, 3♀, Mākua Valley, S side, 470 m, 26.V.1974, [no host]; 1♂, 3♀, Mākua Valley, 100 m, 26.V.1974, on *R. sandwicensis*; 3♀, Mānoa Valley, Lyon Arboretum, 130 m, 13.II.1976, on ornamental *Munroidendron racemosum*, S.L. Montgomery. **HAWAIʻI**: 10♂, 8♀, Kaʻū District, Manukā Pk, 1700 ft. [518 m], 29.VII.1970, on *Tetraplasandra hawaiiensis*; 2♂, 2♀, 2N, near S Kona, Government Road, 18.VIII.1919, on *Reynoldsia sandwicensis*, P.H. Timberlake; 7♂, 6♀, ʻŌlaʻa Forest Reserve, Wright Road, 28.XI.1968, on *Cheirodendron trigynum* var. *trigynum*; 9♂, 12♀, Kohala Mts, Kupehau Ridge, 800 m, 5.XII.1968, on *T. hawaiiensis* var. *microcarpa*; 5♂, 3♀, Kīpuka Puaulu, [no elevation], 7.I.1971, on *C. trigynum* var. *trigynum*; 2♂, 1♀, Kīpuka Puaulu, 4000 ft. [1219 m], 29.XI.1968, on *C. trigynum*; 2♂, 2♀, Kohala Mts, Kupehau Ridge, 800 m, 13.VI.1970, on *T. hawaiiensis*; 14♂, 12♀, Kūlani Prison Road, 800 ft. [244 m], 25.VII.1970, on *T. hawaiiensis*; 1♀, Hwy 13 at 1955 lava flow, between Kaimū and Pāhoa, 800 ft. [244 m], 15.IX.1970; 1♀, Kaʻū District, Manukā Pk, 1200 ft. [518 m], 6.I.1971, on *T. hawaiiensis*; and 1♀, Kūlani Prison Road, 760 m, 21.II.1971, on *T. hawaiiensis*; 1♂, Kūlani Prison Road, 920 m, 28.II.1971, on *T. hawaiiensis*; 1♀, Manukā Park, 1760 ft. [523 m], 30.III.1971, on *T. hawaiiensis*; 2♂, 1♀, 19N, ʻŌlaʻa Forest Reserve, Wright Road, 1165 m, 28.II.1968, on *C. trigynum* var. *trigynum* (BPBM).

Variation. Although 2 disjunct island populations are represented here, the variation that could be attributed to geographical isolation appears to be minor. The Hawaiʻi Island population has the flabellum of the spiculum longer, the apex of the left and right claspers are slightly blunter. I do not consider these differences sufficient to warrant considering them as separate species, even though they appear to be absent on the intervening islands. There are discrete geographical size differences. Later nymphal instars (4th and 5th) vary in the degree of development of the propleural stripe. In most specimens it is present as an interrupted line, while a few have the stripe developed only on the anterior aspect.

Some males have the upper arm of the left clasper with lumps and small knobs. This feature seems to indicate incipient denticles and teeth that are so evident in the related *N. trinesiotes*.

Discussion. On both islands a wide range of araliaceous hosts are attacked. The peculiar distribution is paralleled, to some extent, in the endemic avifauna by the ʻ*elepaio* (*Chasiempis sandwicensis*) and the meliphagids.

There is also an interesting pattern in host associations. On Oʻahu practically all those endemic Araliaceae occurring in mesic to xeric habitats are attacked, including *Reynoldsia sandwicensis*, while on Hawai'i Island, the situation is essentially the same, with the notable exception of *Reynoldsia* (1 collection by Timberlake). Apparently *N. timberlakei* is mostly excluded from the latter by the presence of *N. secularis*; *N. secularis*, in turn, has a comparatively narrow host range on Hawai'i, but a broader one on Maui.

This species is one of the smallest in the genus; only *N. secularis* and *N. pallidus* are similarly sized in the *elongatus* subgroup. In the remainder of the genus, a few species of the *usingeri* complex, especially those on Maui, approach it in size.

18. *Nesiomiris pallidus* Gagné, **new species** Figs. 20, 25m, 30l, 45e, 57

Living adults and nymphs are only faintly greenish on the dorsum and are otherwise predominantly pallid white, matching closely the white pubescence on the undersides of the leaves of the host, *Tetraplasandra hawaiiensis*, rendering them cryptic on that background. On W Maui it shares this niche with *N. curvipes*. The adults are superficially easily confused, since both have the same color pattern. However, the latter has the strongly incrassate hind femora (Fig. 46b) as contrasted with the normal, straight femora of *N. pallidus*; also *N. curvipes* is notably larger, is subglabrous dorsally, and in profile has a pronotal shape much different from *N. pallidus* (Fig. 20). The genitalia (Figs. 56, 57) of *N. timberlakei* and *N. pallidus*, although differing in details, have the same basic structure, *viz*., a strongly reduced right clasper, a spiculum with a small, subterminal flabellum, a relatively unmodified aedeagus, and an unarmed pygofer.

Description. **MALE** (n = 6). Body: 5.77 (5.45–5.78) long; 1.20 (1.20–1.35) wide. Head: 0.85 (0.85–0.89) wide; apex broadly rounded; basal carina extending across base of head between eyes, very shallowly impressed along middle; pubescence pale. Antennal segment lengths: I, 0.95 (0.81–1.00); II, 2.80 (2.49–2.80); III, 1.25 (1.06–1.33); IV, broken (0.81–1.06); I and II pale excepting brown apex of II; III and IV weakly infuscate; pubescence on I much denser; scattered erect, longer setae on basal 1/2 of II that exceed the thickest part of this segment in length; I longer than head width (ratio 7.5: 6.75). Rostrum: 1.52 (1.45–1.61) long; just surpassing apex of mesosternum. Pronotum: 0.90 (0.90–1.00) long; 1.20 (1.20–1.35) wide; surface dull; hind lobe very shallowly rugose-punctate; propleural stripe interrupted above and posterior to coxal cleft. Differing from *N. timberlakei* as follows: broader and more robust; dorsum at most with scattered, irregular faint greenish areas, especially on hind lobe of pronotum and on hemelytra; pubescence long, dense, erect, and pale (Fig. 20); otherwise wholly pallid. Legs: erect pubescence of hind femora and tibiae about equal to thickness of base of femora, slightly shorter and not so predominantly erect on other legs; hind femora weakly, reddish annulate near apex. Hemelytra: membrane creamy infumate adjacent to anal ridge only; areolar veins yellowish; large areole narrowly coriaceous along inner border; small areole greenish, coriaceous. Mesosternum creamy, of a different texture than remainder of venter. Abdomen: pallid; pygofer, claspers, and aedeagus as illustrated (Fig. 57); pygofer unarmed.

FEMALE (n = 6). Body: 6.11 (5.57–6.11) long; 1.33 (1.22–1.35) wide. Head: 0.89 (0.85–0.89) wide. Antennal lengths: I, 0.92 (0.81–1.00); II, 2.65 (2.49–2.84); III, 1.10 (1.10–1.65); IV, 1.09 (0.89–1.09). Rostrum: 1.55 (1.53–1.61) long. Pronotum: 0.92 (0.90–1.00) long; 1.33 (1.22–1.35) wide. Differing from male as follows: antennal segment I weakly annulate rufescent at apex, ratio of length to antennal segment I to head width as 7: 6.75. Structure of bursa copulatrix as illustrated (Fig. 25m).

NYMPHS. *1st instar*. Unknown. *2nd instar* (n = 1). Body: 1.70 long; 0.46 wide. Head: 0.44 wide; apex evenly rounded; basal carina absent; setae brown, a row of 6 setae along hind margin, the pair astride the ecdysial line strongest and longest, a few more setae near and on apex, and 1 between antennal socket and eye. Antennal lengths: I, 0.18; II, 0.26; III, 0.20; IV, 0.56; III strikingly shorter and IV strikingly longer than for same instar of other species examined; II and III reddish annulate near their apices; IV infuscate at apex, otherwise antennae pallid, the scattered, stronger erect setae about equal in length to the thickness of antennal segment I on all segments, only imperceptibly shortening on distal segments. Rostrum: 0.89 long; reaching apices of hind coxae. Pronotum: 0.22 long; 0.39 wide; propleural stripe strongly evident at anterior humeral angle of pronotum, faint at posterior humeral angle, similarly appearing on posterior 2/3 of mesopleuron, where it is beset with 3 strong, evenly-spaced setae. Dorsum and legs beset with erect and suberect setae. Legs: entirely pallid. *3rd instar* (n = 1). Body: 2.14 long; 0.69 wide. Head: 0.61 wide. Antennal segment lengths: I, 0.39; II, 0.95; III, 0.95; IV, broken. Rostrum: 0.76 long. Pronotum: 0.27 long; 0.56 wide. Differing from 2nd instar as follows: a few setae on dorsum brown; pair of setae astride the ecdysial line not strikingly shorter, antennal setae as described for 3rd instar of *N. timberlakei*; rostrum reaching apices of mesocoxae; propleural line weakly interrupted only at middle, with a strong seta at posterior humeral angle, with area about base infuscate and reaching posterior margin of pronotum; no stripe on mesopleuron. *4th instar*. Unknown. *5th instar* (n = 16). Body: 3.20 long; 1.17 wide. Head: 0.82 wide. Antennal segment lengths: I, 0.56; II, 1.64; III, 1.03; IV, 1.23. Rostrum: 1.45 long. Pronotum: 0.56 long; 0.85 wide. Differing from 3rd instar as follows: most setae on dorsum brown; propleural stripe broadly interrupted at middle; setae bordering posterior margin of pronotum sometimes infuscate about their bases; wing bud reaching apex of 5th abdominal segment; legs hirsute, posterior pair more densely so; posterior femora with reddish annulate near apices.

Distribution. E and W Maui.

Host. *Tetraplasandra hawaiiensis* and its var. *gracilis*.

Etymology. Named for the predominantly pallid, only faintly greenish color.

Type data. Holotype ♂ (BPBM 14,843), allotype ♀, paratypes 9♂, 6♀, **W MAUI**: Makamaka‘ole Stream, 420 m, 31.III.1970, on *Tetraplasandra hawaiiensis* var. *gracilis*, W.C. Gagné. Additional paratypes with data differing only as follows: 1♀, S fork of Makamaka‘ole Stream, 420 m, 21.II.1970; 5N, Makamaka‘ole Gulch, 600m, 1.IV.1970; 2♂, 5♀, 2.II.1970; 15♂, 9♀, 1N, 11.IV.1971; 2♀, Kaulalewelewe, 915 m, 27.V.1972, on *T. hawaiiensis*; **E MAUI**: 3♂, 10N, Hana Forest Reserve, Kawaipapa Stream, 1100 ft. [335 m], 14.XII.1971, on *T. hawaiiensis*; 27♂, 22♀, Hana Forest Reserve, Kawaipapa Gulch, 335 m, 14.XII.1971 [no host]; 4N, Waiho‘i Valley, 610 m, 18.VII.1972, on *T. hawaiiensis* (BPBM).

Variation. The males sometimes lack the rufescent annulation near the apex of the hind femora.

Discussion. On W Maui, *N. trinesiotes* and, on E Maui, *N. secularis* sporadically occur on the green, glabrous sucker growth of *T. hawaiiensis*. Both these bugs are predominantly green and are also easily distinguished from *N. pallidus* in details of the male genitalia and the inconspicuous dorsal setae.

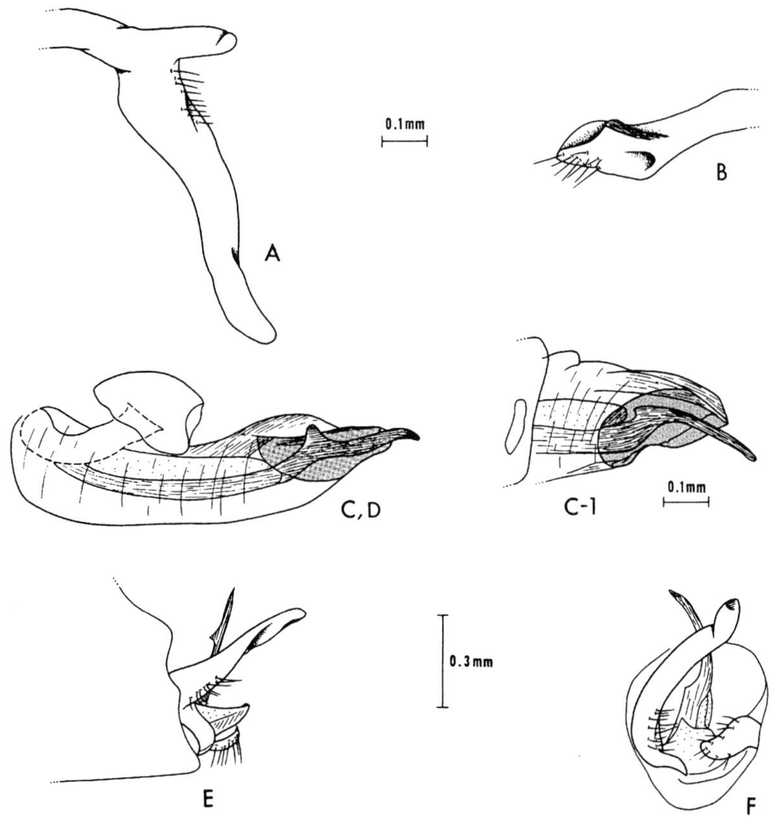

Fig. 57. *Nesiomiris pallidus*, n. sp., male genitalia. **a**, left clasper, internal; **b**, right clasper, internal; **c, d**, aedeagus and spiculum, lateral; **c-1**, apex of aedeagus and spiculum, dorsal; **e**, pygofer, left lateral; **f**, pygofer, terminal.

The specific differences in the hind femora, body pubescence and pronotal profile of *N. pallidus* and *N. curvipes* would appear to offer important clues in the morphological isolating mechanism of these closely related species. The 2 are convergent in morphology and in their specificity to the host, and the feeding and probably breeding sites are identical. The similar color patterns render these insects cryptic on the host. Hence, with quite similar male genitalia, the selection for the isolating mechanisms has appeared to center especially on body pubescence and pronotal profile. Antennal "patting" may provide the necessary information as to the identity of the other individual to avoid hybridization.

The *lihuensis* Group

***Description*.** **ADULTS**. Not sexually dimorphic for color; areolar veins rufescent sometimes yellowish (greenish in *N. bifurcatus*), not partially coriaceous nor infuscate; head usually rufescent; membrane never reduced; apex of venter of male without tuft of dense, short setae; aedeagus with apical or lateral tooth; propleural stripe present or absent; spiculum usually subapically flabellate.

NYMPHS. Propleural stripe absent; legs pallid; apices of antennae II and III at most infuscate at apices, never reddish.

***Distribution*.** All main islands, except Hawai'i I.

***Host*.** Species of *Cheirodendron* and *Tetraplasandra*.

***Included species*.** Nos. 19–39.

The *lihuensis* Subgroup

***Description*.** **ADULTS**. Apex of head bifid; antennal segment I and II yellowish, II becoming brown on apical 1/2; pubescence on dorsum and appendages erect, hirsute, golden brown; propleural stripe absent; pygofer toothed on dorsum; left clasper with upper arm uniramous; aedeagus with a right lateral tooth, spiculum strongly subapically flabellate; area about gonopore not denticulate.

***Distribution*.** Kaua'i I.

***Host*.** *Tetraplasandra lihuensis*.

***Included species*.** No. 19.

***Discussion*.** This subgroup is the most plesiomorphic in the *lihuensis* group because of the strongly flabellate spiculum and toothed aedeagus, characters which are otherwise widespread and shared with some species in both the *trifurcatus* and *lanaiensis* species groups.

19. *Nesiomiris lihuensis* Gagné, **new species** Fig. 58

This comparatively large and robust species lacks any indication of a propleural stripe, a character it shares in the *lihuensis* group with *N. waimeae*. It differs from the latter in having the dorsum as densely pilose as the legs and antennal segment I; the pronotum, like that of *N. tetraplasandra* (Fig. 45g), is strongly bilobed in profile, with the basal lobe strongly convex.

In details of the male genitalia, *N. lihuensis* approaches the general conformation of species in *uncinatus* complex. Since the spiculum is strongly flabellate, but not uncinate, I am placing *N. lihuensis* in a separate subgroup.

***Description*.** **MALE** (n = 6). Body: 7.27 (6.95–7.27) long; 1.54 (1.54–1.62) wide. Head: 1.06 (1.06–1.09) wide. Antennal segment lengths: I, 1.19 (1.17–1.28); II, 3.43 (3.20–3.45); III, 2.21 (1.93–2.21); IV, 1.95 (1.78–1.95). Rostrum: 1.51 (1.45–1.55) long. Pronotum: 1.10 (1.09–1.17) long; 1.54 (1.54–1.62) wide. Differing from *N. waimeae* as follows: head reddish; antennal segment II becoming brown on apical 1/2; III and IV, blackish; pubescence on I predominantly inclined rather than erect; longer setae not extending onto II; rostrum reaching apex of mesosternum. Dorsum of pronotum, mesonotum, scutellum, and hemelytra with long, slightly inclined, dense golden setae, these becoming shorter and paler on cuneus. Apex of scutellum green. Setae on legs still longer than that on dorsum, erect. Abdomen: teeth on pygofer dorsally located (Fig. 58e, f); right claspers larger than left; aedeagus with toothlike prominence on right side (Fig. 58c); spiculum with a subapical flabellum (Fig. 58d).

FEMALE (n = 6). Body: 7.58 (7.21–7.58) long; 1.62 (1.54–1.62) wide. Head: 1.06 (1.06–1.09) wide. Antennal segment lengths: I, 1.23 (1.17–1.28); II, 3.39 (3.20–3.45); III, 2.11 (1.93–3.21); IV, 1.95 (1.78–1.95). Rostrum: 1.52 (1.45–1.51) long. Pronotum: 1.17 (1.09–1.17) long; 1.62 (1.61–1.71) wide. Differing from male as follows: areolar veins more deeply rufescent.

NYMPHS. *1st instar* (n = 1). Body: 1.61 long; 0.46 wide. Head: 0.44 wide apex rounded; a row of 6 setae across base, 3 across the apex anterior to the eye and a pair on the apex; antennae and legs immaculate, former sparsely setaceous on I-III, IV glabrous, each setae about equal in length to thickness of I, setae inclined. Antennal lengths: I, 0.20; II, 0.34; III, 0.24; IV, 0.64. Rostrum: 0.69 long; reaching apices of hind coxa. Setae on head and thoracic segments longest, strongest, and darkest, each subequal in length to width of antennal segment I. Pronotum: 0.20 long; 0.46 wide; anterior margin convex, hind margin concave, lateral margin strongly convex, beset with an anterior row of 4 setae and a posterior row of 8 setae, the middle pair offset, with a patch of 8 setae on each disc, metanota with 6 setae on each disc; each abdominal tergite with a posterior row of setae; femora with shorter inclined setae, tibiae. *2nd instar*. Unknown. *3rd instar* (n = 1). Body: 2.84 long; 0.76 wide. Head: 0.66 wide. Antennal segment lengths: I, 0.43; II, 0.92; III, 0.69; IV, 1.28. Rostrum: 0.90 long. Pronotum: 0.37 long; 0.73 wide. Differing from 1st instar as follows: apex of head striolate; dorsal setae proportionately shorter; antennal segments II and III apically infuscate, IV becoming infuscate on apical 1/2, latter with short, inconspicuous setae, length of longest antennal setae about 1/2 that of thickness of I; rostrum reaching middle of mesocoxae; dorsal setae not longer, stronger, or darker than those on appendages; anterior margin of pronotum almost straight; lateral margin weakly convex; patch of setae on mesonotum more concentrated near anterior humeral angle; abdomen with spiracular openings infuscate; setae on femora mostly erect. *4th instar* (n = 1). Body: 3.23 long; 1.07 wide. Head: 0.82 wide. Antennal segment lengths: I, 0.64; II, 1.43; III, 1.17; IV, 1.49. Rostrum: 1.09 long. Pronotum: 0.46 long; 0.89 wide. Differing from 3rd instar as follows: antennal segment IV mostly infuscate; setae on appendages longer, paler, and pilose; thoracic tergites more densely setose and setae evenly distributed. *5th instar* (n = 1). Body: 5.45 long; 1.75 wide. Head: 1.12 wide. Antennal segment lengths: I, 0.81; II, 2.30; III, 1.65; IV, 1.78. Rostrum: 1.24 long. Pronotum: 0.61 long; 1.17 wide. Differing from 4th instar as follows: apex of head weakly bifid; pronotal collar present; rostrum reaching bases of mesocoxae.

Distribution. Kaua'i I.

Host. *Tetraplasandra lihuensis*.

Etymology. Named for the species epithet of its host.

Type data. Holotype ♂ (BPBM 14,844), allotype ♀, paratypes 7♂, 9♀. **KAUA'I**: Hā'upu, 480 m, 24.I.1973, on *Tetraplasandra lihuensis*, W.C. Gagné. Paratypes: 1♂, 1♀, 9N, Hoary Head Mts, N side of Hā'upu, 21.I.1973, on *Tetraplasandra lihuensis*, W.C. Gagné (BPBM).

Variation. The areolar veins appear to darken to red from yellows in older specimens; and in the latter, the small areolar vein sometimes remains brownish yellow on the basal half.

The *uncinatus* Subgroup

Description. **ADULTS.** Apex of head bifid to rounded; head and antennal segments I and II rufescent in fresh specimens; pubescence on dorsum and appendages reclined, brown, inconspicuous; propleural stripe absent (obsolescent in *N. uncinatus*); pygofer with a tooth on dorsum or above left clasper uniramous; aedeagus with a terminal "lip," or tooth, on right apex; spiculum apically uncinate, and sometimes subapically flabellate; area about gonopore sometimes denticulate.

Distribution. Kaua'i, Oahu, Moloka'i, and Maui Is.

Host. *Cheirodendron Fauriei, C. trigynum* and its varieties.

Included species. Nos. 20–26.

Discussion. This is the only subgroup of several existing on Kaua'i in the *lihuensis* group that has managed to colonize another island. This has happened only once, since the 3 species on O'ahu (*uncinatus* complex) are clearly monophyletic. Those on Moloka'i and Maui form another natural group (*usingeri* complex) of 3 species, which are most parsimoniously derived from an *uncinatus*-like species from O'ahu.

The *uncinatus* Complex

Description. **ADULTS.** Apex of head bifid; spiculum subapically flabellate and apically uncinate; tooth on pygofer lobate or acute; right clasper chelate, biramous; area ventral to gonopore usually denticulate.

Distribution. Kaua'i and O'ahu Is.

Host. *Cheirodendron* spp.

Included species. Nos. 20–23.

Discussion. I consider the subapically flabellate and apically uncinate spiculum to be apomorphic.

20. ***Nesiomiris cheirodendronae*** Gagné, **new species** Figs. 26f, 59

I am placing this species in the *uncinatus* complex because of the pubescence, color, the incipient lobate tooth on the pygofer, the similar claspers, aedeagus, uncinate spiculum, and host plant (*Cheirodendron*). The *K* structure is most similar to *N. trifurcatus*.

Description. **MALE** (n = 3). Body: 6.28 (6.10–6.28) long; 1.41 (1.39–1.41) wide. Head: 0.97 (0.97–1.03) wide; apex bifid, weakly impressed along middle of dorsum; weakly carinate across middle; dorsum and tylus rufescent; subglabrous; callosities shining. Antennal segment lengths: I, 0.94 (0.93–1.09), rufescent, brown; setae mostly reclined with a few erect longer ones on the inner aspects that are about 1/2 as long as thickness of segment, subequal to head width; II, 2.54 (2.54–3.01), brownish; III, 1.92 (1.92–2.37), dark brown; IV, 2.19 (2.12–2.19), dark brown. Rostrum: 1.35 (1.35–1.45) long. Pronotum: 0.96 (0.95–0.96) long; 1.41 (1.39–1.41) wide; basal lobe laterally sinuate, greenish across base, remainder of pronotum yellowish, sparsely pubescent. Mesonotum and scutellum yellowish, latter green only at extreme apex. Hemelytra: pale greenish; extreme apex of cuneus rufescent, large areolar vein rounded at apex. Legs and abdomen entirely pale. Pygofer (Fig.

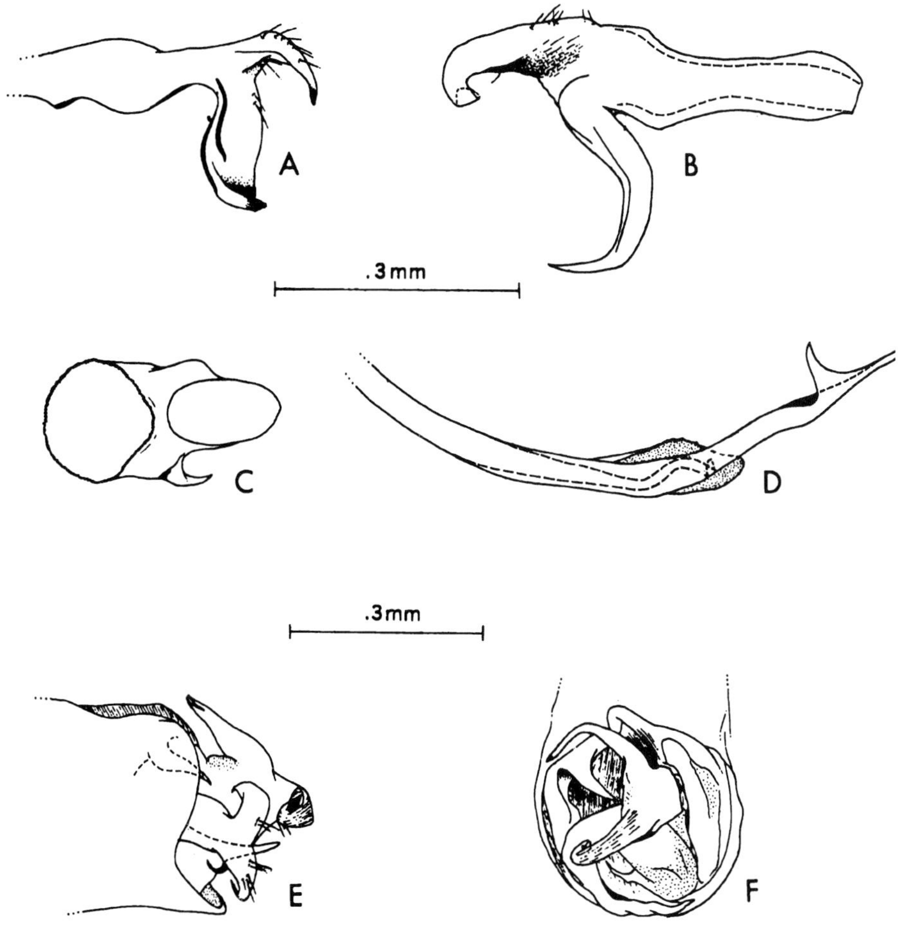

Fig. 58. *Nesiomiris lihuensis,* n. sp., male genitalia. **a**, left clasper, internal; **b**, right clasper, internal; **c**, **d**, aedeagus and spiculum, lateral; **c-1**, apex of aedeagus and spiculum, dorsal; **e**, pygofer, left lateral; **f**, pygofer, terminal.

59e) with an incipient tooth dorsally; upper arm of left clasper without denticles (Fig. 59a, f); right clasper biramous (Fig. 59b); aedeagus with a terminal lip (Fig. 59c) and a stout projection laterally on both sides (Fig. 59c–1); aedeagus uncinate, hooked at the flabellum (Fig. 59d); lacking denticles on fleshy lobe ventral to gonopore.

FEMALE (n = 4). Body: 6.80 (6.70–6.92) long; 1.44 (1.40–1.53) wide. Head: 0.95 (0.95–1.09) wide. Antennal segment lengths: I, 1.06 (1.05–1.17); II, 3.05 (3.01–3.37); III, 2.50 (2.14–2.50); IV, 2.37 (2.14–2.53). Rostrum: 1.56 (1.56–1.64) long. Pronotum: 1.00 (1.00–1.09) long; 1.44

(1.40–1.53) wide. Differing from male as follows: collar greenish; apical 1/2 of scutellum greenish; internal genitalia as illustrated (Fig. 26f); *K* structure weakly lobed, outer lobe shorter than inner lobe, and with few denticles, shape similar to *N. trifurcatus* (Fig. 25a).

NYMPHS. *1st to 3rd instars*. Unknown. *4th instar* (n = 1). Body: 2.35 long; 0.92 wide. Head: 0.80 wide. Antennal lengths: I, 0.64; II, 1.50; III, 1.28; IV, 1.44; extreme apices of II and III and all of IV infumate, remainder pallid. Rostrum: 1.10 long, reaching base of abdomen. Pronotum: 0.39 long; 0.73 wide. Appendages and dorsum clothed with sparse, erect, brown setae, those on mesothorax and segments beyond arranged in horizontal rows. *5th instar* (n = 1). Body: 4.32 long; 1.21 wide. Head: 0.91 wide. Antennal lengths: I, 0.81; II, 2.28; III, 1.95; IV, 2.14. Rostrum: 1.34 long. Pronotum: 0.58 long; 0.92 wide. Anterior wing buds with a row of lateral setae, others irregularly arranged on lateral aspects of wing bud; rostrum not exceeding apices of mesocoxae.

Distribution. Kaua'i I.

Host. *Cheirodendron Fauriei, C. kavaiensis* var. *kavaiensis, C. Helleri*.

Etymology. Named for the genus of the host plant, *Cheirodendron*.

Type data. Holotype ♂ (BPBM 14,845), allotype ♀, paratype 1♂, 2♀, 3N, **KAUA'I**: Hoary Mts, Hā'upu, 670 m, 30.VIII.1970, on *Cheirodendron Fauriei*, W.C. Gagné. Additional paratypes with data differing only as follows: 1♂, Pole Line Trail near Kualapa, 28.VIII.1970, on *C. kavaiensis* var. *kavaiensis*; 1♀, Pole Line Trail near Kualapa, 28.VIII.1970, on *C. Helleri*; 1♀, 20.XII.1968, on *Cheirodendron* (BPBM).

Discussion. This is a most systematically difficult area of the genus because the lineages that appear so distinct on the other islands, begin to merge on Kaua'i, and there are about a dozen species to contend with in the *bifurcatus* and related complexes on Kaua'i. This is not to say, however, that the species, individually, are not easily distinguished; if they were not, I would have to assume that the populations represented 1 quite variable species. What we appear to have is a great proliferation of species on a single island, and little simultaneous extinction to create more striking gaps between species.

21. ***Nesiomiris uncinatus*** Gagné, **new species** Fig. 60

Details of the male genitalia relate this species to *N. cheirodendronae*, that is, the apically downwardly directed, reduced upper arm of left clasper (Fig. 60e), comparatively weakly developed dorsal tooth on pygofer (Fig. 60f), internally facing blunt tooth between the arms of right clasper (Fig. 60b), a lobe at base of atrium on aedeagus (Fig. 60c), and uncinate spiculum (Fig. 60d). These characters could be considered plesiomorphic to those of *N. limitatus* and *N. beardsleyi*, which have the upper arm of the left clasper dorsally directed and facing the strongly developed dorsal tooth on the pygofer (Figs. 61e, 62e) and lack an internally facing blunt tooth between the arms of the right clasper (Figs. 61b, 62b). Synapomorphic characters here that make *N. uncinatus*, *N. limitatus*, and *N. beardsleyi* a natural group and a sister group of *N. cheirodendronae* are: presence of denticles on fleshy lobe beneath gonopore (Figs. 61d, 62d), subapical flabellum on spiculum (Figs. 61d, 62d), 2 unreduced arms on right clasper (Figs. 61b, 62b), and incipient vein at apex of the large areolar vein. Other differences from *N. cheirodendronae* are outlined below.

The obsolescent propleural stripe, which extends from behind the pronotal collar to the posterior humeral angle, at once distinguishes this species from other members of the *N. uncinatus* complex.

***Description*.** **MALE** (n = 3). Body: 6.29 (5.67–6.29) long; 1.32 (1.28–1.34) wide. Head: 0.94 (0.92–0.96) wide. Antennal segment lengths: I, 1.26 (1.15–1.28); II, 3.40 (2.84–3.40); III, 1.18 (1.81–1.85), IV, 0.75 (0.71–0.89). Rostrum: 1.56 (1.54–1.56) long. Pronotum: 1.00 (0.90–1.00) long; 1.32 (1.28–1.34) wide. Differing from *N. cheirodendronae* as follows: head more deeply rufescent, its apex reddish striolate; ratio of length of antennal segment I to head width as 10: 7.5; pronotum with apical half of hind lobe only with some green; propleuron with a very weakly indicated propleural stripe extending from behind collar to posterior humeral angle; setae on femora pale, not light brownish; apex of clavus rufescent. Abdomen: differences as noted above; pygofer, claspers, spiculum, and aedeagus as illustrated (Fig. 60).

FEMALE (n = 3). Body: 6.44 (6.40–6.50) long; 1.35 (1.35–1.40) wide. Head: 0.96 (0.92–0.96) wide. Antennal segment lengths: I, 1.21 (1.15–1.28); II, 3.32 (2.84–3.40); III, 1.85 (1.81–1.85); IV, 0.78 (0.71–0.89). Rostrum: 1.56 (1.54–1.56) long. Pronotum: 1.03 (0.90–1.00) long; 1.35 (1.28–1.35) wide. Apex of ovipositor rufescent, not exceeding apex of abdomen.

NYMPHS. Unknown.

***Distribution*.** O'ahu I.

***Host*.** *Cheirodendron trigynum* var. *hawawanum.*

***Etymology*.** Named for the uncinate or hooked apex of the spiculum.

***Type data*.** Holotype ♂ (BPBM 14,846), allotype ♀, paratypes 1♂, 2♀, **O'AHU**: W side Ka'au Crater, 600 m, 27.XII.1968, on *Cheirodendron trigynum* var. *hawawanum*, W.C. Gagné. Additional paratype with data differing only as follows: 1♂, Mt Tantalus, 1800 ft. [549 m], 19.IX.1968, on *Cheirodendron trigynum* var. *hawawanum*, W.C. Gagné (BPBM).

22. *Nesiomiris limitatus* Gagné, **new species** Fig. 61

This species is very close to *N. beardsleyi* and *N. uncinatus*, although there are constant differences in the genitalia of the population of the 4 males available. These differences are such that I consider it to occupy a position somewhat intermediate to *N. beardsleyi* and *N. uncinatus*. These characters are: dorsal sinuate tooth of pygofer with some small subsidiary teeth (Fig. 61e), apex of upper arm of left clasper also facing tooth on pygofer and just failing to meet it, but lacking denticles (Figs. 61a, e); spiculum is quite similar to *N. beardsleyi* but with more denticles on lobe below the gonopore; and subapical flabellum longer (Fig. 61d). The apical lip on the aedeagus is prolonged into a unique hornlike process (Fig. 61c). The right clasper is quite similar in these 3 species. Other differences from *N. beardsleyi* are noted below.

***Description*.** **MALE** (n = 4). Body: 5.96 (5.96–6.39) long; 1.22 (1.16–1.28) wide. Head: 0.91 (0.91–0.93) wide. Antennal segment lengths: I, 1.15 (1.15–1.26); II, 3.25 (2.95–3.38); III, 1.78 (1.78–2.08); IV, 0.91 (0.91–0.92). Rostrum: 1.55 (1.41–1.55) long. Pronotum: 1.00 (0.94–1.00) long; 1.22 (1.16–1.28) wide. Differing from *N. beardsleyi* as follows: ratio of length of antennal segment I to head width as 1.15: 0.91; pronotal callosities yellowish; hemelytra with some yellowish areas adjacent to base of scutellum. Pygofer, claspers, aedeagus and spiculum as illustrated (Fig. 61).

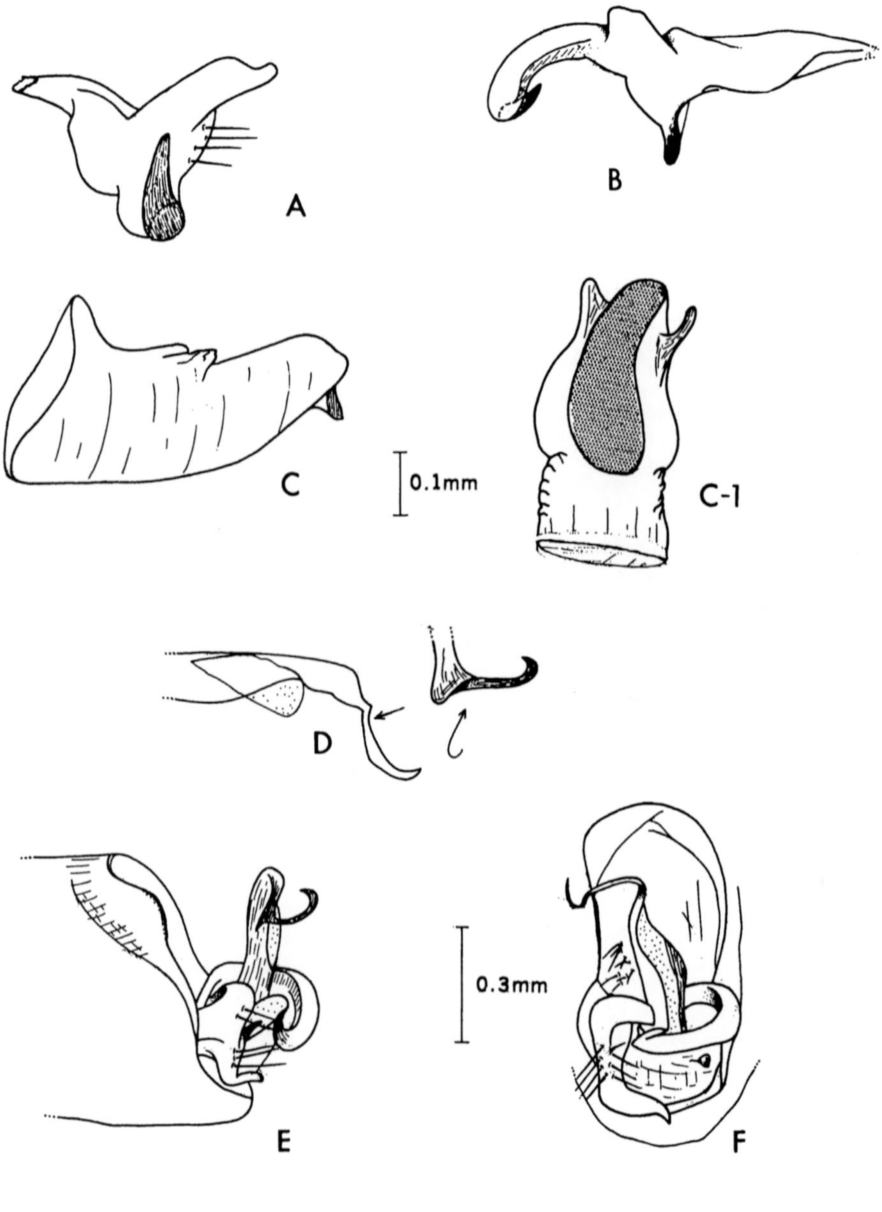

Fig. 59. *Nesiomiris cheirodendronae,* n. sp., male genitalia. **a**, left clasper, internal; **b**, right clasper, internal; **c**, aedeagus, left lateral; **c-1**, aedeagus, dorsal; **d**, spiculum, lateral and dorsal, **e**, pygofer, left lateral; **f**, pygofer, termnial.

FEMALE. Unknown.

NYMPHS. *1st to 3rd instars*. Unknown. *4th instar* (n = 1). Body: 3.55 long; 0.95 wide. Head: 0.76 wide. Antennal lengths: I, 0.75; II, 2.99; III, 1.17; IV, 0.80. Rostrum: 1.07 long. Pronotum: 0.40 long; 0.73 wide. Closely resembling same stage in *N. beardsleyi* with respect to stout, short, sparse setae on dorsum and anterior aspect of legs. *5th instar.* Unknown.

Distribution. O'ahu I.

Host. *Cheirodendron trigynum*.

Etymology. Named for its comparatively limited distribution on O'ahu I.

Type data. Holotype ♂ (BPBM 14,847), paratypes 1♂, 1N, **O'AHU**: Waiāhole Forest Reserve, Turnover Peak, 2000 ft. [605 m], 21.XI.1971, on *Cheirodendron trigynum*, W.C. Gagné. Additional paratypes with data differing only as follows: 2♂, 1N, 31.V.1970.

Discussion. This species has the narrowest distribution on O'ahu of any in this complex. It is apparently limited to the Turnover Peak area of the Waiāhole Forest Reserve, a mountain isolated by only a couple of kilometers from the northwestern end of the main Ko'olau Range. This isolation appears to have prevented gene flow from the more widespread *N. beardsleyi*. It is now quite isolated geographically from *N. uncinatus*.

23. ***Nesiomiris beardsleyi*** Gagné, **new species** Figs. 24c, 26a, 49b, 62

This species appears to form a natural group on O'ahu with *N. limitatus* and *N. uncinatus*. *Nesiomiris beardsleyi* shares characters with the O'ahu species as noted under *N. uncinatus*. From these it differs in having the upper arm of the left clasper apically armed with denticles and the dorsal tooth on the pygofer, itself with several prominent teeth.

Description. **MALE** (n = 7). Body: 5.78 (5.50–5.80) long; 1.21 (1.17–1.31) wide. Head: 0.91 (0.89–0.92) wide. Antennal segment lengths: I, 1.15 (1.05–1.15); II, 3.11 (2.95–3.20); III, 1.69 (1.60–1.85); IV, 0.94 (0.89–1.06). Rostrum: 1.39 (1.28–1.61); reaching base of mesocoxa. Pronotum: 0.94 (0.90–1.00) long; 1.21 (1.17–1.31) wide. Differing from *N. cheirodendronae* as follows: ratio of length of antennal segment I to head width as 1.15: 0.91; pronotum and scutellum entirely greenish, mesonotum greenish yellow; large areolar vein not rounded apically, with short branch at apex (Fig, 49b); membrane reddish infumate; femora greenish yellow, tibiae paler; setae on hemelytra paler; pygofer with dorsal tooth having several subsidiary teeth, upper arm of left clasper with denticles (Fig. 62e), its apex facing apex of dorsal tooth on pygofer (Fig. 62f), aedeagus with a terminal lip only (Fig. 62c); spiculum with fleshy lobe beneath gonopore, lobe with 5 denticles (Fig. 62d).

FEMALE (n = 7). Body: 6.28 (5.78–6.28) long; 1.29 (1.17–1.31) wide. Head: 0.92 (0.90–0.96) wide. Antennal segment lengths: I, 1.14 (1.03–1.14); II, 3.10 (2.84–3.10); III, 1.82 (1.60–1.85); IV, 0.91 (0.89–1.06). Rostrum: 1.35 (1.28–1.61) long. Pronotum: 0.95 (0.90–1.05) long; 1.29 (1.22–1.35) wide. Differing from male as follows: apex of head weakly striolate; ratio of length of antennal segment I to head width as 9: 7; pronotal callosities yellowish; clavus with yellowish areas adjacent to scutellum. *K* structure (Fig. 26a) distinctly bilobed, each lobe with denticles; sclerotized rings highly infolded laterally (Fig. 24c).

NYMPHS. *1st and 2nd instars*. Unknown. *3rd instar* (n = 1). Body: 2.49 long; 0.72 wide. Head: 0.58 wide, apex rounded. Antennal segment lengths: I, 0.39; II, 0.99; III, 0.89; IV, 0.64; ex-

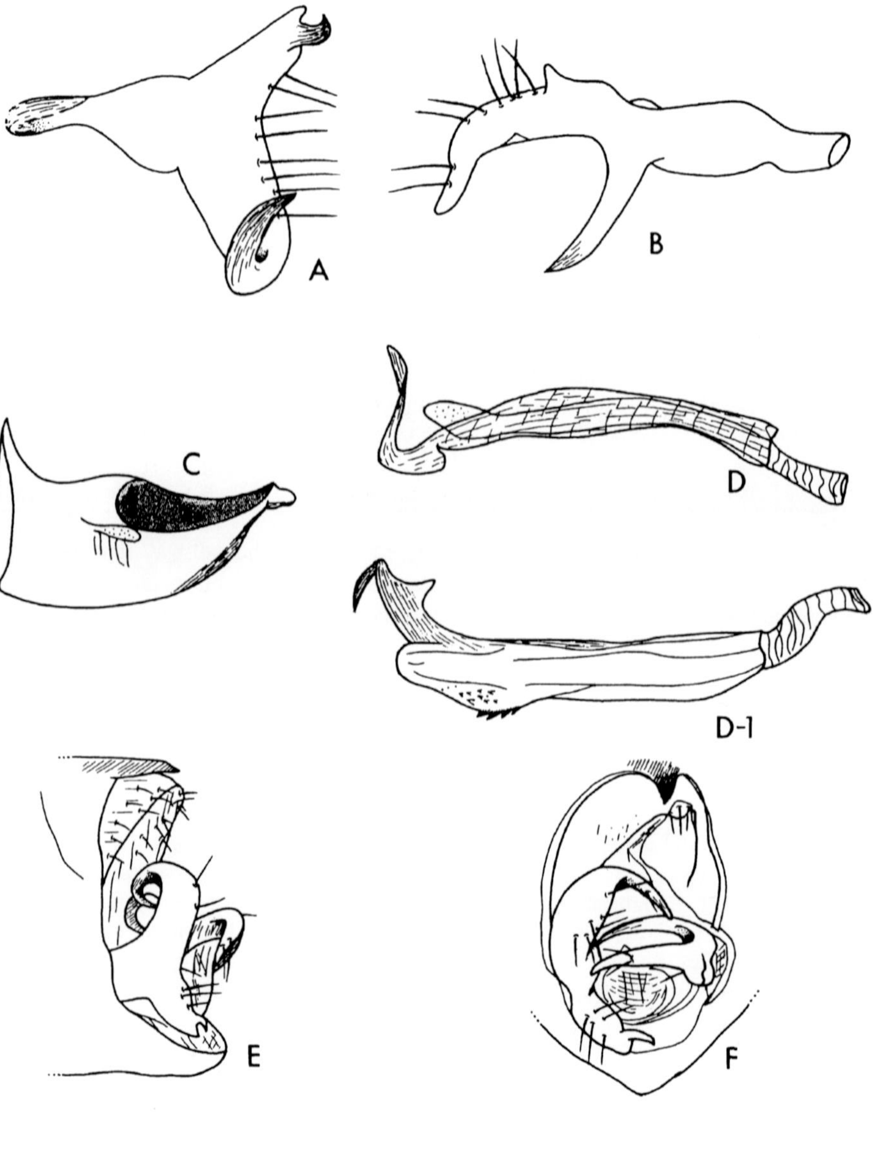

Fig. 60. *Nesiomiris uncinatus,* n. sp., male genitalia. **a**, left clasper, internal; **b**, right clasper, internal; **c,** aedeagus, lateral; **d**, spiculum, dorsal; **d-1**, spiculum, right lateral; **e**, pygofer, left lateral; **f**, pygofer, terminal.

treme apices of II and III weakly infuscate, otherwise pallid. Rostrum: 0.82 long; reaching hind coxae. Pronotum: 0.27 long; 0.60 wide. Setae on antennal segment I, femora, and dorsum short, brown, stout, spinelike, sparse, a row on each side of mid-line and 1 only along lateral margin of each segment, these becoming obsolescent on apical segments of abdomen; legs entirely pallid. *4th instar* (n = 1). Body: 2.84 long; 0.90 wide. Head: 0.70 wide. Antennal segment lengths: I, 0.64; II, 1.39; III, 1.17; IV, 0.69. Rostrum: 1.00 long. Pronotum: 0.37 long; 0.69 wide. Differing from 3rd instar as follows: antennal segment IV infumate; wing buds with a row of strong, stout setae along their lateral margins. *5th instar* (n = 1). Body: 3.01 long; 1.11 wide. Head: 0.85 wide. Antennal segment lengths: I, 0.64; II, 1.39; III, 1.17; IV, 0.69. Rostrum: 1.00 long. Pronotum: 0.37 long; 0.69 wide. Differing from 4th instar as follows: setae on appendages no longer spinose; apex of head indented; rostrum reaching apex of mesosternum; dorsal setae shorter, stouter, sparser.

Distribution. O'ahu I.

Host. *Cheirodendron trigynum* vars. *trigynum*, *Hillebrandii*, and *halawanum*.

Etymology. Named for the 1st collector J.W. Beardsley Jr. of the University of Hawai'i, Honolulu, a specialist on the Coccoidea and Encyrtidae, who has made outstanding contributions to the systematics of these groups on Hawai'i and the Pacific Basin.

Type data. Holotype ♂ (BPBM 14,848), allotype ♀, paratypes 2♂, 6♀, 8N, **O'AHU**: Wiliwilinui Ridge, 1800 ft. [549 m], 6.IX.1968, on *Cheirodendron trigynum* var. *Hillebrandii*, W.C. Gagné. Additional paratypes with data differing only as follows: 1♀, 26.XII.1956, [no host], J.W. Beardsley; 1♂, 19.VI.1964, on *Cheirodendron* sp., J.W. Beardsley; 4♂, 2♀, 3N, 8.XI.1968; 1♂, 1♀, 1N, Lā'ie Trail, 1400 ft. [427 m], 11.XI.1968, on *C. trigynum* var. *Hillebrandii*; 1♂, 3♀, 1N, 9.V.1970; 2♂, 1♀, 4N, Mt Ka'ala, Dupont Trail, 1200 m, 21.V.1970, on *C. trigynum* var. *halawanum*; 7♂, 4♀, Lā'ie Trail, 1800 ft. [550 m], 23.V.1970, on *C. trigynum*; 2♂, 6♀, Mākaha Valley, 800 m, 28.V.1970, on *C. trigynum*; 4♂, 1♀, 3N, Ko'olau Mts, Castle Trail, 790 m, 30.V.1971, on *C. trigynum*; 1♂, 2♀, 2N, 28.XI.1971, on *C. trigynum* (BPBM).

Discussion. This species has the widest distribution of this complex on O'ahu and is the only one found in both the Wai'anae and Ko'olau mountain ranges.

The *usingeri* Complex

Description. **ADULTS**. Apex of head rounded; spiculum apically uncinate, without subapical flabellum; tooth on pygofer located laterally above left clasper; right clasper not chelate, arms much reduced or absent; area ventral to gonopore not denticulate.

Distribution. Moloka'i and Maui Is.

Host. *Cheirodendron trigynum*.

Included species. Nos. 24–26.

Discussion. I consider the shared loss of the subapical flabellum on the spiculum to be synapomorphic.

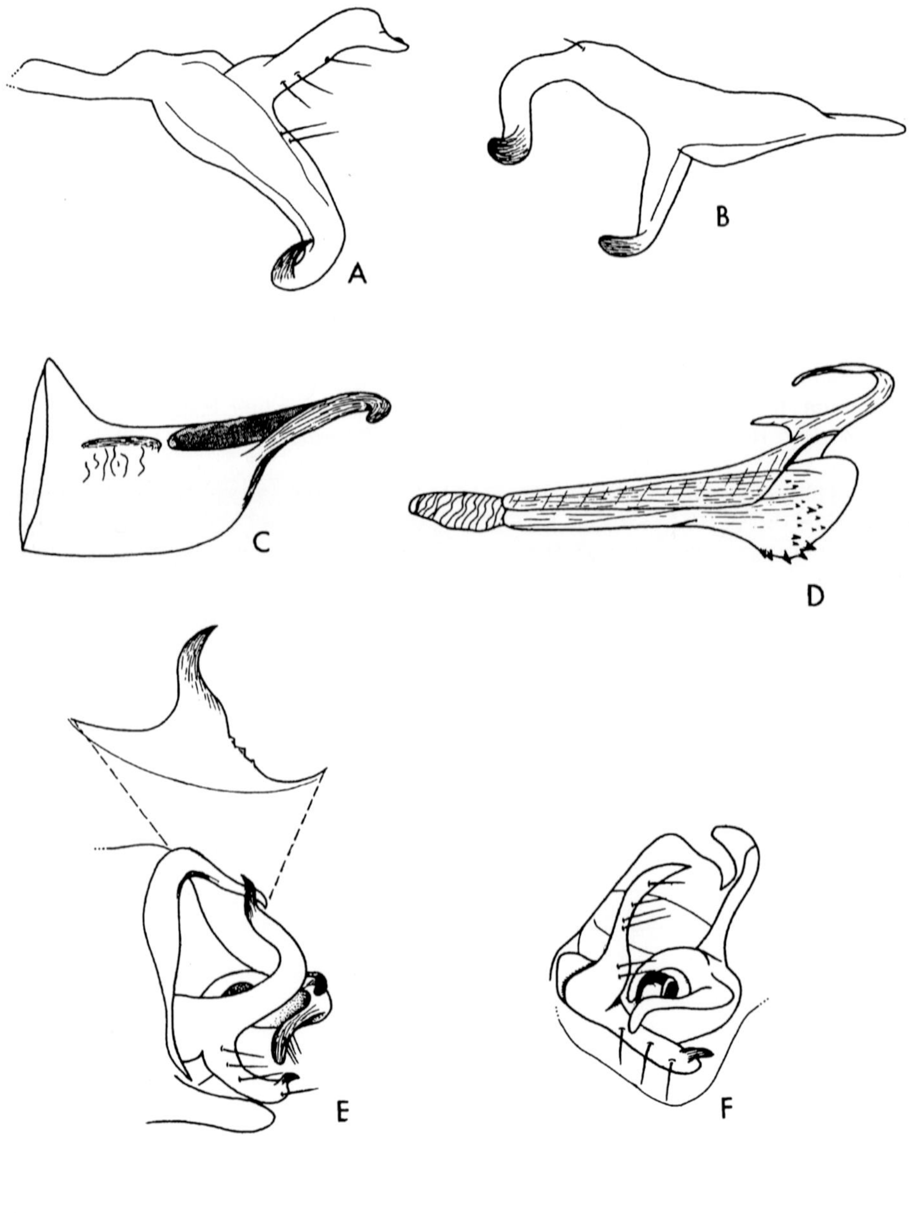

Fig. 61. *Nesiomiris limitatus,* n. sp., male genitalia. **a**, left clasper, internal; **b**, right clasper, internal; **c,** aedeagus, lateral; **d**, spiculum, right lateral; **e**, pygofer, left lateral with dorsal tooth on a pygofer inset; **f**, pygofer, terminal.

24. ***Nesiomiris usingeri*** Gagné, **new species** Figs. 26b, 63

Nesiomiris sp. Usinger, 1972, *Mem. Pac. Coast Entomol. Soc.* 4: 66.

This species is the most closely related to *N. kraussi* and *N. rufonotus.*

***Description*.** **MALE** (n = 6). Body: 4.92 (4.92–5.36) long; 1.19 (1.19–1.30) wide. Head: 0.90 (0.88–0.93) wide; apex bluntly rounded; weakly carinate across base; rufescent; a few brown setae on carina, frons and inner angle of eyes. Antennal segment lengths: I, 1.35 (0.7–0.83), rufus, vestiture dark brown; erect bristles paler; II, 2.23 (2.17–2.33), rufus on proximal 1/4 becoming dark brown distally, vestiture on distal 3/4 dense, a mixture of recumbent pale setae and erect brown setae; III, 1.29 (1.29–1.53), black, vestiture as for II distally; IV, 1.20 (1.20–1.40), color and vestiture as for III, shorter or equal in length to III. Rostrum: 1.20 (1.07–1.20) long; reaching apex of mesosternum. Pronotum: 0.77 (0.77–0.81) long; 1.19 (1.17–1.21) wide; yellow, paler than head; vestiture as for *N. cheirodendronae.* Hemelytra: clavus with green flecks; vestiture dark brown, semierect, comparatively dense. Legs: femora yellowish, vestiture pale, short erect; tibiae and tarsal segments 1 and 2 rufus, 3 brown; vestiture on tibiae erect, brown, long exteriorly, recumbent interiorly. Abdomen: pale and nonsclerotized above pygofer; atrium and genitalia as illustrated (Fig. 63).

FEMALE (n = 6). Body: 5.57 (5.57–6.00) long; 1.35 (1.30–1.40) wide. Head: 0.93 (0.91–0.93) wide. Antennal segment lengths: I, 1.19 (1.19–1.30); II, 2.06 (2.17–2.33); III, 1.42 (1.42–1.57); IV, 1.32 (1.32–1.47). Rostrum: 1.15 (1.15–1.30) long. Pronotum: 0.83 (0.64–0.83) long; 1.35 (1.33–1.40) wide. Differing from male as follows: antennal segment II, rufus on distal 1/2, vestiture on distal 1/2 not so dark, the pale recumbent setae sparser; antennal segment III, pale at base; tip of ovipositor rufus; *K* and *L* structures as illustrated (Fig. 26b).

NYMPHS. Unknown.

***Distribution*.** Moloka‘i I.

***Host*.** *Cheirodendron trigynum.*

***Etymology*.** The late R.L. Usinger (1972: 66) mentioned in his autobiography collecting this species on "*Cheirodendron Gaudichaudii*" (= *C. trigynum*) on Mapulehu-Puna‘ula Ridge, Moloka‘i; I have not recollected it, and no one else has either. It seems fitting to name this species for R.L. Usinger, to whom I owe both my career in entomology and my introduction to the fascination that is Hawai‘i. He has also played a major role in elucidating the systematics and evolution of the Heteroptera of the Hawaiian Islands.

***Type data*.** Holotype ♂ (BPBM 14,849), allotype ♀, paratypes 11♂, 12♀, **MOLOKA‘I**: Mapulehu-Puna‘ula Ridge, [no elevation], 15.VIII.1936, [no host], R.L. Usinger. Additional paratype with data differing only as follows: 1♀, on *Elaeocarpus* (BPBM).

***Discussion*.** The host record "on *Elaeocarpus*" is puzzling in 2 respects. *Elaeocarpus* is not known to occur on Moloka‘i, nor is it known to serve as a host plant for any other *Nesiomiris* species. Hence, this record may be a mislabeling, a misidentification of the plant, and/or a perching record of the insect.

25. ***Nesiomiris kraussi*** Gagné, **new species** Fig. 64

The male claspers and the tooth on the pygofer are closer in detail to *N. usingeri* than to *N. rufonotus*; it differs from the former in having a rufus head and from the latter in

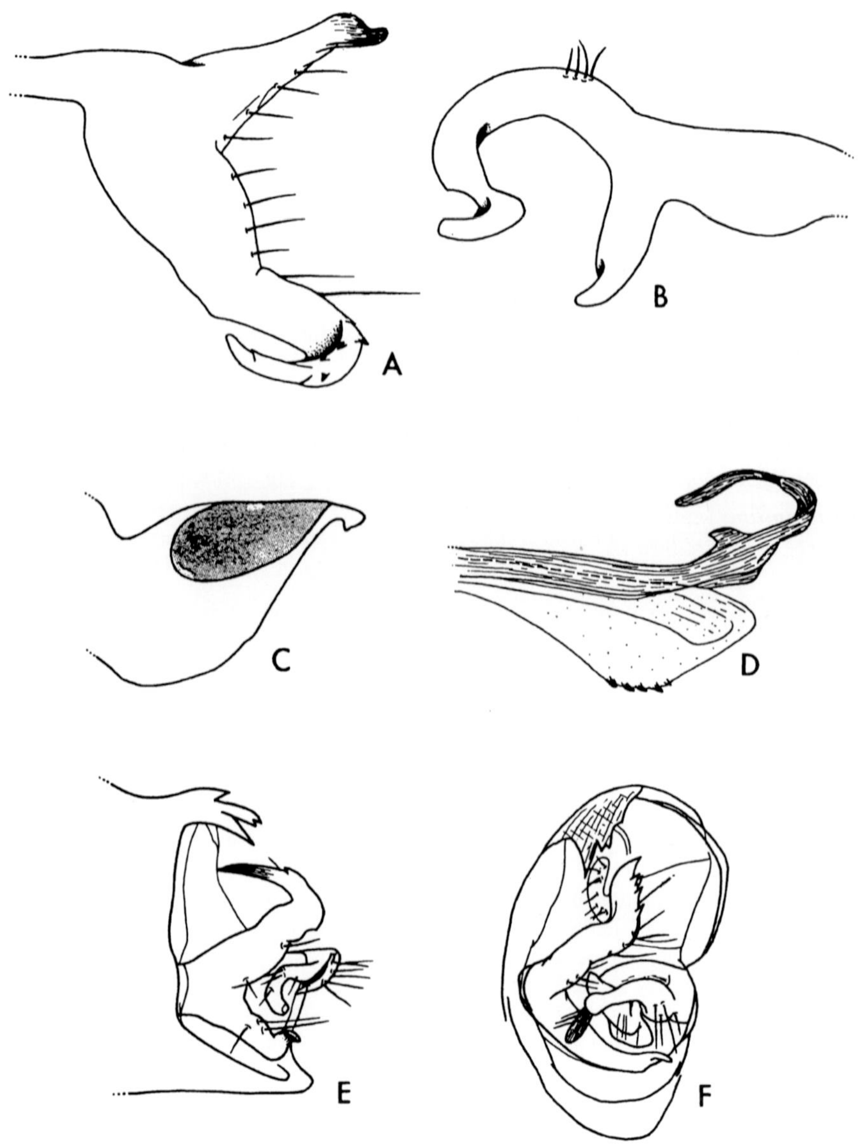

Fig. 62. *Nesiomiris beardsleyi,* n. sp., male genitalia. **a**, left clasper, internal; **b**, right clasper, internal; **c**, aedeagus, dorsal, spiculum removed; **d**, spiculum, right lateral; **e**, pygofer, left lateral; **f**, pygofer, terminal.

having the apical tooth on the pygofer darkened at its tip and in the right clasper having apically acute lobes (Fig. 64b), rather than being entirely rounded (Fig. 65b).

Description. **MALE** (n = 4). Body: 5.25 (5.10–5.30) long; 1.26 (1.17–1.26) wide. Head: 0.91 (0.81–0.91) wide; rufus; apex rounded; carinate; vestiture comparatively dense, especially on apex and inner angles of eyes. Antennal segment lengths: I, 0.97 (0.66–0.97), rufescent, vestiture brown; II, 2.39 (2.03–3.01), color of segment and vestiture as for I; III, 1.70 (1.38–1.70), dark brown with a pale base; IV, 1.45 (1.33–1.50). Rostrum: 1.34 (1.22–1.34) long; reaching apex of mesosternum. Pronotum: 0.82 (0.76–0.82) long; 1.26 (1.17–1.26) wide; paler than head; vestiture as dense as on hemelytra. Hemelytra: apex of cuneus rufus, remainder greenish; legs: vestiture of femora sparse, short, suberect; last tarsal segment darker. Abdomen: nonsclerotized and pale above atrium of pygofer; tooth on pygofer small, spiculate, darkened at tip; genitalia and pygofer as illustrated (Fig. 64).

FEMALE (n = 4). Body: 5.85 (5.45–5.85) long; 1.34 (1.33–1.37) wide. Head: 0.85 (0.85–0.90) wide. Antennal segment lengths: I, 0.94 (0.64–0.94); II, 2.41 (2.20–2.60); III, 1.49 (1.43–1.60); IV, 1.07 (1.07–1.60). Rostrum: 1.30 (1.30–1.45) long. Pronotum: 0.85 (0.76–0.90) long; 1.34 (1.31–1.35) wide. Differing from male as follows: erect bristles on antennal segment I present interiorly only; antennal segment III light brown on proximal 1/2 becoming dark brown distally.

NYMPHS. Unknown.

Distribution. Maui I.

Host. *Cheirodendron trigynum*.

Etymology. Named for the late Noel H. Krauss, longtime entomologist with the Hawaii State Department of Agriculture, who collected this and several other species of *Nesiomiris*, as well as innumerable insects from a multitude of localities in the Pacific Basin and around the world, adding immensely to the Bishop Museum collections.

Type data. Holotype ♂ (BPBM 14,850), allotype ♀, paratypes 4♂, **W MAUI**: Kaulalewelewe-Pu'u Kukui, 4500–5700 ft. [1371–1737 m], 28.V.1972, on *Cheirodendron trigynum*, W.C. Gagné. Additional paratype with data differing only as follows: 1♀, Kaulalewelewe-Pu'u Kukui Trail, [no elevation], 25.VIII.1965, on *Cheirodendron*, P.D. Ashlock; 2♀, [no elevation], 26.X.1966, [no host], P.D. Ashlock; 1♂, 3♀, **E MAUI:** 'Ukulele, 30.VIII.1932 [no host], N.L.H. Krauss; 1♂, NW Haleakalā, 6000–6500 ft. [1830–1980 m], 18.VIII.1937, [no host], E.C. Zimmerman; 1♂, Haleakalā, 'Ukulele Pipe Line, 5000 ft. [1525 m], 12.VII.1919, [no collector, but labels in Timberlake's handwriting] (BPBM).

Variation. Some males from the W Maui population have an inwardly hooked apex of the right clasper. Otherwise, the genitalia are very similar to the other specimens both from W Maui and from E Maui. Since this feature does not appear to be constant, these specimens are considered to be within the range of variation of *N. kraussi*.

26. *Nesiomiris rufonotus* Gagné, **new species** Figs. 26d, 65

This species differs from *N. usingeri* and *N. kraussi* in details of the male genitalia and in having the tooth on the pygofer pale at its tip.

Description. **MALE** (n = 5). Body: 5.05 (5.05–5.15) long; 1.30 (1.21–1.30) wide. Head: 0.84

(0.82–0.85) wide; rufus; apex rounded; weakly carinate across base; subglabrous. Antennal segment lengths: I, 0.89 (0.86–0.90), rufus, vestiture brown, sparse, only 2–3 erect bristles on inner angles; II, 2.50 (2.30–2.64), rufus, becoming brownish toward tip; vestiture entirely short and sparse; III, 1.70 (1.25–1.70), dark brown with a pale base; IV, broken. Rostrum: 1.19 (1.19–1.35), reaching base of procoxae. Pronotum: 0.89 (0.85–0.90) long; 1.30 (1.21–1.30) wide; paler than head; vestiture very short, pale, and sparse. Legs: as for *N. kraussi*. Abdomen: pale and nonsclerotized above atrium; left atrial tooth very small, short, and nonsclerotized; genitalia and pygofer as illustrated (Fig. 65).

FEMALE (n = 5). Body: 5.15 (5.15–5.78) long; 1.34 (1.32–1.36) wide. Head: 0.91 (0.81–0.91) wide. Antennal segment lengths: I, 0.90 (0.81–0.90); II, 2.20 (2.20–2.30); III, 1.60 (1.23–1.78); IV, broken (0.71–0.89). Rostrum: 1.24 (1.24–1.35) long. Pronotum: 0.91 (0.76–0.96) long; 1.34 (1.28–1.36) wide. Differing from male as follows: generally larger and wider; antennal segment II yellowish on proximal 1/2, becoming rufus on distal 1/2; rostrum longer but failing to reach base of mesocoxae. *K* and *L* structure closely resembling *N. usingeri* and *N. beardsleyi* (Fig. 26d).

NYMPHS. Unknown.

Distribution. E Maui I.

Host. *Cheirodendron trigynum*.

Etymology. Named for the reddish or rufus head.

Type data. Holotype ♂ (BPBM 14,851), allotype ♀, paratypes 2♀, **E MAUI**: SW Rift Zone, Pu'u Mākua, 1540 m, 14.IV.1971, on *Cheirodendron trigynum*, W.C. Gagné. Additional paratype with data differing only as follows: 6♂, 6♀, Haleakalā, 9400 ft. [2864 m], 23.III.1932, on *Cyathodes*, O. Bryant (BPBM).

Variation. The only male present in the SW Rift Zone population shows some variation from the claspers of the Bryant series. The upper arm of the left clasper is more inwardly hooked but other aspects of the male genitalia appear to be practically identical, so I interpret this to be within the limits of variation.

Discussion. It would appear from dissections of the male genitalia that the right clasper is held relatively immobile against the apex of the aedeagus. In fact, the male genitalia are among the most reduced and simple in the genus. The right clasper may not function to clasp the female in copulation, but rather may hold the aedeagus firmly while the spiculum is everted.

Bryant's series has evidently been mislabeled, at least as to host, and probably as to altitude as well. I am certain that the whole *usingeri* complex is specific to *Cheirodendron*. I have never taken any *Nesiomiris* species on *Cyathodes* (= *Styphelia tameiameiae*), even as perching individuals. The altitude given, 2764 m, is well above the upper limits of *Cheirodendron*, which drops off quite abruptly at about 2100 m, the upper limit of the orographic cloud belt and associated rain forest.

The *fasciatus* Subgroup

Description. **ADULTS**. Apex of head bifid; antennal segments I and II yellowish; propleural stripe present only on apical area of forelobe of pronotum; pubescence on dorsum and appendages reclined, brown, inconspicuous; pygofer toothed or not, tooth when present, located above base of left clasper; upper arm of left clasper uniramous; aedeagus with a terminal "lip"; spiculum flabellate; area about gonopore and fleshy lobe below it, denticulate.

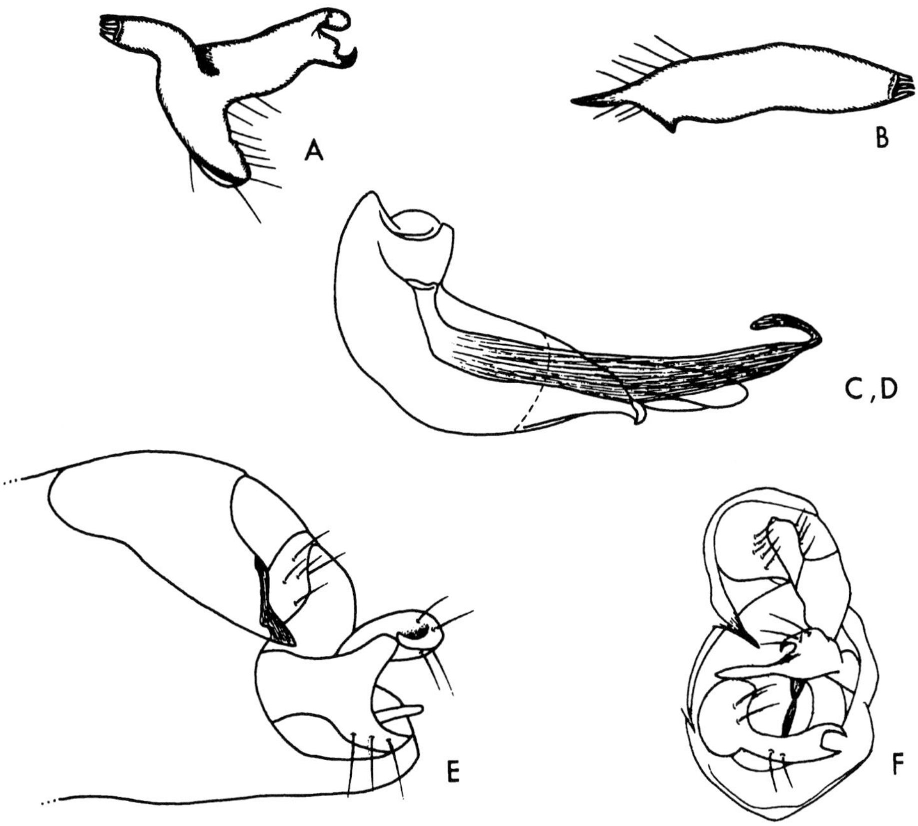

Fig. 63. *Nesiomiris usingeri*, n. sp., male genitalia. **a**, left clasper, internal; **b**, right clasper, internal; **c**, **d**, aedeagus and spiculum, lateral; **e**, pygofer, left lateral; **f**, pygofer, terminal.

Distribution. Kaua'i I.
Host. *Tetraplasandra kavaiensis* and *T. meiandra*.
Included species. Nos. 27–29.

The *fasciatus* Complex

Apart from the unique green spotting on the hemelytra and the yellowish areolar veins, these species are morphologically and ecologically allied with *N. elongatus*.

I find that the females of this complex are practically inseparable morphologically. I hesitantly associate them with the respective males; that is, since I have not found *N. fasciatus* males in the drier Kōke'e area, I am presuming the females found there to be *N.*

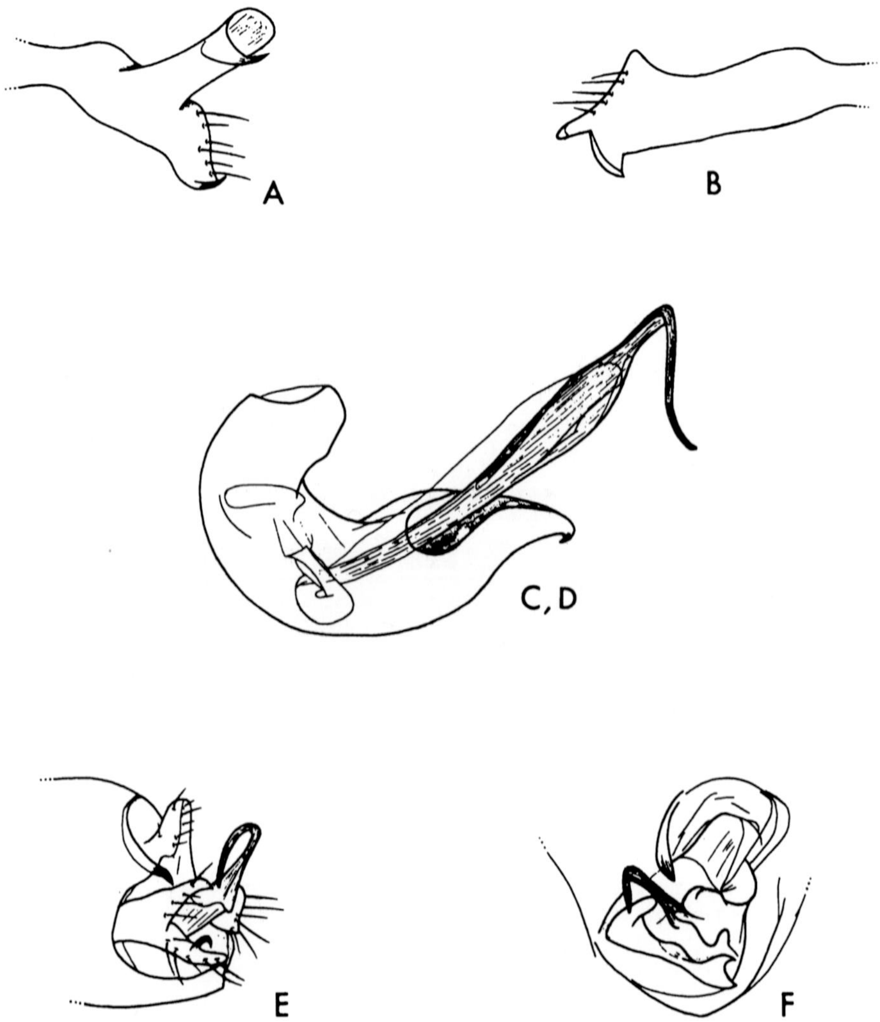

Fig. 64. *Nesiomiris kraussi*, n. sp., male genitalia. **a**, left clasper, internal; **b**, right clasper, internal; **c**, **d**, aedeagus and spiculum, lateral; **e**, pygofer, left lateral; **f**, pygofer, terminal.

maculatus, and since most of the male population at Mt Kāhili were *N. fasciatus* (12 male *fasciatus*, 1 male *N. maculatus*), I am presuming that most of the females but 1, which lacks a propleural stripe, were *N. fasciatus*.

However, I am describing the nymphs here since I do not feel that I can use the rationale expressed above and feel that it would be necessary to rear these out to properly associate them. It seems likely that these will prove to be *N. fasciatus*.

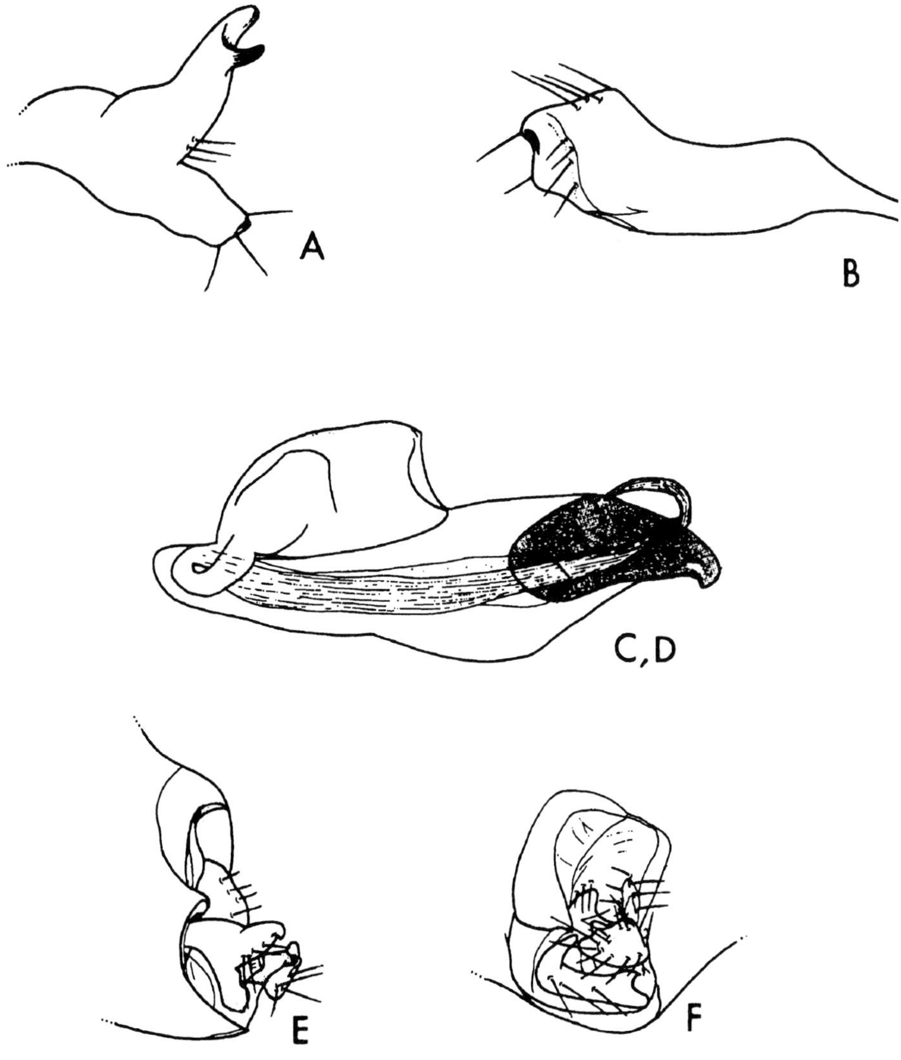

Fig. 65. *Nesiomiris rufonotus*, n. sp., male genitalia. **a**, left clasper, internal; **b**, right clasper, internal; **c**, **d**, aedeagus and spiculum, lateral; **e**, pygofer, left lateral; **f**, pygofer, terminal.

Description. **ADULTS**. Pale yellow species with green spots on the dorsum, excepting the head; rostrum reaching apex of mesosternum; propleural stripe present, at least in males; pygofer with a tooth above base of left clasper; upper arm of left clasper reduced, toothed at apex.

NYMPHS. *1st to 4th instars*. Unknown. *5th instar* (n = 2). (Fig. 27b). Body: 3.01–4.15 long; 1.21–1.39 wide. Head: 0.99–1.00 wide; slightly indented and infuscate at apex; remainder pale; subglabrous; a few pale, inconspicuous setae on periphery between eye and apex; basal carina absent.

Antennal segment lengths: I, 0.90–0.92; II, 2.35–2.50; III, 1.97–2.20; IV, broken; I to IV mostly pale; II slightly infuscate at apex. Rostrum: 0.58–0.74 long; both 0.92 wide, infuscate on posterior humeral angles; subglabrous, the few setae short, pale, inconspicuous excepting a strong 1 on each humeral angle; mesonotal wing buds infuscate adjacent to scutellum and again apically adjacent to the abdominal scent gland; a few pale setae predominantly along the lateral margins. Legs: pale; setae on femora short (about 1/4 thickness of this segment), erect, pale brown, sparser on ventral and posterior aspects, those on tibiae longer, paler, several irregular rows of longer, stouter setae on posterior aspects. Abdomen: segments anterior to scent gland glabrous; segments IV to VII with lateral setae, VIII and IX each with a row of setae across their apices.

My collection notes indicate that the nymphs, when alive also had a pattern of green spots on their dorsa; but the green appears to have faded from those stored in alcohol.

Distribution. Kaua'i I.

Host. *Tetraplasandra kavaiensis*.

Included species. Nos. 27–28.

27. *Nesiomiris fasciatus* Gagné, **new species** Fig. 66

This species and *N. maculatus* are immediately differentiated from all others by the pattern of green spots on the coriaceous parts of the hemelytra. *Nesiomiris fasciatus* differs from *N. maculatus* primarily in having the right claspers less reduced and longer than the left clasper.

The closest relative, other than *N. maculatus*, based on the general conformation of the male genitalia, would appear to be *N. elongatus*. These 3 species apparently attack the same host plant. However, the spiculum lacks a subterminal flabellum. The left clasper is much more reduced and the upper arm is toothed as in many other species of the *fasciatus* subgroup.

Description. **MALE** (n = 7). Body: 6.61 (6.10–6.61) long; 1.42 (1.39–1.45) wide. Head: 1.09 (1.02–1.09) wide; apex brown striate; weakly bifid; carinate across base, weakly impressed along middle; subglabrous; stylus reddish, remainder yellowish, weakly shining. Antennal segment lengths: I, 1.09 (1.02–1.17); II, 3.51 (2.84–3.57); III, 2.46 (1.93–2.49); IV, 2.13 (1.78–2.13); I and most of II yellowish, apex of II and III brownish; IV darker, setae on I brown, inclined with a few stronger, longer, and erect setae on internal surface, becoming progressively shorter and more reclined on subsequent segments, I longer than head width (ratio 9.5: 8.5). Rostrum: 1.43 (1.35–1.57) long, reaching apex of mesosternum. Pronotum: 1.03 (1.00–1.03) long; 1.42 (1.35–1.45) wide; yellowish; subglabrous, the few setae pale, inconspicuous; propleural stripe extending behind coxal cleft to a little beyond the mesal constriction; posterior humeral angles also conspicuously infuscate; posterior lobe rugose wrinkled. Mesosternum and scutellum yellow, subglabrous. Hemelytra: green spotted from apex of scutellum to apex of cuneus; setae mostly golden, inclined, denser along veins and lateral margins; membrane infuscate adjacent to anal ridge and at apex of areoles, remainder smokey; areolar veins yellowish with a little green near base of large areolar vein, large areole narrowly coriaceous along inner border. Legs: pale, setae as described for *N. elongatus*. Abdomen: pygofer (Fig. 66e) pale dorsally, a tooth above the left clasper; claspers and aedeagus as illustrated (Fig. 66a-d).

FEMALE (n = 7). Body: 7.21 (6.70–7.21) long; 1.52 (1.48–1.55) wide. Head: 1.13 (1.09–

1.13) wide. Antennal segment lengths: I, 1.09 (1.09–1.17); II, 3.65 (3.37–3.65); III, 2.62 (2.49–2.65); IV, 2.44 (1.85–2.49). Rostrum: 1.47 (1.47–1.61) long. Pronotum: 1.09 (1.04–1.09) long; 1.52 (1.48–1.55) wide. Differing from male as follows: ratio of antennal segment I to head width as 9.25: 9; base of hind lobe of pronotum pale greenish, green spots from base to apex of scutellum.

NYMPHS. See under description of *fasciatus* complex.

Distribution. Kaua'i I.

Host. *Tetraplasandra kavaiensis* var. *koloana*.

Etymology. Named for the green spots or fascia on the hemelytra.

Type data. Holotype ♂ (BPBM 14,852), allotype ♀, paratypes 13♂, 10♀, 3N, **KAUA'I**: Mt Kāhili, 3000 ft. [914 m], 2.IX.1970, on *Tetraplasandra kavaiensis* var. *koloana*, W.C. Gagné (BPBM).

Variation. The green spotting is sometimes present across the base of the pronotum.

Discussion. This appears to be another instance of a sympatric species pair attacking the same host plant, in which 1 species has reduced claspers compared to the other species. The right clasper of *N. fasciatus* is much longer than that of *N. maculatus* and is even longer and more robust than its own left clasper.

28. *Nesiomiris maculatus* Gagné, **new species** Fig. 67

This species is separated from *N. fasciatus* primarily on the basis of differences in the male genitalia; the claspers, particularly the right, are more reduced, the tooth above the left clasper is much longer and is itself denticulated and is about equal in length to the lower arm of the left clasper (Fig. 67a, b).

Description. **MALE** (n = 3). Body: 5.67 (5.67–6.38) long; 1.31 (1.31–1.34) wide. Head: 1.04 (0.96–1.04) wide. Antennal segment lengths: I, 0.90 (0.90–1.08); II, 2.54 (2.54–3.32); III, 2.58 (2.10–2.58); IV, 2.29 (1.27–2.29). Rostrum: 1.23 (1.20–1.34) long. Pronotum: 0.90 (0.90–1.00) long; 1.37 (1.33–1.43) wide. Differing from *N. fasciatus* as follows: a distinctly smaller species (body less than 1.35 mm wide), apex of head reddish brown; basal carina extending almost to edge of eye; antennal segment I reddish, slightly shorter than head width (ratio 7.5: 7.75); hind lobe of pronotum green spotted across base; propleural stripe shorter, extending only to mesal constriction posterior to coxal cleft; clavus mostly greenish adjacent to scutellum; areolar veins yellowish with irregular reddish portions.

FEMALE (n = 3). Body: 6.26 (6.18–6.26) long; 1.37 (1.33–1.43) wide. Head: 1.01 (1.00–1.03) wide. Antennal segment lengths: I, 0.90 (0.90–1.08); II. 2.54 (2.54–3.32); III, 2.58 (2.10–2.58); IV, 2.29 (1.27–2.29). Rostrum: 1.23 (1.20–1.34) long. Pronotum: 0.90 (0.90–1.00) long; 1.37 (1.33–1.43) wide. Differing from male as follows: ratio of lengths of antennal segment I to head width as 7.5: 8, and I yellowish; collar dorsally, and area of pronotum bordering collar, greenish.

NYMPHS. Unknown.

Distribution. Kaua'i I.

Host. "*Pterotropia*" [now considered a section of *Tetraplasandra* (Sherff 1955) of which only *T. kavaiensis* occurs on Kaua'i] and *T. kavaiensis* var. *koloana*.

Etymology. Named for the pattern of green spots or maculations on the hemelytra.

Type data. Holotype ♂ (BPBM 14,853), allotype ♀, **KAUA'I**: Kōke'e State Park, 22–

25.VII.1968, light trap, W.C. Gagné. Paratypes with data differing only as follows: 1 ♀, Halemanu, 22.VI.1932, on *Pterotropia* sp., O.H. Swezey; 1♂, dead at window, 15–20.XI.1968; 1♂, 1 ♀, Mt Kāhili, 3000 ft. [914 m], 2.IX.1970, on *Tetraplasandra kavaiensis* var. *koloana* (BPBM).

Variation. The male from Mt Kāhili has the propleural stripe represented only by a spot at the mesal constriction, and a female, presumed to belong here, lacks it entirely.

Discussion. Although *N. maculatus* and *N. fasciatus* occur sympatrically on Mt Kāhili, they are believed to be predominantly allopatric and possibly also allochronic in their life cycles. *Nesiomiris maculatus* appears to be more abundant than *N. fasciatus* in drier areas. Despite the differences in male genitalia, they appear to be quite closely related and lack the striking differences in vestiture seen in the sympatric *N. curvipes* and *N. pallidus* on W Maui.

The *seminotatus* Complex

Description. **ADULTS**. Apex of head bifid; species with uniformly greenish hemelytron and hind lobe of pronotum; upper arm of left clasper uniramous, not toothed at apex; pygofer unarmed; propleural stripe present in both sexes, but extending only at most to apex of hind lobe; rostrum reaching base of mesocoxae.

NYMPHS. Entirely greenish in life, not green spotted.

Distribution. Kaua'i I.
Host. *Tetraplasandra meiandra*.
Included species. No. 29.

29. *Nesiomiris seminotatus* Gagné, **new species** Fig. 68

The general conformation of the male genitalia approaches that found in members of the *N. bifurcatus* complex; but the differences—upper arm of the left clasper uniramous, not bifurcate, the lack of a tooth on the pygofer, and the comparatively small size—are such that I am considering it to belong to a sister group of the *fasciatus* complex. Both complexes predominantly feed on *Tetraplasandra* species and have bifid apices of the head.

The *seminotatus* complex shares the character of small denticles about the gonopore with the *fasciatus*, *uncinatus*, and *lineatus* complexes and with many species of the *lihuensis* group, the apical "lip" on the aedeagus. This, in addition to the yellowish, rather than rufescent, head and 1st antennal segment and the weakly developed propleural stripe place this species in a plesiomorphic position with respect to the *bifurcatus* and *lapalapa* subgroups.

Description. **MALE** (n = 6). Body: 5.66 (5.36–5.67) long; 1.19 (1.17–1.55) wide. Head: 0.92 (0.86–0.95); dorsum and tylus yellowish brown; shining; bifid; a few pale, reclined setae nearest eyes and at apex; carinate across base. Antennal segment lengths: I, 0.90 (0.81–1.00), yellowish

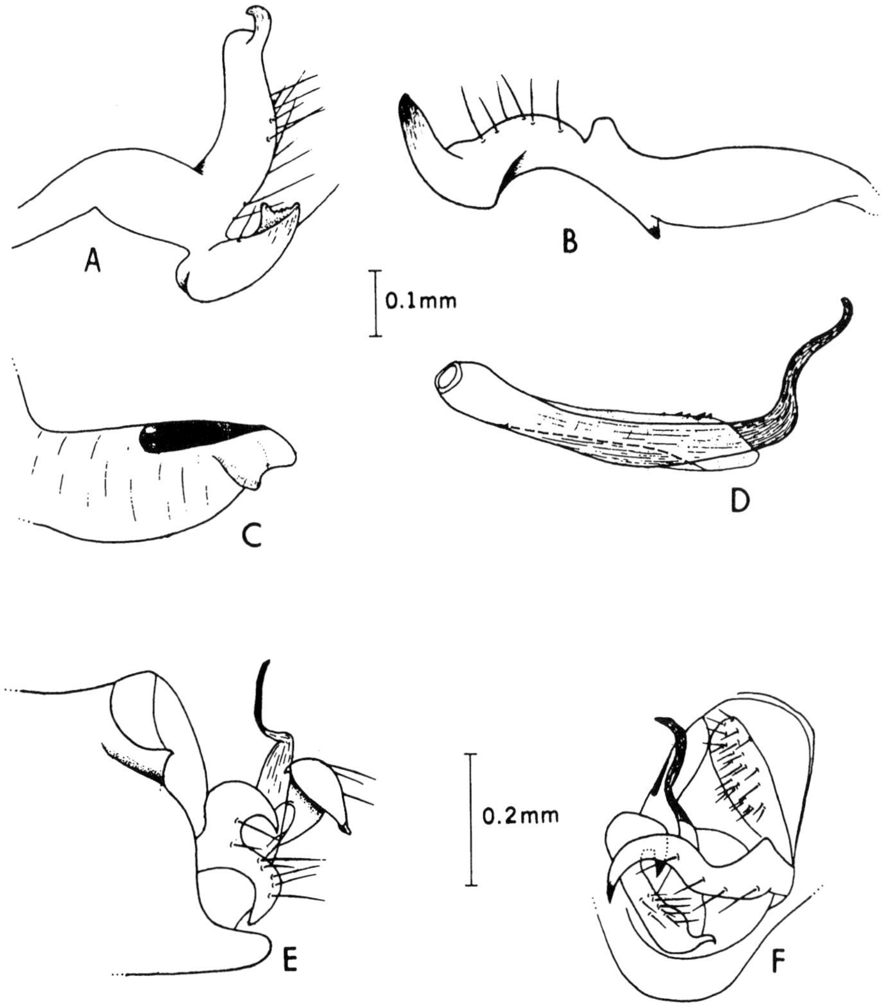

Fig. 66. *Nesiomiris fasciatus,* n. sp., male genitalia. **a**, left clasper, internal; **b**, right clasper, internal; **c**, aedeagus, lateral; **d**, spiculum, right lateral; **e**, pygofer, left lateral; **f**, pygofer, terminal.

brown; setae brown, the stouter, erect setae interspersed with reclined ones, subequal in length to head width (ratio 7: 7); II, 2.93 (2.65–3.20), brownish with blackish apex, setae all short and reclined; III, 2.20 (1.47–2.30); IV, 1.88 (1.25–1.93); III and IV blackish. Rostrum: 1.35 (1.28–1.45) long; reaching bases of mesocoxae. Pronotum: 0.89 (0.85–0.90) long; 1.19 (1.17–1.55) wide; callosities shining; collar and basal punctate lobe dark greenish, remainder yellow; clothed with a few reclined, pale setae. Mesonotum and middle of base of scutellum yellowish, remainder of latter pale greenish, both glabrous. Hemelytra: uniformly greenish; small areolar vein as rufescent as large are-

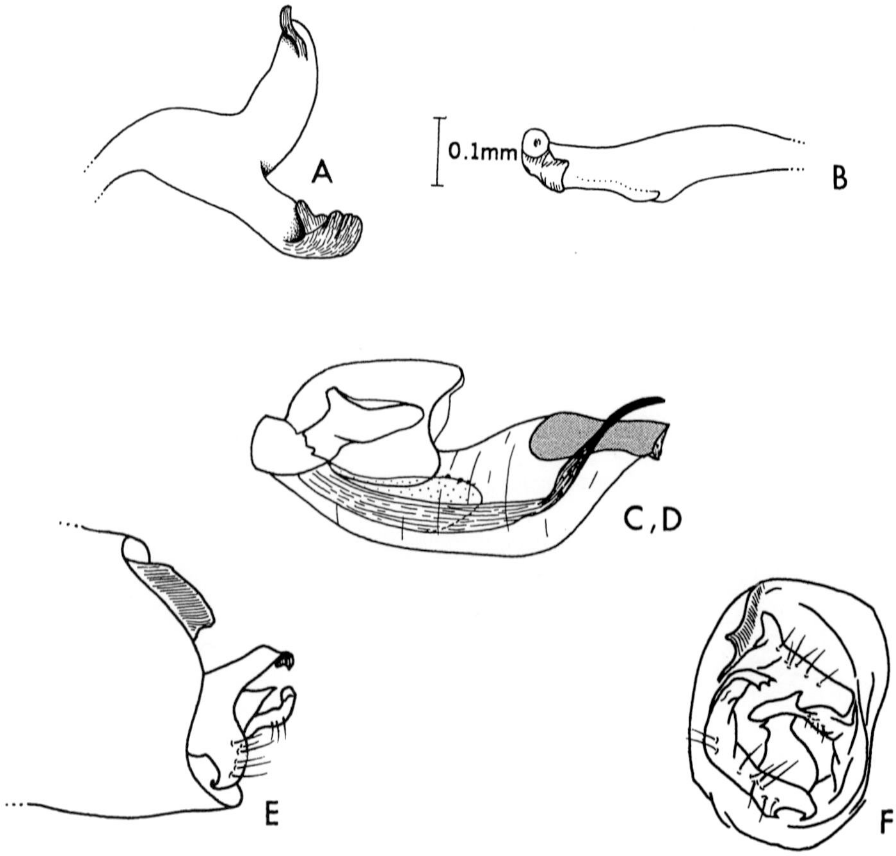

Fig. 67. *Nesiomiris maculatus,* n. sp., male genitalia. **a**, left clasper, internal; **b**, right clasper, internal; **c**, **d**, aedeagus and spiculum, lateral; **e**, pygofer, left lateral; **f**, pygofer, terminal.

olar vein; setae inclined, short, brown; membrane uniformly, weakly infumate; cuneus rufescent at extreme apex. Legs: hind femora pale greenish on apical 1/2; remainder pale yellowish; setae reclined, short, less than 1/2 the thickness of hind femora. Abdomen: posterior margins of tergites, excepting pygofer, infumate, remainder pale yellowish; pygofer, claspers aedeagus and spiculum as illustrated (Fig. 68).

FEMALE (n = 6). Body: 5.96 (5.67–6.10) long; 1.28 (1.24–1.31) wide. Head: 0.95 (0.90–0.96) wide. Antennal segment lengths: I, 0.95 (0.90–1.00); II, 3.08 (2.84–3.13); III, 2.41 (2.18–2.64); IV, 2.34 (1.78–2.34). Rostrum: 1.44 (1.35–1.45) long. Pronotum: 0.90 (all 0.90) long; 1.28 (1.24–1.31) wide. Differing from male as follows: antennal segment I longer than head width (ratio 7.75: 7.5); III, brownish at base; middle femora greenish at apex. Apex of ovipositor rufescent.

NYMPHS. *1st and 2nd instars*. Unknown. *3rd instar* (n = 1). Body: 1.94 long; 0.60 wide. Head: 0.60 wide; apex rounded; a row of 4 evenly spaced setae in a concave arc across base of head between eyes, another pair on vertex on a line drawn 1/2 way between eye and antennal base, a few other pale, scattered setae near vertex. Antennal segment lengths: I, 0.37; II, 1.00; II, 0.85; IV, broken; entirely pale; remainder of setae on dorsum apparently abraded and not readily discernable. Rostrum: 0.90 long; reaching middle of hind coxae. Pronotum: 0.25 long; 0.51 wide. Propleural stripe absent. Legs: entirely pale. *4th instar* (n = 1). Body: 2.80 long; 0.90 wide. Head: 0.76 wide. Antennal segment lengths: I, 0.50; II, 1.71; III, 1.55; IV, broken. Rostrum: 1.08 long. Pronotum: 0.37 long; 0.66 wide. Differing from 3rd instar of same species and 4th instar of *N. lineatus* as follows: antennal segment I with a few sparse, erect short (less than 1/2 thickness of this segment) brown setae; dorsum subglabrous (these possibly abraded); pubescence on legs pale; rostrum reaching middle of mesocoxae. *5th instar* (n = 1). Body: 3.85 long; 1.15 wide. Head: 0.89 wide. Antennal segment lengths: I, 0.80; II, 2.30; III, 2.10; IV, 2.20. Rostrum: 1.28 long. Pronotum: 0.55 long; 0.85 wide. Differing from 4th instar as follows: extreme apex of antennae II and III weakly infuscate.

Distribution. Kaua'i I.

Host. *Tetraplasandra meiandra*.

Etymology. Named for the propleural stripe that extends only 1/2 the length of the forelobe of the pronotum, becoming obsolescent on the basal 1/2 of the forelobe.

Type data. Holotype ♂ (BPBM 14,854), allotype ♀, paratypes 3♂, 5♀, **KAUA'I**: Mt Kāhili, 870 m, 22.VI.1972, on *Tetraplasandra meiandra*, W.C. Gagné. Additional paratypes with data differing only as follows: 9♂, 12♀, 7N, 3000 ft. [915 m], 2.IX.1970 (BPBM).

Variation. Several males lacked the subapical hook on upper arm of the right clasper. Also, some males and females had the propleural stripe clearly reaching the mesal constriction, rather than becoming obsolescent on basal half of the anterior lobe.

The *bifurcatus* Subgroup

Description. **ADULTS**. Apex of head bifid (except *N. lineatus* ♀); antennae I and II pallid or rufescent; propleural stripe present or absent; upper arm of left clasper bifurcate; species with uniformly greenish hemelytra and hind lobes of pronota; dorsal setae brownish, recumbent; spiculum with a subapical flabellum; pygofer with a tooth above base at left clasper; pubescence on dorsum and appendages, suberect to recumbent, golden brown or brownish; aedeagus with a terminal lip or tooth near right apex; spiculum not flabellate; area about gonopore usually not denticulate (except *N. lineatus*).

Distribution. Kaua'i I.

Host. *Cheirodendron* and *Tetraplasandra* spp.

Included species. Nos. 30–33.

The *bifurcatus* Complex

Description. **ADULTS**. Head and antennal segments I and II yellowish; propleural stripe, if present, not extending posteriorly much beyond mesal constriction; right clasper reduced with only 1 arm developed; area about gonopore not denticulate.

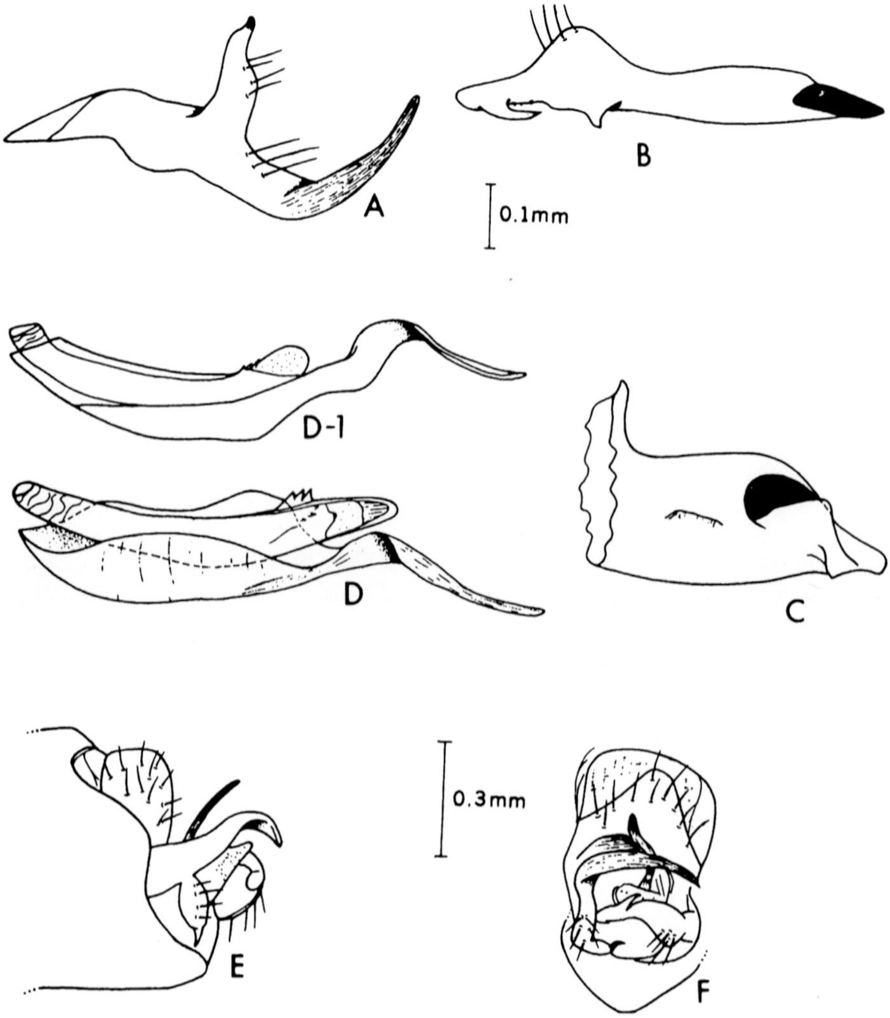

Fig. 68. *Nesiomiris seminotatus,* n. sp., male genitalia. **a**, left clasper, internal; **b**, right clasper, internal; **c**, aedeagus, lateral; **d**, spiculum, dorsal; **d-1**, spiculum, left lateral; **e**, pygofer, left lateral; **f**, pygofer, terminal.

Distribution. Kaua'i I.

Host. *Tetraplasandra meiandra*, *T. waimeae*, *T. lihuensis*.

Included species. Nos. 30–32.

30. ***Nesiomiris bifurcatus*** Gagné, **new species** Fig. 69

This species has the most strongly bifurcate upper arm of the left clasper (Fig. 69a) of any in the *bifurcatus* complex, for which reason it is placed at the base of the subgroup. The similar conformation of the right clasper, the toothed pygofer, and the aedeagus with its terminal lip, relate it to *N. fasciatus* of the *fasciatus* complex. In the *bifurcatus* complex it stands near *N. tetraplasandrae* but with a uniramous upper arm of the left clasper (Figs. 69, 70). Apart from the darker, yellowish areolar veins, the females are similar to those of *N. seminotatus*. It also occurs with an elongate, pale green (when alive) species so far known only from several females. The latter somewhat resemble those of *N. waimeae*, but the association cannot be more firmly established in the absence of additional males. *Nesiomiris bifurcatus* is unique in the *lihuensis* group in possessing green areolar veins when alive.

Description. **MALE** (n = 2). Body: 5.97 (5.57–5.97) long; 1.27 (1.27–1.32) wide. Head: 1.03 (1.00–1.03) wide; apex distinctly bifid; yellowish brown; subglabrous; tylus darker than dorsum; shallowly impressed along middle; very weakly carinate across middle of base. Antennal segment lengths: I, 0.91 (0.91–0.96); II, 2.69 (2.69–2.95); III, 1.80 (1.8–2.03); IV, broken. Rostrum: 1.45 (1.45–1.48) long; reaching apices of mesocoxae. Pronotum: 0.91 (0.90–0.91) long; 1.27 (1.25–1.27) wide; subglabrous, with a few scattered, short, pale, inclined setae and a stronger, darker seta on each humeral angle; collar and forelobe yellowish; hind lobe broadly greenish; propleural stripe extending from behind collar to just beyond mesal constriction. Mesonotum and scutellum yellowish brown with a median longitudinal green line; glabrous. Scutellum greenish except at middle of base, a few short, brown, recumbent setae. Hemelytra: coriaceous parts pale greenish; a few setae as described on scutellum; areolar veins green in life, fading to brownish yellow after death; membrane slightly infumate. Legs: yellowish, setae as described in *N. fasciatus*. Abdomen: pygofer (Fig. 69e) pale dorsally; claspers and aedeagus as illustrated (Fig. 69a-d).

FEMALE (n = 5). Body: 6.40 (6.50–6.40) long; 1.45 (1.39–1.45) wide. Head: 1.07 (1.00–1.07) wide. Antennal segment lengths: I, 1.00 (0.90–1.00); II, 2.94 (2.65–3.01); III, 2.24 (2.06–2.30); IV, 2.72 (2.30–2.72). Rostrum: 1.62 (1.45–1.62) long. Pronotum: 0.92 (0.85–0.92) long; 1.45 (1.39–1.45) wide. Differing from male as follows; mesonotum and scutellum yellowish; apex of latter green; ratio of length of antennal segment I to head width as 7.25: 8.25; collar pale greenish above.

NYMPHS. Unknown.

Distribution. Kaua'i I.

Host. *Tetraplasandra meiandra*.

Etymology. Named for the bifurcate upper arm of the left clasper.

Type data. Holotype ♂ (BPBM 14,855), allotype ♀, paratypes 1♂, 1♀, **KAUA'I**: Mt Kāhili, 3000 ft. [914 m], 2.IX.1970, on *Tetraplasandra meiandra*, W.C. Gagné. Paratypes with data differing only as follows: 3♀, 870 m, 22.VI.1972; 1♀, 19.XI.1968 (BPBM).

Variation. In the more recently collected females, the scutellum is pale green; the collar is also pale dorsally.

Discussion. This species demonstrates another transition from species with green areolar veins and partially coriaceous areolar cells (the most parsimoniously plesiomorphic state), to species with reddish areolar veins and noncoriaceous areolar cells. Living adults of *N. bifurcatus* have the areolar veins greenish, fading to yellowish, but the areoles are noncoriaceous. Its obviously close relatives, *N. tetraplasandra* and *N. waimeae*, judging from close correspondence of details of the male genitalia, have reddish areolar veins and noncoriaceous areoles and so fall comfortably into the *lihuensis* group. With the combination of characters outlined above, this species should be placed near the base of this group.

31. ***Nesiomiris tetraplasandrae*** Gagné, **new species** Figs. 45g, 70, 82a

Nesiomiris hawaiiensis: Swezey, 1954, *B.P. Bishop Mus. Spec. Publ.* 44: 217 (host, misidentification).

This species lies in an intermediate systematic position to *N. bifurcatus* and *N. waimeae*, based on the general conformation of the male genitalia and armature of the pygofer.

Description. **MALE** (n = 1). Body: 7.57 (paratype mutilated) long; 1.67 wide. Head: 1.14 wide; slightly bifid, weakly carinate across base; vestiture as for *N. bifurcatus*, shape as illustrated (Fig. 45g). Antennal segment lengths: I, 1.13, yellow, shorter than head width, vestiture brown, erect bristles present on inner aspect only; II, 3.55, yellow with light brown apex; III, 2.43, brownish on proximal 1/2, black on distal 1/2; IV, 2.65, black. Rostrum: 1.87 long. Pronotum: 1.00 long; 1.67 wide; disc greenish on basal 1/2, marmorate, yellowish green on anterior 1/2, remainder yellow; propleural stripe obsolescent; vestiture very sparse, a prominent bristle on outer anterior angle of callus; borders of calli indistinct interiorly; shape as illustrated (Fig. 45g). Scutellum yellow. Hemelytra: yellowish green anterior to anal ridge, green posteriorly; cubitus rufus. Legs: femora, tibia, and 1st 2 tarsal segments yellow, last tarsal segment brownish; vestiture on femora pale and sparse. Abdomen: pale; pygofer with a prominent left lateral atrial tooth; pygofer and genitalia as illustrated (Fig. 70).

FEMALE (n = 1). Body: 7.53 (paratype mutilated) long; 1.57 wide. Head: 1.15 wide. Antennal segment lengths: I, 1.13; II, 3.40; III and IV, broken. Rostrum: 1.83 long. Pronotum: 1.13 long; 1.57 wide. Differing from male as follows: head more strongly carinate, propleural stripe present on anterior lobe of pronotum; calli with vestiture denser.

NYMPHS. Unknown.

Distribution. Kaua‘i I.

Host. *Tetraplasandra* sp. (probably *waimeae*).

Etymology. Named for the host plant, *Tetraplasandra*.

Type data. Holotype ♂ (BPBM 14,856), allotype ♀, paratypes 1♂, 1♀, **KAUA‘I**: Alaka‘i Swamp, 14.VII.1932, on *Tetraplasandra*, O.H. Swezey (BPBM).

Discussion. This species is represented by a series of only 4 faded specimens (2 male, 2 female) collected in 1932 by Swezey. This probably is the same material listed as "*Nesiomiris hawaiiensis*," and as being "numerous on *Tetraplasandra* leaves on the trail, Kaua‘i, and at Kainalu, Moloka‘i." (Swezey 1954: 217). The Moloka‘i record refers to *N. curvipes*, and I presume this was the "numerous" species. I have been unable to uncover any additional material matching Swezey's Kaua‘i data. I expect that this species will be found to attack *Tetraplasandra waimeae* in the Alaka‘i Swamp region. Its close relative, *N. waimeae*, attacks this host and *T. meiandra*, another glabrous, potential host species, which does not appear to occur at those higher elevations.

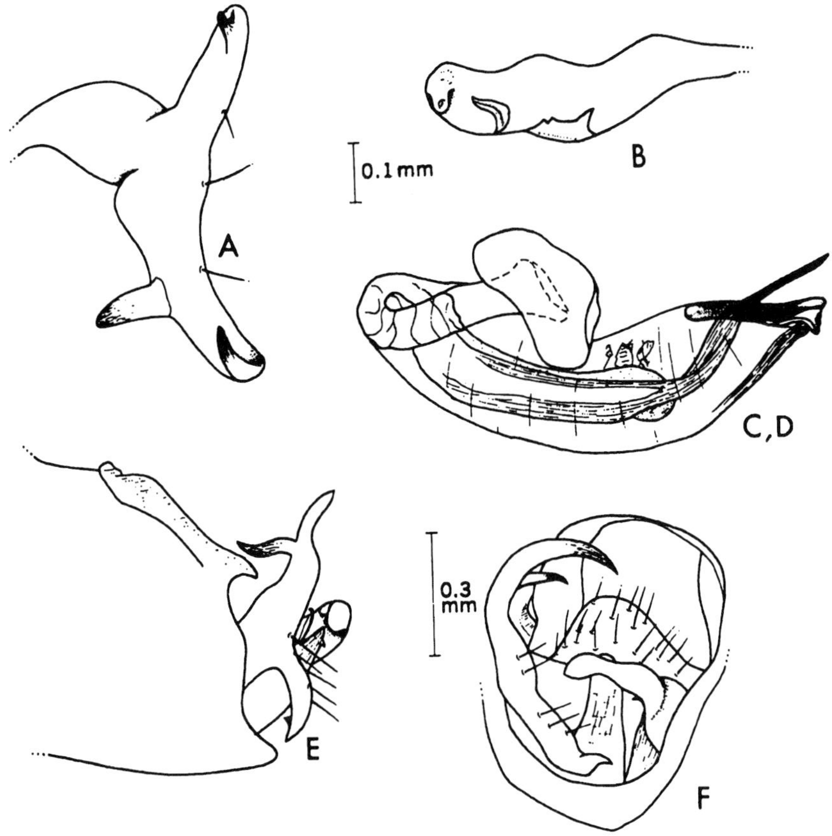

Fig. 69. *Nesiomiris bifurcatus*, n. sp., male genitalia. **a**, left clasper, internal; **b**, right clasper, internal; **c**, **d**, aedeagus and spiculum, lateral; **e**, pygofer, left lateral; **f**, pygofer, terminal.

32. *Nesiomiris waimeae* Gagné, **new species** Fig. 71

This is a pale green species in life. Specimens faded rapidly to pale yellow after death, with any green that remained usually only present along the anterior lateral borders of the wings. Among members of the *bifurcatus* complex, it is readily differentiated by the pilose femora and antennae I, the lack of propleural stripe, and in details of the male genitalia. The latter place it next to *N. lineatus*, *N. tetraplasandrae* and *N. bifurcatus*, perhaps closer to the latter since both species lack the propleural stripe, while the 1 female available of *N. tetraplasandrae* has it. *Nesiomiris lineatus*, by contrast, is named for its strikingly developed propleural stripe, but both have a weakly bilobate, relatively flat pronotum (in profile) not approached in any other species.

Description. **MALE** (n = 7). Body: 6.50 (6.18–6.75) long; 1.41 (1.35–1.41) wide. Head: 1.06 (1.05–1.07) wide; yellowish apex distinctly bifid; a few pale setae on anterior aspects. Antennal segment lengths: I, 1.21 (1.09–1.35); II, 3.41 (3.37–3.59); III, 2.97 (1.93–2.95); IV, broken (1.45–2.30); I pale, longer than head width (ratio 9.5: 8.25); densely clothed with golden brown, erect setae about equal to the thickness of this segment; II, pale except darker apex, reclined golden brown setae only on apical 1/2 of II, on III and IV, recumbent. Rostrum: 1.38 (1.35–1.53) long; reaching middle of mesosternum. Pronotum: 1.07 (1.00–1.09) long; 1.41 (1.35–1.41) wide; flat in profile; hind lobe about same level as forelobe, not strikingly declivous anteriorly; a few pale, scattered setae; propleural stripe absent; pale throughout. Mesosternum and scutellum pale; latter with a few golden brown, suberect setae. Hemelytra: faded to yellowish, excepting some green on anterior lateral edges and inner edge of clavus; beset with golden setae; large areolar vein weakly rufescent; small areolar vein pale, indistinct; membrane infuscate only near anal ridge. Legs: pale; femora and hind tibiae hirsute, setae as long as those on antennal segment I, but paler; fore and middle tibiae less densely hirsute. Abdomen: pale yellow; pygofer toothed on left side (Fig. 71e), pale above; spiculum not flabellate; claspers and aedeagus as illustrated (Fig. 71a-d).

FEMALE (n = 7). Body: 6.60 (6.50–7.05) long; 1.37 (1.36–1.48) wide. Head: 1.06 (1.02–1.12) wide. Antennal segment lengths: I, 1.16 (1.09–1.30); II, 3.23 (3.23–3.57); III, 2.50 (2.30–2.84); IV, 1.78 (1.43–2.30). Rostrum: 1.41 (1.35–1.50) long. Pronotum: 1.16 (1.05–1.16) long; 1.37 (1.36–1.48) wide. Not notably differing from male.

NYMPHS. *1st to 3rd instars*. Unknown. *4th instar* (n = 1). Body: 2.52 long; 0.73 wide. Head: 0.76 wide; apex weakly indented and striolate; 3 pairs of erect, long brown setae, 1 at middle of base, another between eyes and another at apex, as well as 1 between eye and antennal base. Antennal segment lengths: I, 0.61; II, 1.90; III, 1.78; IV, broken; I with sparse, erect, brown setae about equal in length to thickness of segment, these continuing on II increasingly sparse almost to apex and interspersed with shorter reclined increasingly shorter setae on III and IV; II slightly infuscate at apex, remainder yellowish. Rostrum: 1.09 long; reaching apices of hind coxae. Pronotum: 0.37 long; 0.58 wide; subquadrate, slightly narrower across apex; lacking propleural stripe; a sparse row of erect brown setae along lateral edge continuing similarly on mesonotal wing bud, plus 2 pairs along mid-line. Legs: pale; femora with setae as described for those on antennal segment I, those on tibia more reclined. Abdominal segments with a row of setae across the apex of each. *5th instar* (n = 1). Body: 4.05 long; 1.21 wide. Head: 1.00 wide. Antennal segment lengths: I, 1.00; II, 2.74; III, 2.36; IV, 2.14. Rostrum: 1.17 long. Pronotum: 0.55 long; 0.92 wide. Differing from 4th instar as follows: appendages hirsute, as described in adult, similar setae also present on dorsum, especially on mesonotal wing buds and are randomly arranged; less dense on pronotum; head distinctly bifid; rostrum reaching bases of mesocoxae.

Distribution. Kaua'i I.

Host. *Tetraplasandra waimeae* var. *waimeae*.

Etymology. Named for the species of the host plant.

Type data. Holotype ♂ (BPBM 14,857), allotype ♀, paratypes 2♀, **KAUA'I**: Kōke'e State Park, Honopū Trail, 4000 ft. [1219 m], 25.VIII.1970, on *Tetraplasandra waimeae* var. *waimeae*, W.C. Gagné. Additional paratypes with data differing only as follows: 1♀, Kōke'e, 24.VIII.1959, light trap, J.W. Beardsley; 1♂, 2♀, 15.XI.1968; 11♂, 10♀, 9N, Halemanu, Kaunuohua Ridge, 3500 ft. [1066 m], 24.VIII.1970; 3♂, 1♀, Po'omau Canyon Lookout, 1037 m, 25.VIII.1970; 1♂, 1♀, Koai'e-Wai'ale'ale Trail, 3500 ft. [1066 m], 25.VIII.1970, on *Cheirodendron Helleri* var. *multiflorum* (BPBM).

Discussion. I consider the pair collected on *Cheirodendron* to be a perching record of individuals straying from the closely adjacent *Tetraplasandra waimeae* trees. There is no evidence of breeding on *Cheirodendron* since no nymphs were taken.

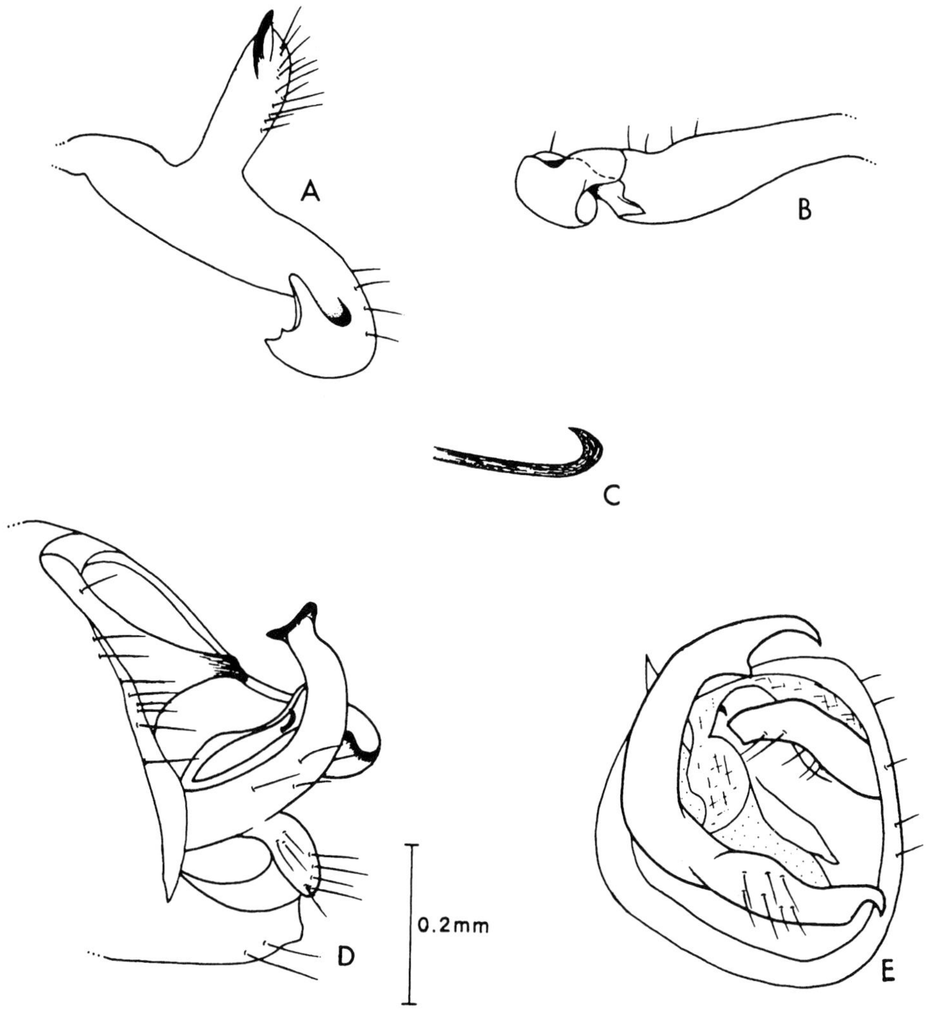

Fig. 70. *Nesiomiris tetraplasandrae,* n. sp., male genitalia. **a**, left clasper, internal; **b**, right clasper, internal; **c**, apex of spiculum; **d**, pygofer, left lateral; **e**, pygofer, terminal.

The *lineatus* Complex

Description. **ADULTS**. Head rufescent apex straight in ♀; antennal segment I brownish, II rufescent; propleural stripe extending from base of head, across collar, to base of pronotum; right clasper reduced to 2 short arms; area ventral to gonopore denticulate.

Distribution. Kaua'i I.

Host. *Cheirodendron Helleri* and *C. kauaiense* and its varieties.

Included species. No. 33.

Discussion. There is a small (5.95mm long) female, which may belong to this complex (Kaua'i I, Kōke'e State Park, Honopū Trail, 1200 m, 25.VIII.1970, on *Tetraplasandra waimeae* var. *waimeae*, W.C. Gagné), that has the propleural stripe extending over almost the entire length of the propleuron. Alternatively, this species may be related to *N. legnotus* and really feeds on *Cheirodendron*. I am refraining from describing it in spite of its distinctiveness until an associated male is collected.

As noted below in the discussion under *N. lineatus*, it is difficult to place this complex in the cladistic analysis. If *N. lineatus* and *N. waimeae* can be considered synapomorphic on the basis of the flat pronotum, then this would make the *bifurcatus* complex polyphyletic. So, I am placing *N. lineatus* as an autapomorphic species.

33. ***Nesiomiris lineatus*** Gagné, **new species** Fig. 72

The strikingly developed propleural stripe immediately separates *N. lineatus* from other members of the *lihuensis* group except the much smaller *N. legnotus*. The flat pronotum approaches *N. waimeae* closely in shape, vestiture, and profile. In other details, particularly of the male genitalia, there is an indication that it could be a sister species of *N. waimeae* and has become host specific on *Cheirodendron*, evolving from a *tetraplasandra*-feeding progenitor.

Description. **MALE** (n = 7). Body: 6.80 (6.70–7.32) long; 1.41 (1.35–1.46) wide. Head: 1.05 (1.02–1.09) wide. Antennal segment lengths: I, 1.26 (1.22–1.30); II, 3.45 (3.34–4.05); III, 2.14 (1.93–2.30); IV, 2.12 (1.78–2.12). Rostrum: 1.71 (1.61–1.71) long. Pronotum: 1.13 (1.03–1.13) long; 1.41 (1.35–1.46) wide. Differing from *N. waimeae* as follows: head and tylus rufescent, base of tylus dark brown, setae present adjacent to eyes and at apex; setae of antennal segment I almost all reclined, brownish, longer than head width (ratio 10.5–8.25); II, rufescent; rostrum reaching apex of mesosternum; pronotum subglabrous, shining, hind lobe green on basal 1/2, propleural stripe extending on an unbroken line from collar to base of hind lobe; mesosternum greenish at apex and with a few inclined, brown setae on disc; hemelytra darker greenish, setae reclined, brownish; large areolar vein strongly rufescent; membrane infumed, most strongly so external to juncture of areolar veins; legs with pubescence as described on antennal segment I; hind femora ventral brownish semiannulate. Abdomen: pygofer with a tooth above left clasper (Fig. 72e); claspers, aedeagus, and spiculum as illustrated (Fig. 72a-d).

FEMALE (n = 7). Body: 6.80 (6.70–7.11) long; 1.47 (1.35–1.50) wide. Head: 1.06 (0.99–1.15) wide. Antennal segment lengths: I, 1.35 (1.19–1.35); II, 3.39 (3.26–3.55); III, 2.26 (1.93–2.26); IV, 2.23 (1.71–2.23). Rostrum: 1.65 (1.61–1.80) long. Pronotum: 1.18 (1.09–1.18) long; 1.47 (1.35–1.50) wide. Differing from male as follows: apex of head straight; collar, mesodorsally behind collar, hind lobe of pronotum and apical 1/2 of mesosternum, greenish.

NYMPHS. *1st to 3rd instars*. Unknown. *4th instar* (n = 2). Body: 2.95 long; 0.86 wide. Head: 0.77–0.80 wide; apex weakly bifid; a row of 4 evenly spaced, erect, brown setae along hind margin between eyes, another row of 4 setae across middle of vertex on a line drawn between eye and antennal socket, and a pair of setae astride apical indentation. Antennal segment lengths: I, 0.70; II, 1.78–1.83; III, 1.38–1.48; IV, 1.30; apices of II and III infuscate, otherwise pale; I, clothed with mostly brown setae that are about equal in length to the thickness of I. Rostrum: 1.07 long; reach-

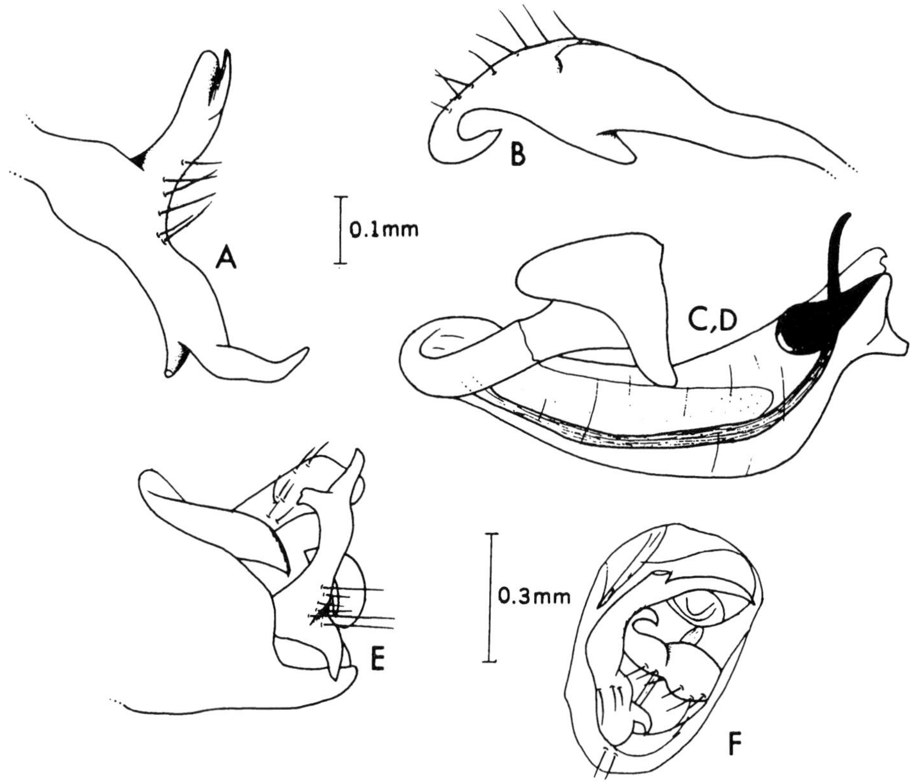

Fig. 71. *Nesiomiris waimeae,* n. sp., male genitalia. **a**, left clasper, internal; **b**, right clasper, internal; **c**, **d**, aedeagus and spiculum, lateral; **e**, pygofer, left lateral; **f**, pygofer, terminal.

ing apices of mesocoxae. Pronotum: 0.40–0.46 long; 0.65–0.73 wide; lacking propleural stripe; setae, size and length as described for antennal segment I, present in a single row about the periphery, the posterior pair astride the median line offset apically from the straight posterior margin. Mesonotal wing bud with reclined setae concentrated along lateral margin and a few others scattered mostly on middle of disc. Legs: pale, setae similar to those on antennal segment I, especially on femora, those on tibiae same length but more reclined. Abdomen: transverse row of setae on each segment about equidistant from posterior and anterior borders on segments 5 to 8, setae on apical segments closer to the posterior border and more inclined. *5th instar*. Unknown.

Distribution. Kaua'i I.

Host. *Cheirodendron Helleri*, *C. kauaiense* and its vars. *Forbesii* and *kauaiense*.

Etymology. Named for the long propleural stripe.

Type data. Holotype ♂ (BPBM 14,858), allotype ♀, paratypes 9♂, 4♀, **KAUA'I**: Pole Line Trail, near Kualapa, 620 m, 28.VIII,1970, on *Cheirodendron Helleri*, W.C. Gagné. Additional

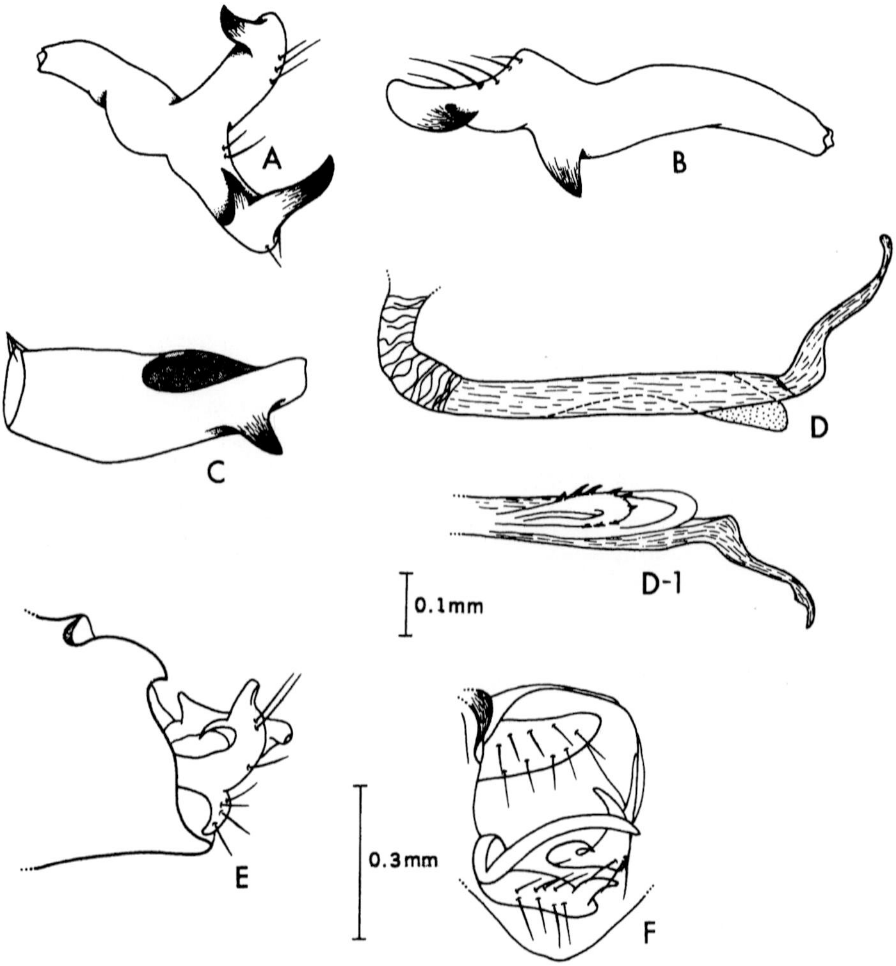

Fig. 72. *Nesiomiris lineatus*, n. sp., male genitalia. **a**, left clasper, internal; **b**, right clasper, internal; **c**, aedeagus, lateral; **d**, spiculum, left lateral; **d-1**, spiculum, ventral; **e**, pygofer, left lateral; **f**, pygofer, terminal.

paratypes with data differing only as follows: 3♂, 2♀, on *C. kavaiensis* var. *kauaiense*; 5♂, 3N, Mt Kāhili, 19.XI.1968, on *C. kavaiensis* var. *Forbesii*; 3N, Mt Kāhili, 17.XI.1968, on *C. kavaiensis* var. *Forbesii*; 2♂, 4N, Kōloa District, Līhu'e-Kōloa Forest Reserve, Kāhili, 2395 ft. [725 m], 2.IX.1970, on *C. kauaiense*; 2♂, Mt Kāhili, 2400 ft. [727 m], 2.IX.1970, on *C. Helleri*; 1♂, 1♀, Mt Kāhili, 870 m, 22.VI.1972, *C. kauaiense* (BPBM).

Discussion. This species is difficult to place because it presents a mosaic of apomorphic and plesiomorphic characters, including the host association, well-developed propleural stripe, lack of flabellum, bifurcate upper arm of left clasper, bifid apex of the head, lack of flabellum on the spiculum, and flat pronotum in profile.

The *lapalapa* Subgroup

***Description*. ADULTS.** Apex of head weakly bifid to straight; antennal segment I and II yellowish to rufescent; propleural stripe usually absent (except *N. legnotus*); pubescence of dorsum and appendages short, recumbent, and brownish; spiculum apically twisted, lacking subapical flabellum; pygofer toothed or not; upper arm of left clasper uniramous (weakly biramous in *N. lapalapa*); aedeagus with a terminal "lip"; spiculum not flabellate; area about gonopore denticulate or not.

***Distribution*.** Kaua'i I.
***Host*.** *Cheirodendron* spp.
***Included species*.** Nos. 34–39.

The *lapalapa* Complex

***Description*. ADULTS.** Propleural stripe absent; right clasper with an apical hook and a subapical, blunt arm.

***Distribution*.** Kaua'i I.
***Host*.** *Cheirodendron Helleri*, *C. Fauriei*, *C. kauaiense* and some of their varieties.
***Included species*.** Nos. 34–38.
***Discussion*.** Females of this complex are largely indistinguishable. I have designated them in series where the males of only 1 species were collected. But, since there is a mixture of the sympatric *N. lapalapa* and *N. kauilamahu*, for example, no females were designated.

Members of this complex are restricted to the higher (above 800 m elevation) rain forest of Kaua'i where they feed on *Cheirodendron* species. They occur in the Makaleha Mountains and in the Alaka'i Swamp, areas separated by less than 10 km, and with suitable hosts occupying the intervening areas. Hosts in these areas below 800 m are, however, occupied by other *Nesiomiris* species such as *N. cheirodendronae* and *N. lineatus*. Thus the isolation of the 2 massifs appears to be sufficient to curtail greatly any gene flow between members of the *lapalapa* complex. Successful colonization of hosts of 1 massif from the other may have been analogous to multiple invasions and subsequent character displacement in male genitalia. The 5 members of the complex are very similar morphologically. I have segregated species with little other than discontinuities in the male genitalia for guidance. Had there been any indication of variation among them, I would not have hesitated to combine several of these species.

I cannot be sure that I have associated the nymphs with the adults properly. Nymphs collected with adults of this complex lacked propleural stripes and reddish markings on the appendages. Those in Alaka'i Swamp fall into 2 groups: (1) infrequently collected with the dorsum clothed with erect, pale brown setae, and (2) more frequently collected with pale, short, partly reclined, and obscure setae. I consider these to be immatures of *N. mahu* and *N. ehu*, respectively, as judged by the distribution and relative abundance of the adults as well as by differences in the pubescence of the dorsa.

34. *Nesiomiris lapalapa* Gagné, **new species** Fig. 73

I am placing this species 1st in this complex of small (less than 6.50 mm long) bugs specific to *Cheirodendron* species. This complex includes also, in presumed phylogenetic order, *N. olapa*, *N. ehu*, *N. kauilamahu*, and *N. mahu*. *Nesiomiris lapalapa* has a number of characters that I consider plesiomorphic to the remaining members of this complex; pygofer with a large, flat tooth above the left clasper; left clasper with upper arm apically bifurcate, and right clasper least reduced of the complex.

This combination of characters relate it to the species of the *fasciatus* subgroup. But, a propleural stripe is completely absent in the *lapalapa* subgroup except *N. legnotus*.

This species occurs sympatrically and synchronously with *N. kauilamahu*, which has more reduced claspers and a ventrally acute apex on the tooth of the pygofer that in *N. lapalapa* is ventrally rounded. I cannot separate the females of these 2 species and so have refrained from designating any as allo- or paratypes.

Description. **MALE** (n = 3). Body: 5.15 (5.05–5.15) long; 1.19 (1.18–1.19) wide. Head: 0.93 (0.92–0.93) wide; apex weakly bifid; greenish striolate, remainder of dorsum weakly rufescent; shining; carinate across base almost to eye margins; subglabrous. Antennal segment lengths: I, 0.71 (all 0.71); II, 2.51 (2.33–2.55); III, 1.28 (1.81–1.88); IV, 2.13 (2.12–2.13); I rufescent at base, becoming greenish apically, sparsely clothed with recumbent brown setae and a few stronger, erect setae on inner aspects, shorter than on I and becoming still stronger apically, weakly rufescent, becoming darker apically; III and IV blackish. Rostrum: 1.41 (1.36–1.46) long; reaching apex of mesosternum. Pronotum: 0.74 (0.71–0.75) long; 1.19 (1.18–1.19) wide; shining; basal lobe and collar greenish, calli yellowish green; propleural stripe absent, subglabrous. Mesoscutellum yellow. Scutellum pale greenish with a few recumbent, short brown setae. Hemelytra: mostly greenish with some yellow areas adjacent to scutellum, setae sparse as described for scutellum; present mostly on veins; membrane clear adjacent to extreme apex of cuneus. Legs: mostly pale with some greenish remaining in femora and tibiae; setae as described for scutellum, but paler. Abdomen: shining totally pale yellow; pygofer, claspers and aedeagus as illustrated (Fig. 73).

FEMALE. Unknown.

NYMPHS. Unknown.

Distribution. Kaua'i I.

Host. *Cheirodendron kauaiense* var. *kauaiense* and *Forbesii*.

Etymology. The Hawaiian name for *Cheirodendron kauaiense* is *lapalapa*.

Type data. Holotype ♂ (BPBM 14,859), paratype 1♂, **KAUA'I**: Kawaihau Dist, Makaleha Mts, 2700 ft. [818 m], 4.IX.1970, on *Cheirodendron kauaiense* var. *Forbesii*, W.C. Gagné. Additional paratype with data differing only as follows: 1♂, on *C. kauaiense* var. *kauaiense* (BPBM).

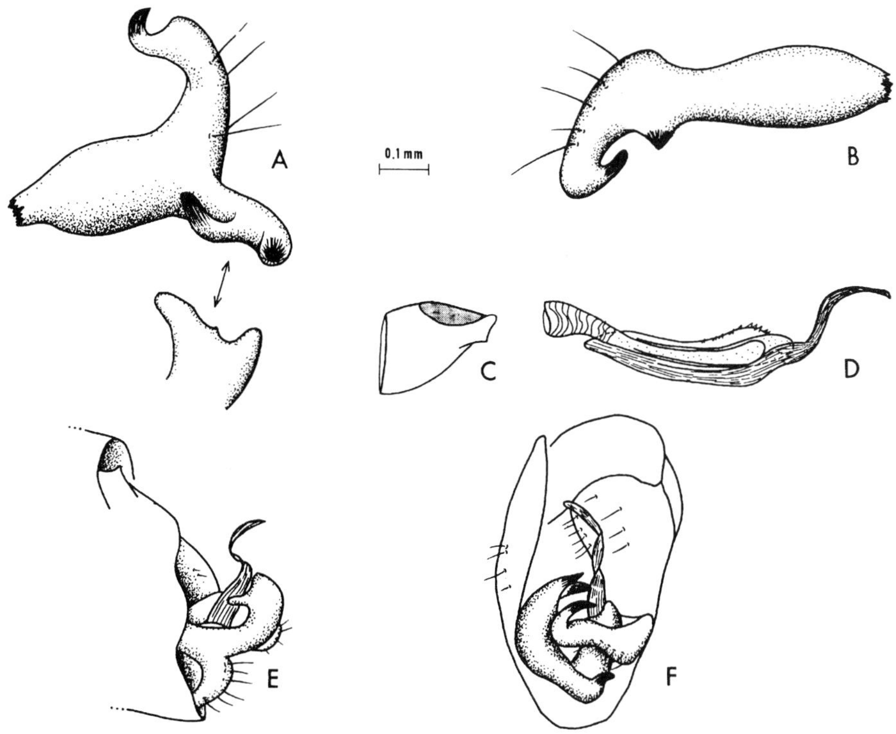

Fig. 73. *Nesiomiris lapalapa,* n. sp., male genitalia. **a**, left clasper, internal; **b**, right clasper, internal; **c**, aedeagus, lateral; **d**, spiculum, lateral; **e**, pygofer, left lateral; **f**, pygofer, terminal.

35. *Nesiomiris olapa* Gagné, **new species** Figs. 26e, 74

Other than having a uniramous upper arm of the left clasper, this species is very close to *N. lapalapa* of the Makaleha Mountains. The rostrum is slightly longer, reaching the apices of the mesocoxae, in addition to other differences noted below.

Description. **MALE** (n = 3). Body: 5.45 (5.45–5.70) long; 1.26 (1.26–1.28) wide. Head: 0.94 (0.94–0.95) wide. Antennal segment lengths: I, 0.80 (0.73–0.80); II, 2.59 (2.59–2.90); III, 2.01 (1.93–2.06); IV, 2.20 (2.14–2.20). Rostrum: 1.46 (1.28–1.48) long. Pronotum: 0.79 (0.79–0.85) long; 1.26 (1.26–1.28) wide. Differing from *N. lapalapa* as follows: apex of head straight, greenish yellow, indistinctly carinate across base; antennal segment I greenish yellow; ratio of length of antennal segment I to head width as 6.5: 7.5; basal lobe of pronotum darker green on basal 1/2, hemelytra subglabrous, the few setae present pale. Abdomen: pygofer, claspers, aedeagus, and spiculum as illustrated (Fig. 74).

FEMALE (n = 3). Body: 5.55 (5.55–5.90) long; 1.26 (1.26–1.31) wide. Head: 0.92 (0.92–0.95) wide. Antennal segment lengths: I, 1.25 (0.70–1.25); II, 2.24 (2.24–2.49); III, 1.67 (1.67–

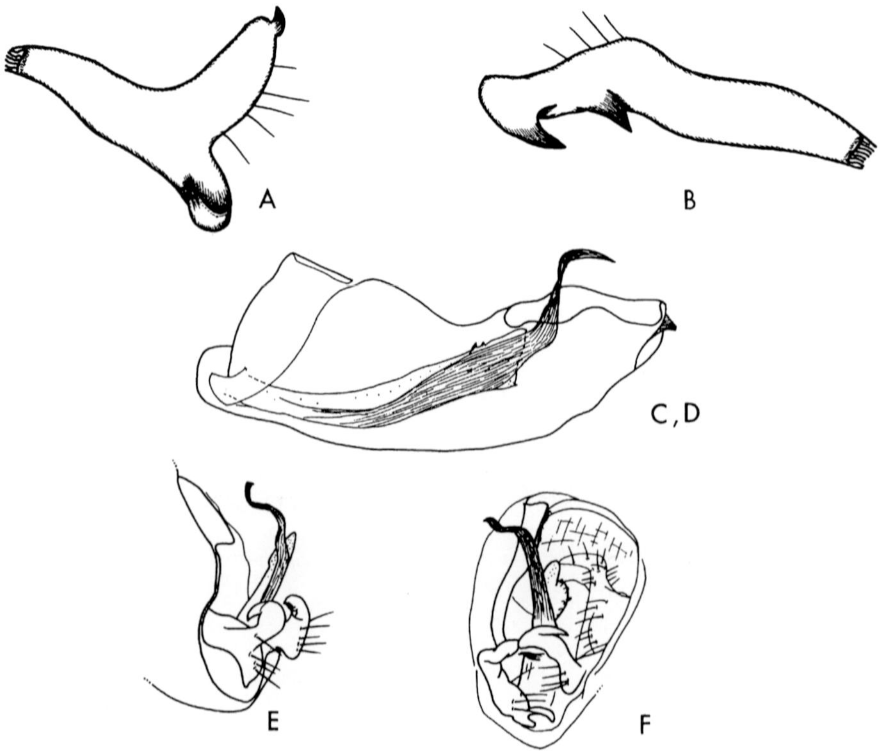

Fig. 74. *Nesiomiris olapa,* n. sp., male genitalia. **a**, left clasper, internal; **b**, right clasper, internal; **c, d**, aedeagus and spiculum, lateral; **e**, pygofer, left lateral; **f**, pygofer, terminal.

1.78); IV, 1.80 (1.80–1.93). Rostrum: 1.29 (1.28–1.45) long. Pronotum: 0.81 (0.81–0.85) long; 1.26 (1.26–1.31) wide. Differing from male as follows: apex of head weakly bifid, weakly rufescent; antennal segment I shorter, yellowish red; basal lobe of pronotum evenly pale as is scutellum and hemelytra. Apex of ovipositor rufescent; *K* and *L* structure as illustrated (Fig. 26e).

NYMPHS. Unknown.

Distribution. Kaua'i I.

Host. *Cheirodendron Fauriei*.

Etymology. A Hawaiian name for *Cheirodendron* is *olapa*.

Type data. Holotype ♂ (BPBM 14,860), allotype ♀, paratypes 3♂, 2♀, **KAUA'I**: Alaka'i Swamp, [no elevation], 24.VII.1968, on *Cheirodendron Fauriei*, W.C. Gagné. Additional paratypes with data differing only as follows: 4♂, 2♀, 10.VII.1928, [no host or elevation], E.H. Bryan, Jr. (BPBM).

Discussion. Since males of no other *Nesiomiris* were collected in the 2 collections comprising the type series, I am designating the females of this species.

36. ***Nesiomiris ehu*** Gagné, **new species** Figs. 26g, 75, 82b

This species possesses a small acute tooth on the pygofer above the left clasper. Related species in the *lapalapa* complex either have the tooth much larger: *N. lapalapa* (Fig. 73e), *N. olapa* (Fig. 74e), and *N. kauilamahu* (Fig. 76e); or entirely absent: *N. mahu* (Fig. 77e). Also, the uniramous upper arm of the left clasper is much longer and erect than other members of the complex, where it is uniformly more robust and curved inwardly toward the right clasper.

The area about the gonopore lacks evident denticles, a character shared with *N. kauilamahu* and *N. mahu*. The spiculum is comparatively short, and does not form an "S" shape beyond the gonopore (Fig. 75c).

Description. **MALE** (n = 7). Body: 5.97 (5.67–5.97) long; 1.29 (1.28–1.34) wide. Head: 0.94 (0.89–0.96). Antennal lengths: I, 1.00 (0.90–1.00); II, 3.39 (3.01–3.39); III, 1.84 (1.78–1.93); IV, broken (1.25–1.78). Rostrum: 1.68 (1.68–1.77) long. Pronotum: 0.91 (0.86–0.91) long; 1.29 (1.28–1.34) wide. Differing from *N. lapalapa* as follows: head and tylus strongly rufescent, a few brown setae between eye and apex; pronotum greenish only on collar and center of basal lobe, remainder of its dorsum weakly rufescent; rostrum reaching apices of mesocoxae; scutellum greenish on disc; hind femora also with a few erect setae; areoles, cuneus, pygofer, claspers, aedeagus, and apex of hemelytron as illustrated (Figs.75, 82b); area about gonopore lacking denticles.

FEMALE (n = 7). Body: 6.19 (6.10–6.45) long; 1.27 (1.27–1.39) wide. Head: 0.96 (0.89–0.96) wide. Antennal segment lengths: I, 0.96 (0.922–1.00); II, 2.90 (2.65–3.20); III, 1.88 (1.78–2.06); IV, 1.78 (1.60–1.93). Rostrum: 1.70 (1.61–1.95) long. Pronotum: 0.92 (0.90–0.96) long; 1.27 (1.27–1.39) wide. Differing from male as follows: antennal segment I and 1/2 of II rufescent, I shorter than head width; basal lobe of pronotum mostly pale greenish; inner apex of cuneus from areoles to apex rufescent. *K* structure sparsely denticulate (Fig. 26g).

NYMPHS. *1st instar*. Unknown. *2nd instar* (n = 1). Body: 1.55 long; 0.46 wide. Head: 0.50 wide; with a few erect, pale setae. Antennal segment lengths: I, 0.27; II, 0.56; III, 0.48; IV, 0.82; antennae totally pallid; setae short, inconspicuous. Rostrum: 0.75 long; reaching base of abdomen. Pronotum: 0.21 long; 0.46 wide; propleural stripe absent. Thoracic tergites and legs with a few erect, pale setae, the latter totally pallid. Abdomen: glabrous. *3rd instar* (n = 1). Body: 1.78 long; 0.58 wide. Head: 0.56 wide. Antennal segment lengths: I, 0.34; II, 1.02; III, 0.82; IV, 1.17. Rostrum: 0.82 long. Pronotum: 0.21 long; 0.55 wide. Not otherwise differing from 2nd instar. *4th instar* (n = 1). Body: 2.65 long; 0.52 wide. Head: 0.73 wide. Antennal segment lengths: I, 0.46; II, 1.55; III, 1.15; IV, 1.47. Rostrum: 1.09 long. Pronotum: 0.37 long; 0.73 wide. Differing from 3rd instar as follows: rostrum reaching between base and apex of hind coxae; pronotum with a stronger seta at each humeral angle; mesonotal wing bud with about 3 stronger, evenly spaced setae along lateral edge, other setae barely discernable; setae denser on appendages. *5th instar* (n = 1). Body: 3.81 long; 1.31 wide. Head: 0.83 wide. Antennal segment lengths: I, 0.81; II, 2.24; III, 1.81; IV, 1.50. Rostrum: 1.55 long. Pronotum: 0.58 long; 0.80 wide. Differing from 4th instar as follows: wing buds glabrous; a few sparse, pale, inconspicuous setae on apical abdominal tergites only; rostrum reaching meso- to metacoxae; setae on tibiae mostly erect and longer than any others.

Distribution. Kaua'i I.

Host. *Cheirodendron Helleri* var. *multiflorum*, *C. kauaiense* var. *kauaiense*, (*C.* "*Gaudichaudii*" is a *nomen nudum* (*fide* Sherff 1954)).

Etymology. A Hawaiian name for *Cheirodendron* is *ehu*.

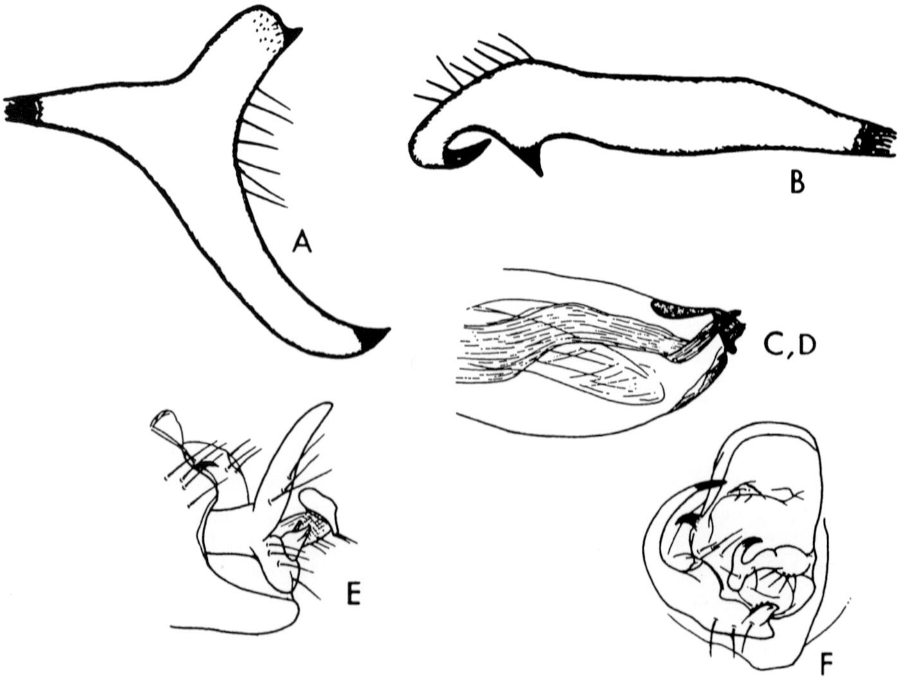

Fig. 75. *Nesiomiris ehu*, n. sp., male genitalia. **a**, left clasper, internal; **b**, right clasper, internal; **c**, **d**, aedeagus and spiculum, lateral; **e**, pygofer, left lateral; **f**, pygofer, terminal.

Type data. Holotype ♂ (BPBM 14,861), allotype ♀, paratypes 2♂, 5♀, **KAUA'I**: Kōke'e, Alaka'i Swamp Trail, Kawaikōī Ridge, [no elevation], 18.VIII.1961, on *Cheirodendron*, R.L. Usinger. Additional paratypes with data differing only as follows: 1♂, Wai'alae Riv, 4000 ft. [1219 m], 14.IV.1920, J.A. Kusche & W.M. Giffard; 1♂, 4♀, 9.VIII.1925, O.H. Swezey; 1♂, 1♀, Kumuwela, 18.VI.1932, O.H. Swezey; 1♀, Kumuwela, 27.VI.1932, O.H. Swezey; 6♂, 1♀, Alaka'i Swamp, 14.VII.1932, O.H. Swezey; 1♂, 1♀, NW Alaka'i Swamp, 9.VII.1937, E.C. Zimmerman; 3♂, 5♀, S Mōhihi Ridge, 15.VII.1937, E.C. Zimmerman; 1♂, Alaka'i Swamp, 1158 m, 29.VII.1963, D.E. Hardy; 4♂, 1♀, Kōke'e, Mōhihi Stream, 3.IX.1965, on *C. Gaudichaudii*, D.E. Hardy; 1♂, 3♀, 13–17.IX.1965, J.W. Beardsley; 1♂, 11–17.IX.1965, C.M. Yoshimoto; 3♀, Kōke'e, 12.IX.1965, C.M. Yoshimoto; 1♂, Alaka'i Swamp, 14.IX.1965, C.M. Yoshimoto; 1♂, 3♀, 14.IX.1965, on *Cheirodendron* sp. P.D. Ashlock; 4♂, 5♀, Koai'e Valley, 3600 ft. [1097 m], 21.VI.1966, D.E . Hardy; 3♀, Koai'e-Wai'ale'ale Trail, 21.VI.1966, K.Y. Kaneshiro; 3♂, 1N, Koai'e-Wai'ale'ale Trail, 3600 ft. [1097 m], 18.XI.1968, on *C. Helleri* var. *multiflorum*, W.C. Gagné; 1♂, 3♀, (1 pair *in copula*), 19N, Alaka'i Swamp at Waineke Swamp, 22.VII.1968, on *C. Helleri* var. *multiflorum*, W.C. Gagné; 2♀, Alaka'i Swamp Trail, at Waineke Swamp, 1062 m, 22–25.VII.1968; 2♂, 3♀, near Kalalau Lookout, 23.VII.1968, on *Cheirodendron* sp., W.C. Gagné; 5♂, 11♀, N Alaka'i Swamp, 24.VII.1968, on *C. Helleri* var. *multiflorum*, W.C. Gagné; 2♀, N Alaka'i Swamp, 24.VII.1968, on *C. kauaiense* var. *kauaiense*, W.C. Gagné; 4♀, Alaka'i Swamp

Trail at Waineke Swamp, 1062 m, 25.VII.1968, on *C. Helleri* var. *multiflorum*, W.C. Gagné; 2♂, 3♀, Koai'e-Wai'ale'ale Trail, 25.VII.1968, W.C. Gagné; 4♂, 3N, Alaka'i Swamp, 2nd bog on trail, 24.VII.1968, on *C. kauaiense* var. *kauaiense*, W.C. Gagné; 1♀, Koai'e-Wai'ale'ale Trail, near Koai'e Creek, 1066 m, 25.VII.1968, W.C. Gagné; 3♂, 7♀, near Kalalau Lookout, 19.XI.1968, on *C. Helleri* var. *multiflorum*, W.C. Gagné; 1♂, 2♀, Alaka'i Swamp Trail at Waineke Swamp, 1062 m, 19.XI.1968, on *C. Helleri* var. *multiflorum*, W.C. Gagné; 1♂, 1♀, 6N, near Kalalau Lookout, 20.XI.1968, on *C. Helleri* var. *multiflorum*, W.C. Gagné; 1♂, 2N, Alaka'i Swamp Trail, near Kawaikōī Stream, 20.XI.1968, on *C. Helleri* var. *multiflorum*, W.C. Gagné; 4♂, 2♀, Kōke'e State Park, Kaunuohua Ridge, 3500 ft. [1066 m], 24.VII.1970, on *C. Helleri* var. *multiflorum*, W.C. Gagné; 3♂, 2♀, Kōke'e State Park, Honopū Trail, 4000 ft. [1219 m], 25.VIII.1970, on *C. Helleri* var. *multiflorum*, W.C. Gagné; 1♂, 1♀, Koai'e-Wai'ale'ale Trail, 3500 ft. [1066 m], 25.VIII.1970, on *C. Helleri* var. *multiflorum*, W.C. Gagné; 3♂, Hanapēpē Dist near Pe'ape'a, 914 m, 7.IX.1970, on *C. Helleri*, W.C. Gagné; 2♂, 8♀, Alaka'i Swamp, 2.5 mi. W of Wai'ale'ale, 1375 m, 27–31.XII.1973, on *C. Helleri*, W.C. Gagné (BPBM).

Variation. The right clasper sometimes has the terminal hook almost fused apically to the clasper in which case the base is thickened. The tooth on the pygofer is sometimes reduced and hard to discern among the setae. The apex of the head is very weakly bifid in some males.

Discussion. This is by far the most abundant *Nesiomiris* species in the Alaka'i Swamp massif and predominates on varieties of *Cheirodendron Helleri*, where also occurs the closely related *N. mahu* and the similarly sized but more distantly related *N. legnotus*.

Sherff (1954), in the only monograph of Hawaiian *Cheirodendron*, unfortunately made no analysis of either the possible phylogeny or relationships of the 4 species of this genus of trees in the Alaka'i Swamp. Sherff based his separation of species primarily on the numbers of stigmas and pyrenae as well as the leaf shape, none of which would appear to me to offer great ecological isolating mechanisms for the *Nesiomiris* feeding on the leaves. If the host relationships indicated above are significant on further analysis, there might be differences in the chemical aspects of these species.

Most collections made on *Cheirodendron Helleri* and *C. kauaiense* have immatures with subglabrous dorsa. Since the majority of the adults are of *N. ehu*, I presume these nymphs belong to *N. ehu*, although they may possibly belong to the partially sympatric *N. mahu*.

Only 1 collection of this species was made off the Alaka'i Swamp massif, that near Pe'ape'a at 914 m elevation.

37. ***Nesiomiris kauilamahu*** Gagné, **new species** Fig. 76

This species appears to be closely related to *N. lapalapa* with which it shares its host plant and occurs sympatrically and synchronously. The males are separated from *N. lapalapa* as noted under that species. I find that the females, however, are inseparable and so I have not designated any as paratypes.

N. kauilamahu shares with *N. ehu* the acute tooth on the pygofer above the left clasper, but in the latter the acute tooth is positioned directly on the wall of the pygofer

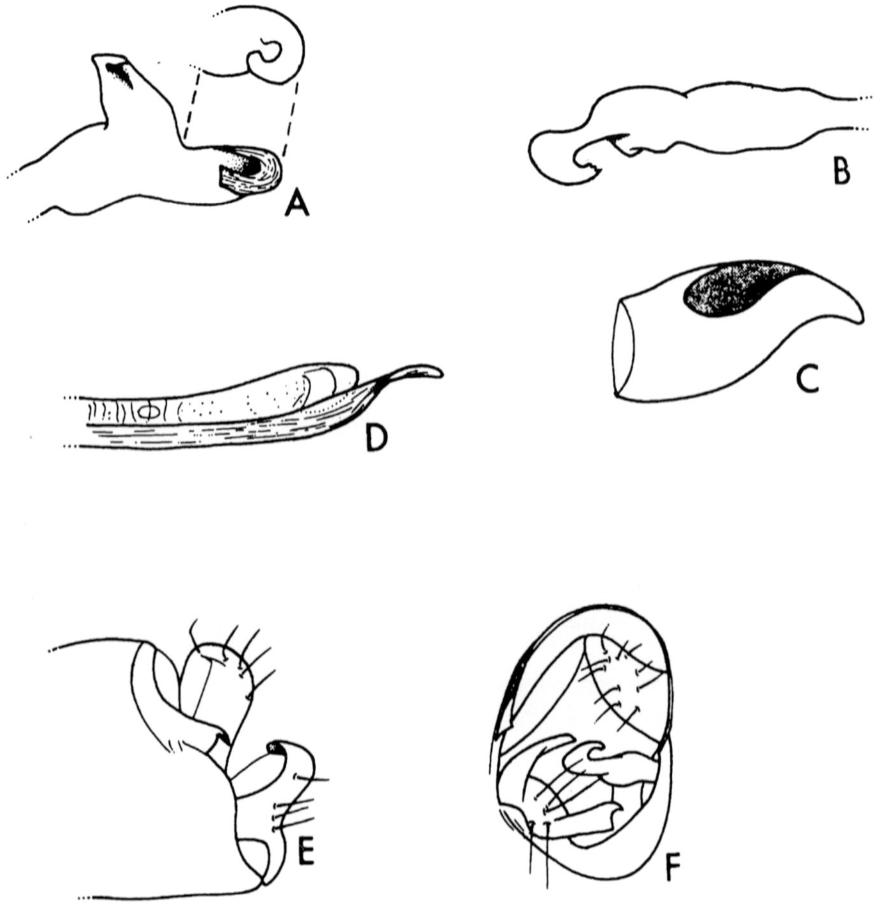

Fig.76. *Nesiomiris kaulimahu,* n. sp., male genitalia. **a**, left clasper, internal; **b**, right clasper, internal; **c**, aedeagus, lateral; **d**, spiculum, right lateral; **e**, pygofer, left lateral; **f**, pygofer, terminal.

and not on a subsidiary tooth as in *N. kauilamahu*. The claspers are more reduced than *N. ehu* (Fig. 75), as is the spiculum (Fig. 75c).

Description. **MALE** (n = 7). Body: 4.93 (4.70–4.94) long; 1.15 (1.09–1.21) wide. Head: 0.90 (0.85–0.90) wide. Antennal segment lengths: I, 0.63 (0.56–0.64); II, 2.10 (2.10–2.45); III, 0.92 (0.92–1.50); IV, broken. Rostrum: 1.18 (1.09–1.28) long. Pronotum: 0.74 (0.73–0.80) long; 1.15 (1.09–1.17) wide. Differing from *N. ehu* as follows: head and tylus yellowish orange but not rufescent, greenish at apex, subglabrous; antennal segment I much shorter than head width (ratio 4.75: 7); hind lobe of pronotum broadly green; scutellum greenish on apical 1/2, remainder yellowish brown, subglabrous; membrane almost entirely infuscate, clear on inner 1/2 of large areole. Pygofer, claspers, and aedeagus as illustrated (Fig. 76); spiculum (Fig. 76c) comparatively short, also failing to form an "S" shape beyond the gonopore and lacking denticles about the gonopore.

FEMALE. Unknown.

NYMPHS. Unknown.

Distribution. Kaua'i I.

Host. *Cheirodendron kauaiense* vars. *kauaiense* and *Forbesii*.

Etymology. A Hawaiian name for *Cheirodendron* on Kaua'i is *kauila māhu*, meaning steam or fog *kauila*.

Type data. Holotype ♂ (BPBM 14,862), paratypes 6♂, **KAUA'I**: Kawaihau Dist, Makaleha Mts, 2700 ft. [818 m], 4.IX.1970, on *Cheirodendron kauaiense* var. *kauaiense*, W.C. Gagné. Additional paratype with data differing only as follows: 1♂, on *Cheirodendron kauaiense* var. *Forbesii* (BPBM).

Variation. Sometimes the mesoscutum is rufescent.

38. ***Nesiomiris mahu*** Gagné, **new species** Figs. 26h, 77

I place this species last in the *lapalapa* complex because a tooth is completely lacking on the pygofer, a character I consider to be apomorphic in the complex. Other aspects, particularly the unreduced right clasper (as compared to *N. kauilamahu*, for example), tempted me to place it systematically lower in the complex. The rufescent head leads me to consider it a sister species of *N. ehu*. The left clasper, however, is inwardly hooked as in other members of the complex, except *N. ehu*.

I consider the females of this species to be those possessing the following combination of characters: apex of head rounded, antennal segment I yellowish brown, rostrum reaching only apex of mesosternum, and hemelytra adjacent to the areoles sometimes rufescent.

Description. **MALE** (n = 4). Body: 6.38 (6.10–6.50) long; 1.45 (1.30–1.45) wide. Head: 1.05 (0.96–1.05) wide. Antennal segment lengths: I, 1.00 (0.95–1.09); II, 2.65 (2.65–3.20); III, 1.78 (1.78–2.14); IV, 1.45 (1.45–1.78). Rostrum: 1.48 (1.45–1.53) long. Pronotum: 1.00 (0.90–1.00) long; 1.45 (1.30–1.45) wide. Differing from *N. ehu* as follows: basal carina of head not reaching edge of eye; head subglabrous, distinctly impressed on middle of apex; antennal segment I and basal 1/2 of II rufescent, apical 1/2 of latter brownish; I shorter than head width; basal lobe of pronotum greenish excepting yellowish brown area adjacent to mesoscutellum; scutellum greenish; hemelytra evenly clothed with brownish, recumbent setae, apex adjacent to areoles rufescent. Membrane deeply infumate; rostrum reaching apex of mesosternum; abdomen yellowish excepting a lateral rufescent area on segments 3–5; pygofer, claspers, and aedeagus as illustrated (Fig. 77).

FEMALE (n = 4). Body: 6.18 (5.87–6.45) long; 1.34 (1.28–1.34) wide. Head: 0.93 (0.89–0.96) wide. Antennal segment lengths: I, 1.00 (0.81–1.00); II, 2.83 (2.83–3.10); III, 1.81 (1.81–1.93); IV, 1.79 (1.78–1.79). Rostrum: 1.90 (1.45–1.90) long. Pronotum: 1.00 (0.81–1.00) long; 1.34 (1.28–1.34) wide. Differing from male as follows: apex of head rounded; only weakly impressed on vertex; antennal segment II yellowish brown (green areas more faded with age and setae apparently considerably abraded), *K* structure as illustrated (Fig. 26h), more denticulate than *N. ehu*.

NYMPHS. *1st and 2nd instars*. Unknown. *3rd instar* (n = 1). Body: 2.30 long; 0.70 wide. Head: 0.65 wide, a row of erect setae encircling the periphery of the vertex; apex rounded. Antennal segment lengths: I, 0.46; II, 1.03; III, 0.95; IV, 1.28; setae same length as those on head on I, II, and basal 1/2 of III; antennae totally pallid. Rostrum: 0.76 long; reaching apex of mesocoxae. Pronotum: 0.30 long; 0.65 wide; subquadrate; erect brown setae distributed in an ill-defined row about its periphery. Wing buds with a row of stouter setae along lateral margins and other setae of varying lengths on their discs. Legs: totally pallid, setae on femora and tibiae similar to those on dorsum. Abdomen: 1st visible tergite with a row of larger setae along lateral border and a row of shorter setae across posterior border, remainder of abdominal tergites each with a posterior row of setae but not becoming abruptly shorter on 2 apical segments. *4th instar* (n = 1). Body: 2.55 long; 1.12 wide. Head: 0.82 wide. Antennal segment lengths: I, 0.51; II, 1.40; III, 1.22; IV, 1.40. Rostrum: 1.12 long. Pronotum: 0.55 long; 0.96 wide. Differing from 3rd instar as follows: mesonotal wing bud with a row of setae along lateral border; rostrum reaching base of abdomen. *5th instar* (n = 1). Body: 4.64 long; 1.39 wide. Head: 0.95 wide. Antennal segment lengths: I, 0.90; II, 2.30; III, 1.65; IV, 1.65. Rostrum: 1.28 long. Pronotum: 1.46 long; 1.00 wide. Differing from 4th instar as follows: rostrum reaching apices of mesocoxae; antennae IV more infumate. No differences apparent between the penultimates of each sex.

Distribution. Kaua'i I.

Host. *Cheirodendron Helleri* var. *multiflorum*, *C.* sp.

Etymology. A Hawaiian name for *Cheirodendron* species is *māhu*.

Type data. Holotype ♂ (BPBM) 14,863), allotype ♀, paratypes 2♂, 4♀. **KAUA'I**: Mt Kāhili, [no elevation], 19.XI.1968, on *Cheirodendron Helleri* var. *multiflorum*, W.C. Gagné. Additional paratypes with data differing only as follows: 2♂, 4♀, Kumuwela, 27.VI.1932, on *Cheirodendron*, O.H. Swezey; 2N, Alaka'i Swamp, 2nd bog on trail, 24.VII.1968, on *C. kauaiense* var. *kauaiense*; 2♂, 2♀, Alaka'i Swamp, 14.VII.1932, on *Cheirodendron*, O.H. Swezey; 1♂, Koai'e-Wai'ale'ale Trail, [no elevation], 25.VII.1968; 1♂, 9N, Alaka'i Swamp Trail at Waineke Swamp, 1062 m, 22–25.VII.1968, on *Cheirodendron*; 1♂, Koai'e-Wai'ale'ale Trail, 3600 ft. [1097 m], 18.XI.1968, on *Tetraplasandra meiandra*; 1♂, Kōke'e State Park, Honopū Trail, 4000 ft. [1219 m], 25.VIII.1970; 6N, Hanapēpē Dist., Mt Pe'ape'a, 3000 ft. [914 m], 7.IX.1970 (BPBM).

Variation. Although both series collected by Swezey are faded, the greenish areas are in the same places, as in material I obtained subsequently.

Discussion. This species has a broader distribution than other members of the *lapalapa* complex, extending from the Alaka'i Swamp massif to Mt Kāhili. However, it skirts the fringes of Alaka'i Swamp and appears not to penetrate the higher and wetter areas as does *N. ehu*, with which it is otherwise sympatric.

I consider the male collected on *Tetraplasandra meiandra* to be a perching record and not indicative of a true host record. Since this tree species is often surrounded by *Cheirodendron Helleri*, the record is understandable.

Because the hemelytra are more densely setose in adults of this species than in *N. ehu*, I should expect the nymphs to be more setose also. Two collections of 7 from Alaka'i Swamp and environs contain strikingly more setose nymphs, as well as adults of *N. mahu*. The other 5 collections have subglabrous nymphs. Both types are represented in collections from Waineke Swamp, where adults of both *N. ehu* and *N. mahu* occur. I consider these setose nymphs to be immatures of *N. mahu*.

A female (Kōke'e, Kaua'i, 22–25.VIII.1968, at light, W.C. Gagné), appears to belong to this species, but I have not designated it as a paratype.

The *legnotus* Complex

Description. **ADULTS**. Head rufescent, apex rounded; propleural stripe present, extending across collar to base of propleuron, right clasper uniramous.

Distribution. Kaua'i I.

Host. *Cheirodendron Helleri* var. *multiflorum*.

Included species. No. 39.

Discussion. This is the sister group of the *lapalapa* complex. There is no indication that *N. legnotus* has reached any island other than Kaua'i. Thus, I consider a single female, with a prominent propleural stripe, bearing the following data "Forest above Honoka'a, Hawaii, 2000', VIII.63, D.E. Hardy, Collector," to be mislabeled. This specimen appears to belong to *N. legnotus*. Several extensive collections from *Cheirodendron trigynum* in the Honoka'a area failed to produce any evidence of this or of any member of the *lihuensis* group.

39. *Nesiomiris legnotus* Gagné, **new species** Figs. 78, 79

Except for the prominent propleural stripe (Fig. 78), this species would be placed in the *lapalapa* complex, since the claspers and pronotum more strongly resemble species there, such as *N. kauilamahu*. The propleural stripe is considered a plesiomorphic character that necessitates placement of this species in its own complex, as a sister complex to the *lapalapa* complex. In the *lihuensis* group only *N. lineatus* shares the character of a fully developed (across collar to base of pronotum) propleural stripe, from which *N. legnotus* differs as outlined below.

Description. **MALE** (n = 2). Body: 4.93 (4.93–5.10) long; 1.18 (1.17–1.18) wide. Head: 0.89 (0.86–0.89) wide. Antennal segment lengths: I, 0.72 (0.72–0.81); II 2.84 (2.84–2.90); III, 1.92 (1.78–1.92); IV, 1.29 (1.29–1.68). Rostrum: 1.49 (1.45–1.49) long. Pronotum: 0.72 (0.72–0.78) long; 1.18 (1.17–1.18) wide. Differing from *N. lineatus* as follows: apex of head rounded; length of antennal segment I subequal to head width, I with only a few scattered recumbent and erect setae; rostrum reaching middle of mesocoxae; pronotum yellowish, glabrous; propleural stripe extending from base of pronotum across the collar and then dorsally along the collar to meet rostrum reaching middle of mesocoxae; scutellum yellowish with lateral carinate shelf, infuscate-like propleural stripe; short, brown, recumbent setae generally distributed on hemelytra; clavus rufescent, corium pale greenish, basal margin infumate, apex rufescent; cuneus pale greenish, its apex and margin adjacent to areoles, rufescent; anal lobe of hind wing broadly infumate; legs yellowish; abdominal tergites infumate, pygofer, claspers, and aedeagus as illustrated (Fig. 79).

FEMALE (n = 3). Body: 5.55 (4.32–5.67) long; 1.27 (1.22–1.22) wide. Head: 0.93 (0.86–0.93) wide. Antennal segment lengths: I, 0.81 (0.73–0.81); II, 3.14 (2.49–3.14); III, 2.08 (1.78–2.08); IV, 1.53 (1.43–1.53). Rostrum: 1.61 (all 1.61) long. Pronotum: 0.81 (0.73–0.81) long; 1.27 (1.22–1.27) wide. Differing from male as follows: antennal segment I shorter than head width, yellowish on basal 1/2 becoming brownish apically; propleural stripe not extending dorsally along collar; inner margins of clavus and cuneus faded to greenish yellow.

NYMPHS. Unknown.

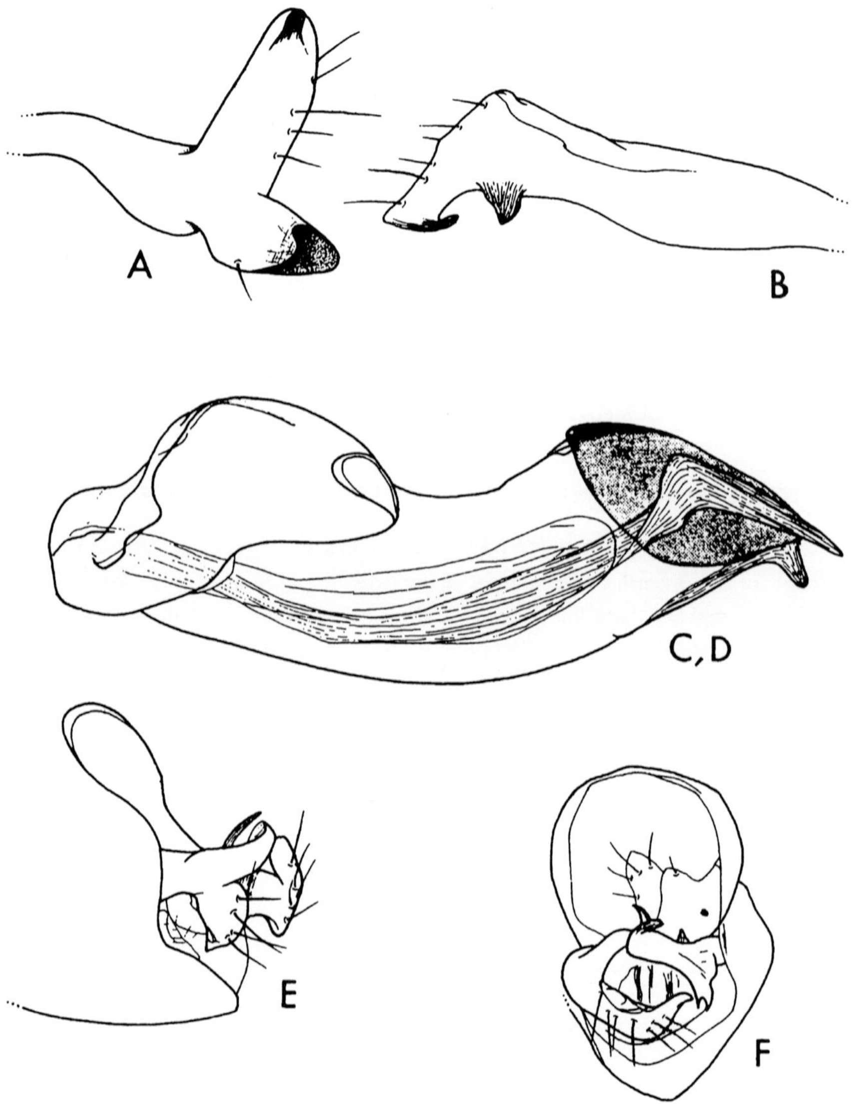

Fig. 77. *Nesiomiris mahu,* n. sp., male genitalia. **a**, left clasper, internal; **b**, right clasper, internal; **c**, **d**, aedeagus and spiculum, lateral; **e**, pygofer, left lateral; **f**, pygofer, terminal.

Distribution. Kaua'i I.

Host. *Cheirodendron Helleri* var. *multiflorum.*

Etymology. Named for the prominent propleural stripe.

Type data. Holotype ♂ (BPBM 14,864), allotype ♀, paratype ♀, **KAUA'I**: Koai'e Valley, 1096 m, 21.VI.1966, D.E. Hardy. Additional paratypes with data differing only as follows: Koai'e-Wai'ale'ale trail, K.Y. Kaneshiro; 1 ♀, S Mōhihi Ridge, beating, 15.VII.1937, E.C. Zimmerman; 1 ♀, N Alaka'i Swamp, 24.VII.1968, on *Cheirodendron Helleri* var. *multiflorum*, W.C. Gagné.

Variation. On 1 of the females (Zimmerman specimen), which has the wings broken off, the abdominal tergites are broadly brownish and the 8th tergite has a blackish apex. This is presumably the case in the allotype as well, although obscured by the wings.

Discussion. The mesoscutum has apparently collapsed in the type, and the wings are partially spread and shriveled apically; otherwise, it is in good condition although apparently somewhat faded, which seems to accentuate the rufescent areas.

The spiculum of the single paratype male available was mutilated when dissection was attempted, so no illustrations can be provided.

The *lanaiensis* Group

Description. **ADULTS**. Sexes strongly to weakly dimorphic for color, males darker green; areolar veins greenish when alive, fading to yellow and yellowish browns after death, never rufescent; propleural stripe absent (excepting some males of *N. lanaiensis*); membrane of forewing variously infuscate, sometimes reduced; pygofer usually with sclerotized projection or tooth on its dorsum, or on left side above base of clasper; and a tuft of short, dense setae on apex of venter; spiculum sinuate but never flabellate; aedeagus with a left apically thickened wall or apical tooth; areoles partially to wholly coriaceous; areolar veins greenish.

NYMPHS. Apices of antennal segments II and III only, infuscate, but never reddish; propleural stripe absent.

Distribution. O'ahu, Lāna'i, Moloka'i, Maui, and Hawai'i Is.

Host. *Ilex anomala*.

Included species. Nos. 40–50.

Discussion. This monophyletic group appears to be the result of a single shift by an *N. elongatus*-like precursor on the Maui group of islands from an araliad, probably *Tetraplasandra*, to *Ilex anomala* of the Aquifoliaceae. Since representatives of both host families grow in close association in the rain forest, this shift may have been due to chance. The *lanaiensis* group does not occur on Kaua'i, where the host plant is abundant.

Swezey (1954), recognizing that a complex of species might be associated with *Ilex anomala* remarked (p. 110), "It is possible that the *Nesiomiris* on that host tree consist of different species on the different islands." This suspicion has now been confirmed.

The nymphs of the species in this group are so closely similar in size in their respective instars that detailed measurements did not seem fruitful. Only the 5th instar of *N. hawaiiensis* was measured.

The *lanaiensis* Subgroup

Description. **ADULTS**. Species with lateral margins linear. Sometimes widest across base of pronotum or pronotal width subequal to body width; apex of head weakly bifid to rounded; heme-

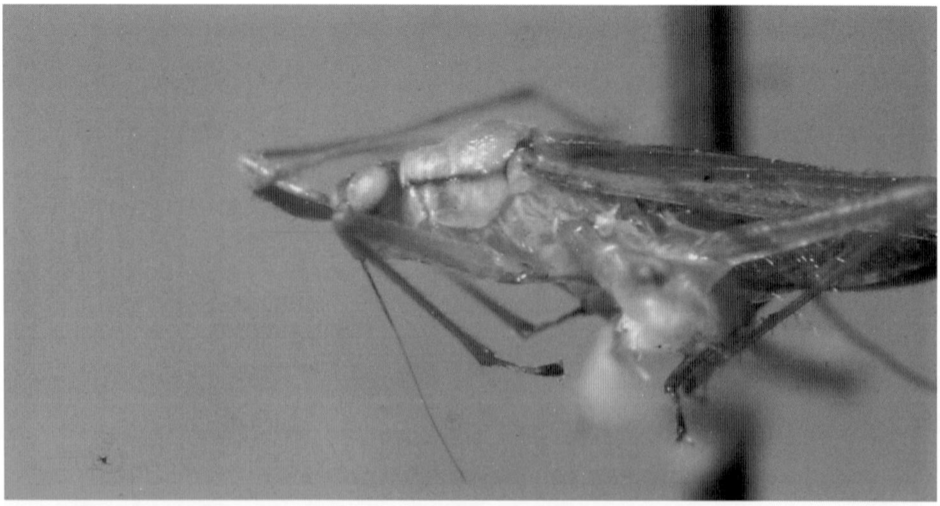

Fig. 78. *Nesiomiris legnotus,* n. sp., male, left lateral. Photo by Wayne C. Gagné.

lytral setae brownish, recumbent, distributed primarily along veins; setae of antennal segment I not pilose, mostly reclined, brown; shorter than greatest diameter of antennal segment I; areoles partially coriaceous; membrane of forewing not reduced, exceeding apex of cuneus by a length greater than length of small areolar cell, infuscate, infuscation extending into large areole; tooth on pygofer, if present, usually dorsally located; upper arm of left clasper usually reduced, thickened, and its apex facing into genital atrium; dorsum of pygofer blackish; lower arm of right clasper not curling over upper arm of left clasper when viewed terminally; tooth on aedeagus, when present, located ventrally at apex or on right side of apex.

NYMPHS. Appendages in all instars not pilose; setae on appendages short and pale, shorter than greatest diameter of antennal segment I.

Distribution. O'ahu, Moloka'i, Lāna'i, Maui, and Hawai'i Is.

Host. *Ilex anomala*.

Included species. Nos. 40–45.

Discussion. The species pairs of this subgroup, which are sympatric or have closely adjacent allopatric distributions on Maui, generally have 1 species with more strongly reduced claspers than the other. This fact might be interpreted as the result of double invasions and subsequent character displacement, as has been postulated for psittirostrine drepanidine birds (see Bock 1970).

Timberlake's (1921) general statement that "the males are usually dark green and the females paler or more yellowish green" refers specifically to this subgroup and is not to be construed as applying to the genus as a whole.

The females of the subgroup are very uniform, and I have found little to distinguish the species. Hence, specimens designated as ♀ allotypes and ♀ paratypes are represented only by those that were collected in association with a male or males that all belonged

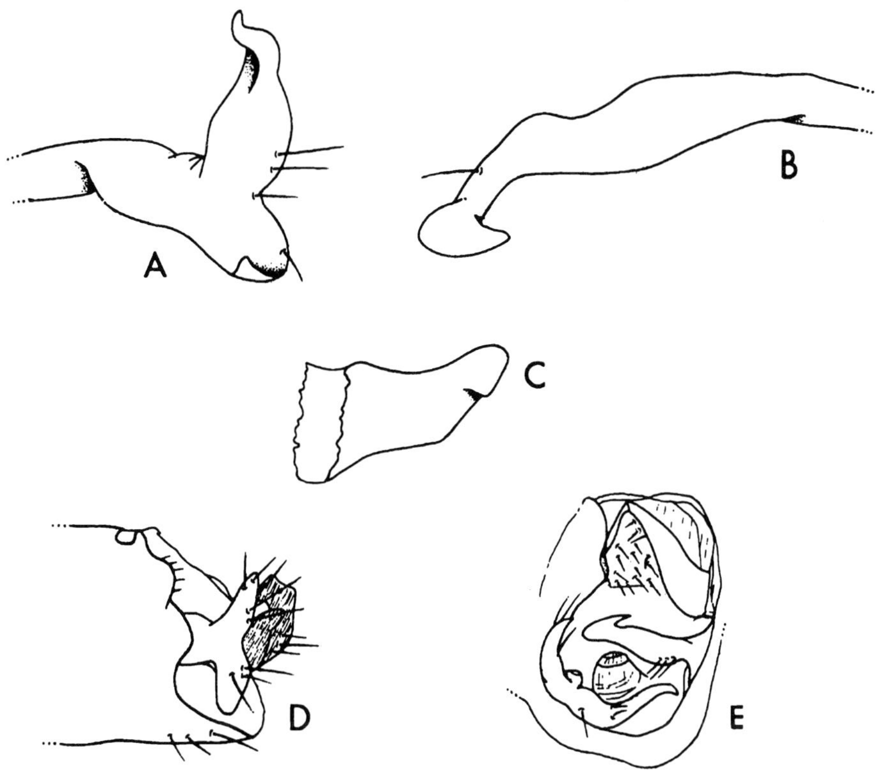

Fig. 79. *Nesiomiris legnotus*, n. sp., male genitalia. **a**, left clasper, internal; **b**, right clasper, internal; **c**, aedeagus, left lateral; **d**, pygofer, left lateral; **e**, pygofer, terminal.

to 1 species, or in cases where monoinsular endemics were present.

The following transformations are apparent in this subgroup: in general, the claspers, particularly the right claspers, are becoming reduced; in the left clasper the upper arm shows a reduction series while the lower arm remains unchanged; in the right clasper the shape is roughly chelate in *N. lanaiensis* and *N. hawaiiensis* (Figs. 81b, 84b, 85b), then the apically acute "thumb" is lost in some populations of *N. hawaiiensis* and in *N. byroniae* (Figs. 85b–1, b–2; 86b–2) and both arms begin to anastomose; the aedeagus is initially without a ventral apical tooth in *N. lanaiensis* and *N. hawaiiensis* (Figs. 81c, 85c), appears weakly at first in *N. mauiensis* (Fig. 87c) and is prominent in the remainder of the complex; the spiculum is initially weakly, apically sinuate in *N. lanaiensis*, *N. hawaiiensis*, *N. byroniae*, and *N. kawau* (Figs. 81d, 84d, 86d, 88d), or also apically twisted in *N. hawaiiensis*, *N. mauiensis*, and *N. oahuensis* (Figs. 85d, 87d, 89d); the pygofer is initially strongly toothed dorsally in *N. lanaiensis*, *N. hawaiiensis*, and *N. byroniae* (Figs. 81e, 85e, 86f), then the tooth becomes situated left laterally in *N. oahuensis* (Fig. 89e) and is weakly defined or absent in *N. kawau* and *N. mauiensis*.

Fig. 80. *Nesiomiris lanaiensis,* n. sp., male, left lateral. Photo by Wayne C. Gagné.

This speciation pattern suggests that a *N. lanaiensis*-like ancestor might have quickly spread over all of Lāna'i, Moloka'i, and Maui when they were joined. At the same time it reached Hawai'i I before isolation, on Maui Nui, began to interrupt gene flow between these populations. *Nesiomiris oahuensis* appears to have been derived from a species on Maui Nui.

40. *Nesiomiris lanaiensis* Gagné, new species — Figs. 26i, 80, 81

Nesiomiris hawaiiensis, Giffard, 1908, *Proc. Hawaii. Entomol. Soc.* 1: 183 (misidentification).

This species is unique in that the propleural stripe, when present, occurs only in males, and then to varying degrees; when present it does not extend posteriorly beyond the mesal constriction of the pronotum (Fig. 80). An apodeme on the sclerotized dorsal wall of the atrium has extended to form a serrate toothlike extension on the dorsum of the pygofer (Fig. 81e-f). Other distinct features are the evenly rounded apex of the head and transverse brownish fascia on the venter of the hind femora. Mature males have the abdominal tergites brownish, while these are pallid in the female.

Description. **MALE** (n = 29). Body: 6.18 (6.18–6.86) long; 1.54 (1.54–1.74) wide. Head: 1.13 (1.13–1.19) wide; broadly rounded; carinate; vestiture present on frons, inner angles of eyes, and evenly spaced along carina. Antennal segment lengths: I, 1.03 (1.03–1.13), yellowish, vestiture brown; II, 3.31 (2.70–3.93), yellowish on proximal 1/2, becoming browner on distal 1/2, tip dark brown; III, 2.05 (1.97–2.20), dark brown; IV, 1.70 (1.33–1.83), dark brown; setae at base only, these as dense as those on III, a few longer, erect setae extending the length of IV. Rostrum: 1.36 (1.36–1.50) long, reaching tip of mesosternum. Pronotum: 1.04 (0.91–1.04) long; 1.54 (1.54–1.74) wide;

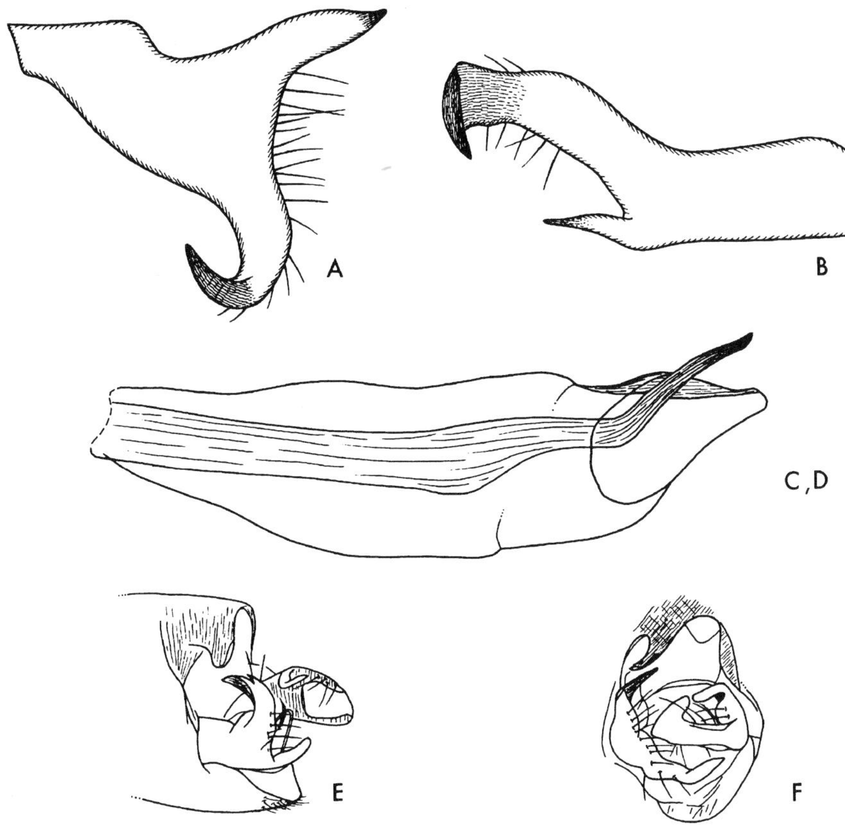

Fig. 81. *Nesiomiris lanaiensis,* n. sp., male genitalia. **a**, left clasper, internal; **b**, right clasper, internal; **c**, **d**, aedeagus and spiculum, lateral; **e**, pygofer, left lateral with dorsal tooth on a pygofer inset; **f**, pygofer, terminal.

disc and collar greenish dorsally and laterally; calli yellowish, brown on inner posterior angles; weak propleural stripe extending from behind collar to mesal constriction. Mesonotum and scutellum yellowish, greenish along borders on distal 1/2. Hemelytra: greenish, marmorate yellow and green on anterior 1/2 of clavus; large areole coriaceous on inner 1/2, membrane infuscate, darker between the anal ridges. Hind femur with ventral brownish annulations, brown at apex, remainder yellowish; other femora totally yellow, vestiture recumbent and dark above the genital atrium, with a dorsal toothlike serrate extension on atrial wall; area about ventral, apical tuft of setae brownish; genitalia pygofer as illustrated (Fig. 81).

FEMALE (n = 29). Body: 6.47 (6.47–7.21) long; 1.59 (1.59–1.83) wide. Head: 1.14 (1.13–1.19) wide. Antennal segment lengths: I, 1.03 (1.00–1.16); II, 3.30 (3.18–3.60); III, 1.55 (1.55–2.27); IV, 1.18 (1.18–1.87). Rostrum: 1.29 (1.29–1.57) long. Pronotum: 1.05 (0.83–1.05) long; 1.59 (1.59–1.83) wide. Differing from male as follows: hind femora lacking brown annulations; tibiae lightly greenish; propleural stripe lacking; body light green; *K* and *L* structures as illustrated (Fig. 26i).

NYMPHS. *1st and 2nd instars*. Unknown. *3rd instar* (n = 6). Longest setae on antennal segment I longer than the greatest thickness of this segment; anterior margin of pronotum convex at middle. *4th instar* (n = 1). Dorsal thoracic setae brown, short, and stout; rostrum reaching apices of mesocoxae; lateral margins of pronotum sublinear; abdomen with sparse recumbent setae posteriorly; setae on legs brownish. *5th instar* (n = 1). Dorsal setae brownish, those on lateral margins of mesonotal wing buds concentrated in patches, densest patch at the anterior humeral angle; lateral margins of pronotum concave; apex of head rounded; apices of mesonotal wing buds rounded; rostrum reaching apex of mesocoxa; hind tibia with erect setae only on anterior aspect; antennal segment II with inclined setae only.

Distribution. Lāna‘i I.

Host. *Ilex anomala*.

Etymology. A patronym for Lāna‘i I.

Type data. Holotype ♂ (BPBM 14,865), allotype ♀, paratypes 4♂, 1♀, 16N, **LĀNA‘I**: Lāna‘ihale, 3000 ft. [914 m], 12.VII.1968, on *Ilex anomala*, W.C. Gagné. Additional paratypes with data differing only as follows: 7♂, 4♀, 7N, 6.VI.1971; 5♂, 6♀, 3300 ft. [1006 m]; 1♂, 2000 ft. [605 m], II.1894, [no host], R.C.L. Perkins; 1♀, 3000 ft. [914 m], II.1894, [no host], R.C.L. Perkins; 4♂, 2♀, Ha‘alelepa‘akai, [no host or elevation], VII.1894, R.C.L. Perkins; 1♂, 1♀, Lāna‘i Mts, [no elevation], 29.X.1947, N.L.H. Krauss; 1♂, 1♀, 1N, 3000 ft. [914 m], 4.I.1962, J.W. Beardsley; 1♂, 1♀, 3000 ft. [914 m], 27.IX.1963, on *Ilex* sp., J.W. Beardsley; 3♂, 2♀, 3000 ft. [914 m], 27.IX.1964, on *Ilex* sp., J.W. Beardsley; 4♂, 5♀, 2N, 1005 m, 25.III.1966, J.W. Beardsley (BPBM).

Variation. The propleural stripe, when present, is evident only in male. The number and location of the denticles on the upper arms of the left claspers also vary, as does the degree of curvature. Sometimes the distal 1/2 of the scutellum is greenish.

Discussion. I consider this to be the most plesiomorphic species in the *lanaiensis* group because it possesses the following combination of characters: a propleural stripe sometimes present in some males only (Fig. 80); brownish annulations present on the venter of the hind femora (Fig. 80), a character shared in other species complexes only with *N. lineatus*, and which is possibly derived from the reddish annulate hind femora of *N. elongatus* of Kaua‘i or *N. trifurcatus* of O‘ahu; the dorsal tooth (Fig. 81e-f) on the pygofer the most strongly developed; the aedeagus lacking a subterminal tooth or projection and more strongly sclerotized on the left side (Fig. 81c). The characters listed for the pygofer and aedeagus are symplesiomorphically shared with *N. hawaiiensis* and *N. byroniae* of Hawai‘i and Maui Is.

41. *Nesiomiris hawaiiensis* Kirkaldy — Figs. 22b, 24b, 26j, 46a, 82c, 83, 84, 85

Nesiomiris hawaiiensis Kirkaldy, 1902, *Fauna Hawaiiensis* 3(2): 150, pl. 5, fig. 50 (original descrip.). Kirkaldy, 1906b, *Trans. Am. Entomol Soc.* 32: 144 (cat.). Giffard, 1908, *Proc. Hawaii. Entomol. Soc.* 1: 183 (misidentification, probably = *N. lanaiensis*, n. sp.). Kirkaldy, 1908, *Proc. Hawaii. Entomol. Soc.* 1: 198 (note on type specimen). Kirkaldy, 1910, *Fauna Hawaiiensis* 2(6): 554 (note). Reuter, 1910, *Acta Soc. Sci. Fenn.* 37: 155 (cat.). Timberlake, 1921 *Proc. Hawaii. Entomol. Soc.* 4: 472 (in part). Swezey & Bryan, 1929, *Proc. Hawaii. Entomol. Soc.* 7: 303, 311 (misidentification, = *N. sinuatus*, n. sp.). Zimmerman, 1948b, *Insects of Hawaii* 3: 217 (in part; "plesiotype," fig. 96 = *N. williamsi*, n. sp.). Carvalho, 1952, *An. Acad.*

Bras. Cienc. 24: 84 (as type of genus). Swezey, 1954, *B.P. Bishop Mus. Spec. Publ.* 44: 50, 110, 217 (in part, also included *N. curvipes*, n. sp.; *N. tetraplasandrae*, n. sp.; and others not presently determinable). Carvalho, 1959, *Arq. Mus. Nac.* (Rio J.) 48: 295 (cat.). Bianchi, 1967, *Proc. Hawaii. Entomol. Soc.* 19: 132 (comp. note).

This species shares with *N. lanaiensis* the strongly developed dorsal tooth on the pygofer (Figs. 81e, 85e), the unreduced, sometimes chelate right clasper, and the lack of a tooth on the aedeagus. It differs from *N. lanaiensis* in having the longer, lower arm of the right clasper apically acute and straight rather than blunt and apically uncinate. The propleural stripe is always absent.

Redescription. **MALE** (n = 53). Body: 6.72–7.50 long; 1.67–1.77 wide, yellowish brown; carinate across the base, vestiture very sparse and pale. Antennal segment lengths: I, 1.10–1.50, pale green dorsally, darker green ventrally, vestiture short, dark brown, with both erect and recumbent setae; II, 3.73–4.44, pale green on proximal 1/2 becoming brown distally, apex dark brown; III, 2.30–2.68, dark brown; IV, 1.60–2.33, dark brown; vestiture on last 3 segments short, brown, and recumbent. Rostrum: 1.35–1.50 long. Pronotum: 0.83–0.97 long; 1.60–1.70 wide; basal lobe, and base and apex of propleuron dark green, remainder of pronotum light brown; vestiture brown; callus dark green. Mesonotum and scutellum light brown, apically greenish; basal 1/2 of claval vein, clavus, and embolium variegate blotched greenish and yellow; vestiture as described for *N. lanaiensis*; cuneus with darker green borders internally; small areole coriaceous, large areole coriaceous on inner 1/2 and darker green than hemelytra. Legs: femora with setae longer, paler, more erect on posterior margin than on anterior margin, yellowish; setae of tibiae brownish, semierect, sparse, yellowish. Abdomen: yellowish, pygofer dorsally sclerotized with 2 indistinct, blunt teeth; genitalia and pygofer as illustrated (Figs. 84, 85).

FEMALE (n = 62). Body: 7.00–7.07 long; 1.70–1.80 wide. Head: 1.10–1.16 wide. Antennal segment lengths: I, 1.10–1.13; II, 3.53–3.73; III, 2.10–2.32; IV, 1.67–1.87. Rostrum: 1.50–1.57 long. Pronotum: 0.83–0.93 long; 1.63–1.70 wide. Differing from male as noted under *N. lanaiensis* subgroup and as follows; dorsum dull, not shining; antennal segment I, head, mesoscutellum, and calli pale yellow; collar pale green; hind lobe of pronotum weakly, longitudinally carinate light brown apically, apex blackish; antennae III and IV dark brown; hind femora immaculate ventrally.

NYMPHS. *1st to 4th instars.* Structurally identical to respective instar of *N. lanaiensis*. *5th instar* (n = 3). Dorsum and appendages uniformly clothed with pale, delicate pilosity resembling that in femora of adult. Body: 4.05–4.52 long; 1.39–1.48 wide. Head: 0.99–1.03 wide. Antennal segment lengths: I, 0.81–1.03, yellow, length of vestiture subequal to diameter of segment; II, 2.80–3.01, yellowish with light brown tips; III, 1.85–2.14, light brown; IV, 1.78–2.30, dark brown; vestiture of last 3 segments shorter and more recumbent distally. Rostrum: 1.28–1.30 long. Pronotum: 0.55–0.66 long; 0.95–1.03 wide; subrectangular, not incised mesally as in adult; dorsal surface flat; calli not evident. Length of mesonotal wing buds: 1.87–1.93; embolar margin now well defined. Legs: pale yellow, last tarsal segment light brown; vestiture of femora more dense than that on body; tibial vestiture recumbent, as dense as that on femora, with a few erect hairs in rows resembling the tibial spines of adult.

Distribution. Hawai'i I.

Host. *Ilex anomala* (= *Byronia sandwicensis*).

Etymology. Apparently a patronym for Hawai'i Island or Hawaiian Archipelago.

Material examined. HAWAIIAN IS: **HAWAI'I**: 1♂, 605 m, IX.1892, [no host], R.C.L. Perkins, Fauna Hawaiiensis Collection; 1♂, Kona, 914 m, IX.1892, [no host], R.C.L. Perkins,

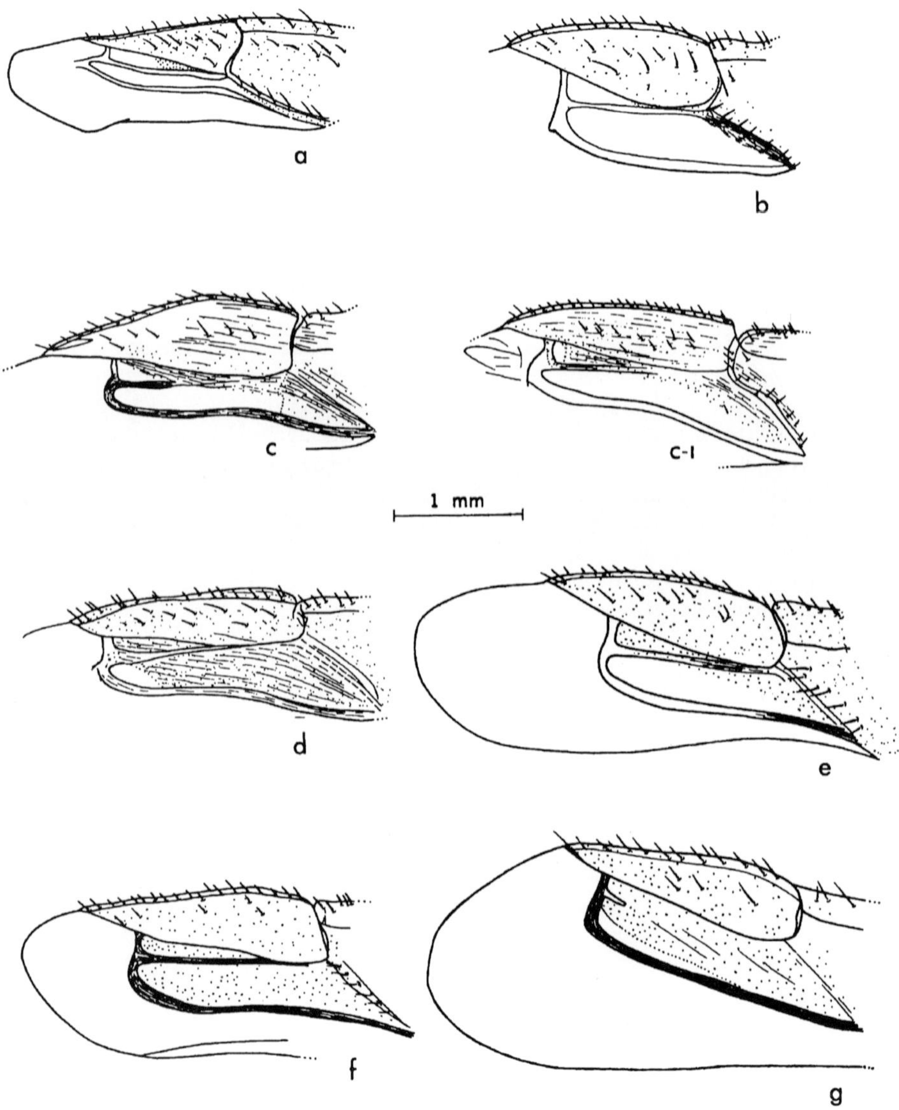

Fig. 82. *Nesiomiris* species, apex of hemelytron and membrane. **a**, *tetraplasandrae,* n. sp.; **b**, *ehu,* n. sp.; **c**, *hawaiiensis* Kirkaldy ('Ōla'a, Hawai'i I); c-1, *hawaiiensis* Kirkaldy (Kealakekua, Hawai'i I); **d**, *byroniae,* n. sp.; **e**, *oahuensis,* n. sp.; **f**, *ilicivorus,* n. sp.; **g**, *planatus,* n. sp.

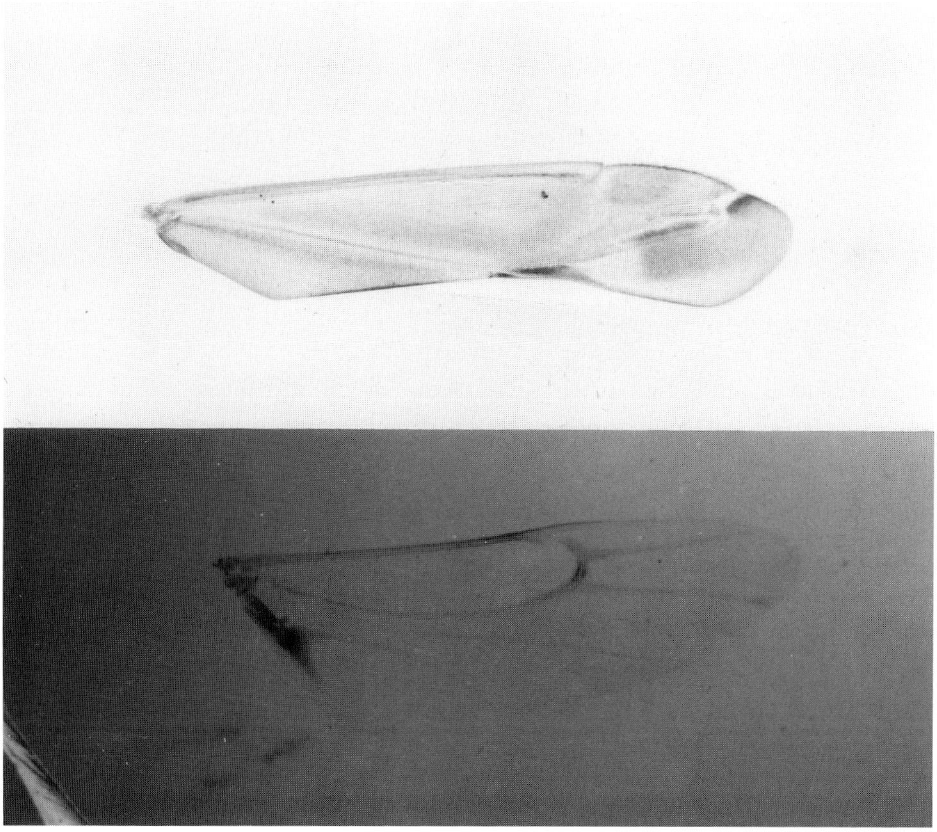

Fig. 83. *Nesiomiris hawaiiensis* Kirkaldy, right wings, dorsal; hemelytron above, wing below.

Fauna Hawaiiensis Collection; 1♂, "572," [Olaa, HAW. I., VI.1895, Fauna Hawaiiensis Collection (label data from Perkins's field notes on file in Bishop Museum)]; 3♂, 1♀, "688," [Olaa, HAW. I., VI.1895, Fauna Hawaiiensis Collection]; 8♂, 6♀, 3N, Kona, Kealakekua, 1066 m, 16.VII.1919, on *Byronia sandwicensis*, P.H. Timberlake; 1♀, Kona, Kealakekua, 1066 m, on *Clermontia coerulea*, P.H. Timberlake; 1♀, 'Ōla'a Hwy, 23 mi, 696 m, 19.IX.1919, on *Antidesma*, D. T. Fullaway; 1♂, Upper Hāmākua Ditch Trail, 15.VIII.1935, on *Ilex anomala*, R.L. Usinger; 1♀, Kīlauea National Park, VIII.1948, C. E. Pemberton; 3♀, Mauna Loa Truck Trail, 1371 m, 14.VIII.1958, on *Byronia*, J.W. Beardsley; 1♀, Hilo Forest Reserve, 19.VIII.1958, on *Byronia*, J.W. Beardsley; 1♀, Kīlauea Iki, Rim Trail, 1157 m, 23.VI.1966, J.W. Beardsley; 1♀, above Waimea, Waikoloa Stream, 975 m, 2.XII.1968, on *Ilex anomala*, W.C. Gagné; 9♂, 4♀, 'Ōla'a Forest Reserve, Wright Road, 28.XI.1968, on *I. anomala*, W.C. Gagné; 2♀, 4N, Saddle Road, Kīpuka #7, 18.4 mi W of Hilo, 1.XII.1968, on *I. anomala*, W.C. Gagné; 2♂, 2♀, Saddle Road, Kīpuka #9, 19.5 mi W of Hilo, 30.XI.1968, on *I. anomala*, W.C. Gagné; 1♂, 6♀, Hawai'i Volcanoes National Park, near headquarters, 1220 m, 17.VI.1970, on *I. anomala*, E.F. Drake; 16♂, 9♀, 7N, Kohala Mts, Upper Hāmākua Ditch Trail, 24.I.1971, on *I. anomala*, W.C. Gagné; 4♂, 1♀, 2N, Mt Hualālai, W

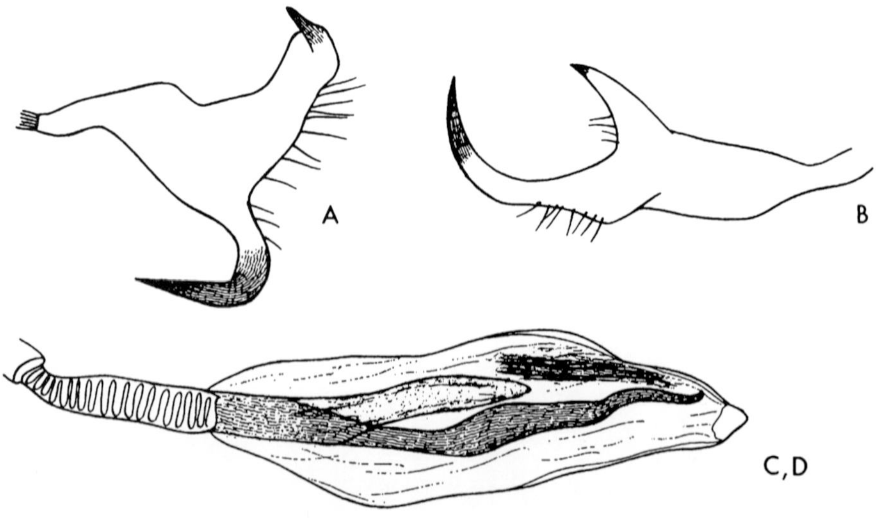

Fig. 84. *Nesiomiris hawaiiensis* Kirkaldy, male genitalia. **a**, left clasper, internal; **b**, right clasper, external; **c**, **d**, aedeagus and spiculum, dorolateral. Drawn by J.C.M. Carvalho from a male paratype in the Natural history Museum labeled "635 Sandwich Is." [= Hawai'i I: Ōla'a, XI.1896, R.C.L. Perkins collector].

side, 600 m, 23.I.1971, on *I anomala*, W.C. Gagné; 1♂, 2♀, 2N, Hilo Forest Reserve, Laupāhoehoe, 1440 m, 25.II.1971, on *I. anomala*, W.C. Gagné; 2♂, 6♀, Hawai'i Volcanoes National Park, Thurston Lava Tube, 1200 m, 11.III.1971, on *I. anomala*, W.C. Gagné; 10♀, Ka'ū, Kaiholena Ridge, 757 m, 8.I.1971, on *I. anomala*, W.C. Gagné; 1♀, 2N, Mt Hualālai, Kahuku, near Kīpuka Nēnē, 1830 m, 23.IV.1972, on *I. anomala*, W.C. Gagné; 4♂, Ka'ū District, Kahuku, near Kīpuka Nēnē, 1830 m, 23.IV.1972, on *I. anomala*, W.C. Gagné; 1♂, 3♀, Kohala Mts, Kehena Ditch Trail, 1159 m, 27.II.1973, on *I. anomala*, W.C. Gagné (BPBM).

Variation. In addition to color variations, the following are noted: the apex of the head varies from rounded to bifid; the inner or upper arm of the right clasper is more truncated (Fig. 85b–1, b–2) in populations primarily on the west and north sides of the island and becomes more chelate (Figs. 84b, 85b) on the east side.

Discussion. The type species of *Nesiomiris hawaiiensis* was said by Kirkaldy (1902) to occur on all islands but O'ahu and Kaua'i. This seems to have been in error, because the carded specimens labeled "specimen figured" in the British Museum are from O'ahu, although no specific locality is given (only "Sandwich Islands"). Later (1908) Kirkaldy specified that the type specimen was from "Olaa, Hawaii I."

A cotype in the BMNH was dissected and figured by Carvalho (Fig. 84). The drawing closely agrees with specimens recently collected from *Ilex anomala* at 'Ōla'a and vicinity, on Hawai'i I. This specimen is labeled "635 Sandwich Is. [Hawaii I, Olaa, XI.96, R.C.L. Perkins, Collector, on basis of Perkins's field notes]."

The host records from "*Antidesma*," "*Clermontia coerulea*," and "*Cheirodendron*" appear attributable to misidentification of host plants and/or perching individuals. *Anti-*

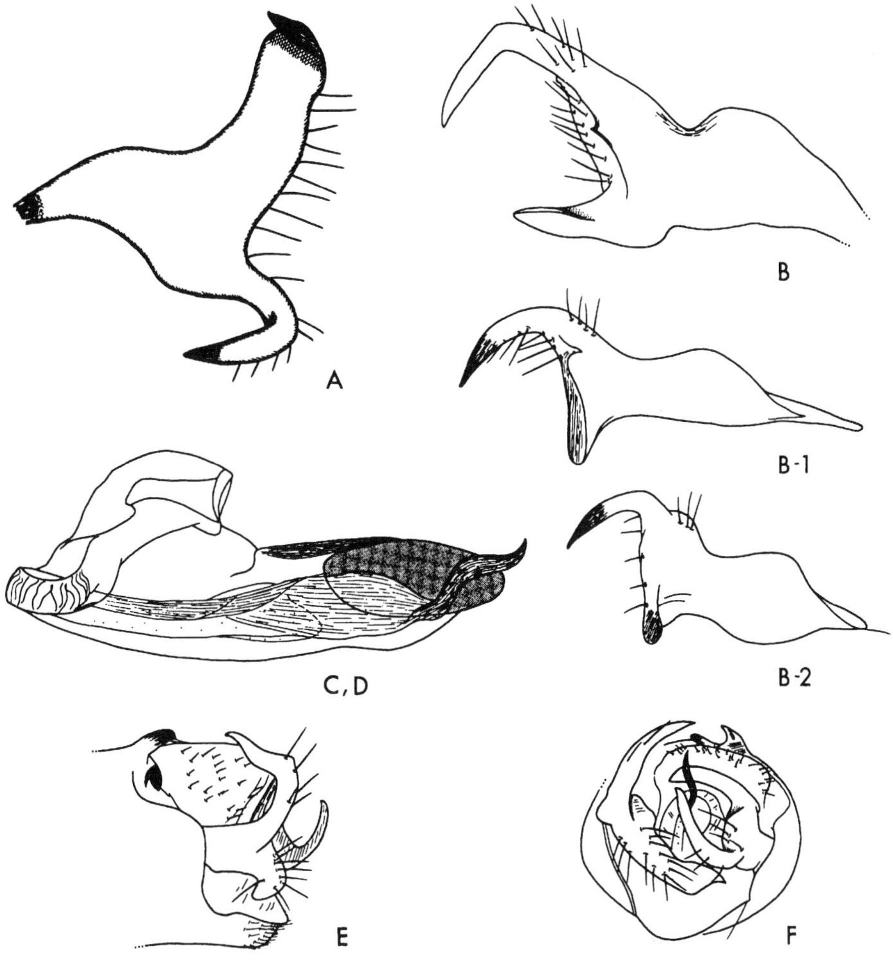

Fig. 85. *Nesiomiris hawaiiensis* Kirkaldy, male genitalia. **a**, left clasper, internal; **b, b-1, b-2**, right clasper, internal, showing clinal type variation (see Fig.. 84b); **c, d**, aedeagus and spiculum, lateral; **e**, pygofer, left lateral; **f**, pygofer, terminal.

desma species, at least, bear a superficial resemblance to *Ilex anomala* and the 2 species are broadly sympatric. I have found *N. hawaiiensis* breeding on *Ilex* only, as shown by restriction of the immature stages to this host.

The redescribed male and female were from material I collected on Wright Road, Hawai'i I. The internal male genitalia were drawn from a specimen labeled: 29 miles 'Ōla'a, Haw., 2300 ft, 17.IX.1919, D. T. Fullaway. These 3 specimens I have illustrated to show the variation of the right clasper (Fig. 85b, b–1, b–2).

42. *Nesiomiris byroniae* Gagné, **new species** Figs. 26k, 82d, 86

This species appears to be the sister species of *N. mauiensis* as judged from the following combination of characters: ventral apex of aedeagus with a strong tooth, spiculum strongly sinuate and twisted apically, upper arm of the left clasper not reduced, dorsum of pygofer strongly sclerotized, and right clasper chelate.

The teeth on the lobes of the pygofer consist of 1 to 2 apically rounded lobes projecting from about the mid-line of the dorsum and another closely adjacent, apically acute, minutely spiculate tooth on the side of the pygofer above the left clasper.

This species occurs on E Maui with *N. kawau*, but the 2 are essentially allopatric because *N. byroniae* dominates the north and west sides of Haleakalā massif, while *N. kawau* occurs on the south and east sides, the 2 overlapping only in the vicinity of ‘Ukulele on the northwest side of the mountain. Too few specimens of either species have been collected on Moloka‘i to see if they are sympatric or allopatric.

Description. **MALE** (n = 31). Body: 7.02 (7.02–8.20) long; 1.53 (1.53–1.83) wide. Head: 1.17 (1.16–1.24) wide, yellowish; bifid; carinate across base; vestiture dark brown, sparse, not striolate. Antennal segment lengths: I, 1.18 (1.13–1.37), vestiture dense, dark brown; II, 3.80 (3.80–4.20), yellow on proximal 1/3, brownish on distal 2/3, dark brown at tip; vestiture dark brown; as long as that on proximal 1/3, becoming more recumbent and shorter distally; III, 2.24 (2.24–2.40), brown, vestiture dark brown; IV. 1.11 (1.91–2.36), color and vestiture as for III. Rostrum: 1.52 (1.52–1.70) long, reaching middle of mesosternum. Pronotum: 0.89 (0.88–1.03) long; 1.53 (1.53–1.83) wide; brownish, vestiture sparse, dark brown; calli weakly swollen, median impression hardly discernable. Mesonotum and scutellum yellowish green. Hemelytra: green, embolar margin paler; vestiture blackish, more recumbent than on pronotum; large and small areole almost completely coriaceous, only their apices clear on the inner 2/3 (Fig. 82d); membrane more deeply infumate between anal ridges; areolar veins green. Legs: yellowish green, vestiture of femora brown, paler posteriorly; vestiture of tibiae pale spines darker; last tarsal segment brown, others yellowish. Abdomen: 3 pygofer teeth present, 2 apically rounded, indistinct on dorsal mid-line, and 1 minutely spiculate and apically acute on left side; genitalia and spiculum as illustrated (Fig. 86).

FEMALE (n = 21). Body: 7.02 (7.02–8.06) long; 1.55 (1.55–1.94) wide. Head: 1.19 (1.19–1.24) wide. Antennal segment lengths: I, 1.18 (1.18–1.43); II, 3.11 (3.11–3.94); III, 2.13 (2.13–2.36); IV, 1.66 (1.66–2.20). Rostrum: 1.71 (1.53–1.71) long. Pronotum: 1.00 (0.93–1.00) long; 1.55 (1.55–1.94) wide. Differing from male as follows: more robust; large areole completely coriaceous; K and L structures as illustrated (Fig. 26k).

NYMPHS. *1st to 5th instars*. Essentially identical to those of *N. lanaiensis*; 5th instar distinctly more pilose.

Distribution. E Maui and Moloka‘i Is.

Host. *Ilex anomala* (= *Byronia sandwicensis*).

Etymology. The host of this species, *Ilex anomala*, was once known as *Byronia sandwicensis*, and its generic name was a patronym for Lord Byron, captain of the HMS *Blonde*, which returned the bodies of the Hawaiian king and queen to Hawai‘i in 1824 after they had died of measles, while in England to attend Queen Victoria’s Jubilee. A number of plants were collected then by the naturalists aboard the *Blonde*, among them the tree originally named after the captain.

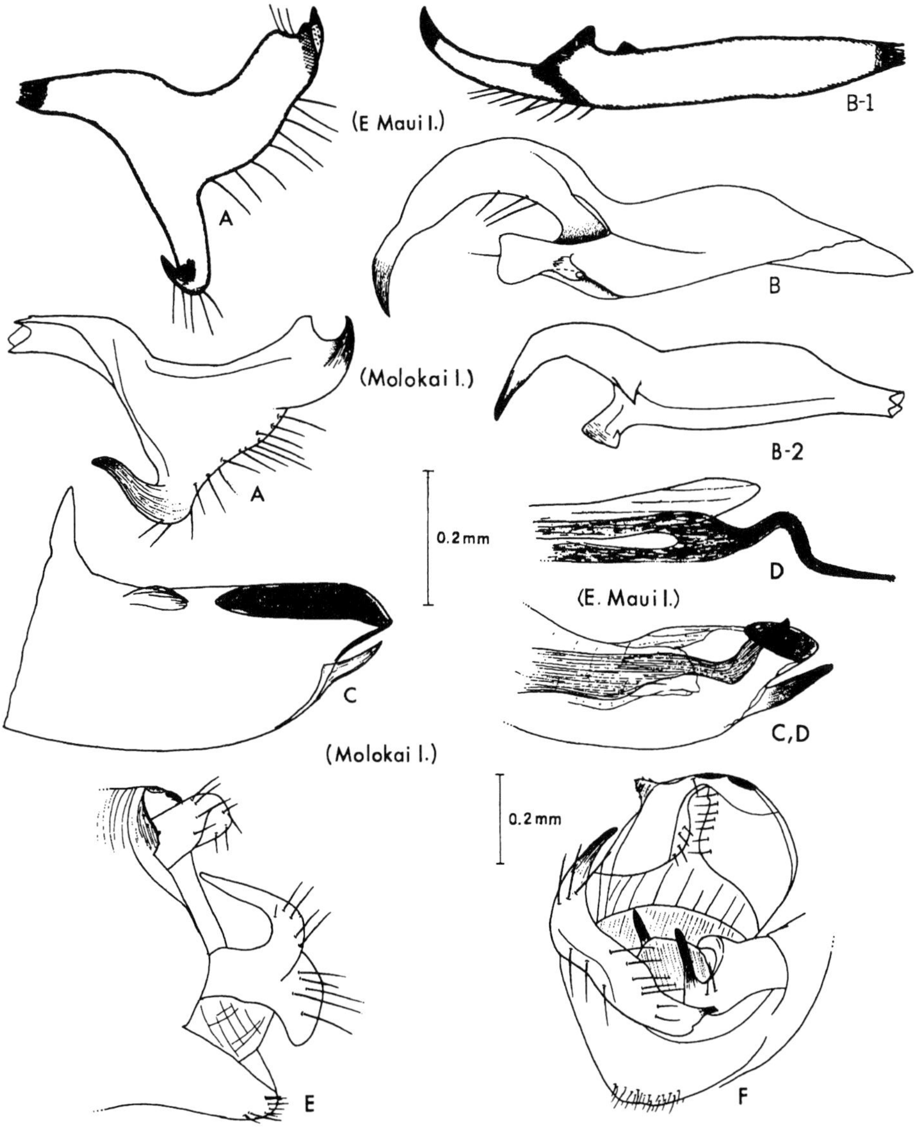

Fig. 86. *Nesiomiris byroniae,* n. sp., male genitalia, showing variation of claspers of aedeagus of E Maui and Moloka'i populations. **a**, left clasper, internal; **b, b-1, b-2**, right clasper, internal; **c, d**, aedeagus and spiculum, lateral; **c**, aedeagus, spiculum removed; **d**, spiculum, dorsal; **e**, pygofer, left lateral; **f**, pygofer, terminal.

Type data. Holotype ♂ (BPBM 14,866), allotype ♀, paratypes 5♂, 2♀, 30N, **E MAUI**: SW Rift Zone at Pu'u Mākua, 1540 m, 14.IV.1971, on *Ilex anomala*, W.C. Gagné. Additional paratypes with data differing only as follows: 3♂, Haleakalā, near 'Ukulele, about 5000 ft. [1524 m], 14.VII.1919, on *Byronia sandwicensis*, P.H. Timberlake; 3♂, Kula Pipeline, [no elevation], 13.VI.1927, on *Byronia*, O.H. Swezey; 1♀, 20N, 'Ukulele Camp,6000 ft. [1830 m], 5.IV.1970, W.C. Gagné. 3♂, 3♀, 'Ukulele, [no host or elevation], 30.VIII.1932, N.L.H. Krauss; 4♂, 3♀, 2N, Haleakalā Crater, Ko'olau Gap, 5600 ft. [1707 m], 29.III.1970; 1♀, 9N, 5.IV.1970; 8♂, 6♀, Hāna Forest Reserve, 2045 m, 3.VIII.1973; 1♂, 1♀, 10N, Upper Hāna Forest Reserve, 2045 m, 4.III.1973; 1♂, 1♀, Hāna Forest Reserve, 1675 m, 4.VIII.1973, F.G. Howarth; 1♂, 1♀, Hāna Forest Reserve, 1740 m, 15.III.1974. **MOLOKA'I**: 4♂, 3♀, 1N, Pēpē'ōpae, 4000 ft. [1219 m], 13.VII.1971 (BPBM).

Variation. In spite of the 2 disjunct populations in the paratype series (E Maui and Moloka'i), I consider them to belong to the same species because of the close similarity of the genitalia and the fact that the variation within the E Maui population encompasses the variation seen in the Moloka'i population. In particular, these variations are: length and inward curvature of the upper arms of the left clasper (Fig. 86a), a degree of curvature and length of the outer arm of the right clasper (Fig. 86b, b–1), the development of a lobe between the inner bases of the arms of the right clasper (Fig. 86b, b–2), and the degree of curvature of the "elbow" near the apex of the spiculum (Fig. 86d). The tooth on the pygofer above left clasper is more distinctly spiculate in the Moloka'i population (Fig. 86e, f). The large areole is sometimes 1/2 membranous on inner aspects, in which case the infumation of the membrane extends into the areole.

Discussion. The development of the tooth on the aedeagus approaches that seen in *N. elongatus* (Fig. 44c, d). The lobe between the bases of the right clasper is probably a remnant of the third arm of the clasper, as seen in *N. trifurcatus* (Fig. 32b, e, f).

43. ***Nesiomiris mauiensis*** Gagné, **new species** Fig. 87

The tooth on the outer, ventral apex of the aedeagus is developed similarly to that of the related *N. kawau*. The arms of the right clasper are less reduced and less closely "pinched" together than in *kawau*. The upper arm of the left clasper is longer and the dorsum of the pygofer is toothed, as in *N. byroniae*.

N. mauiensis appears to be the sister species of *N. byroniae*. On W Maui it is partially sympatric with *N. kawau* but generally occurs at higher elevations. The claspers of the latter (Fig. 88b) are much more reduced, although the aedeagi are quite similar (Figs. 87c, 88c); the spiculum of *N. mauiensis*, however, is more strongly sinuate and S-shaped or elbowed (Fig. 87d), while in *N. kawau* it is only weakly sinuate (Fig. 88d).

This species is placed here because the aedeagus has a tooth on the ventral wall near the apex of the spiculum when it is at rest. The right clasper has the general structure of *N. lanaiensis*, *N. hawaiiensis*, and *N. byroniae* wherein the arms have not anastomosed as in remaining members of the *lanaiensis* subgroup.

Description. **MALE** (n = 22). Body: 5.58 (5.58–6.70) long; 1.46 (1.46–1.93) wide. Head: 1.04 (1.03–1.10) wide; weakly striolate; vestiture brown; apex slightly bifid. Antennal segment lengths:

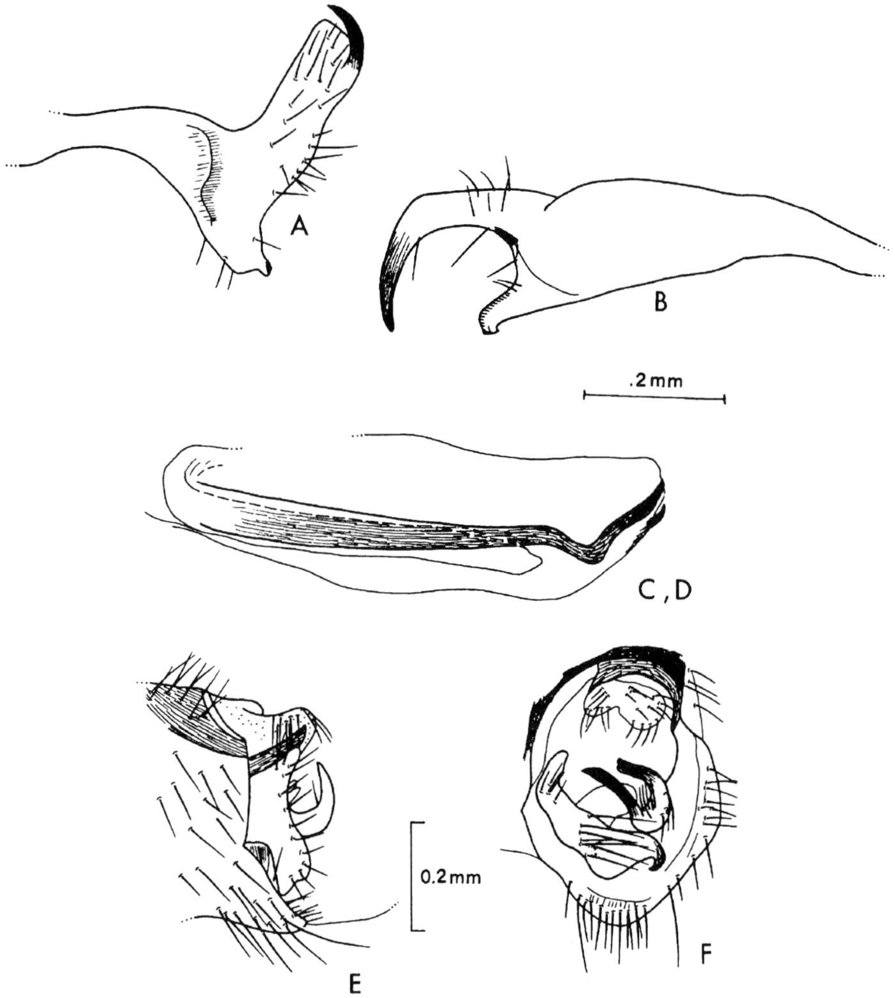

Fig. 87. *Nesiomiris mauiensis,* n. sp., male genitalia. **a**, left clasper, internal; **b,** right clasper, internal; **c, d**, aedeagus and spiculum, lateral; **e**, pygofer, left lateral; **f**, pygofer, terminal.

I, 0.92 (0.90–1.03), same color as head; apex greenish; vestiture brown; II, 2.80 (2.60–3.03), yellow basally, becoming blackish apically; III, 1.85 (1.63–1.93), blackish; IV, 1.70 (1.70–1.93), blackish; vestiture of last 3 segments brown. Rostrum: 1.92 (1.30–1.92) long; reaching apex of mesosternum. Pronotum: 1.05 (0.90–1.05) long; 1.46 (1.40–1.50) wide, basal lobe greenish, remainder yellow; vestiture brown, sparse. Mesoscutellum brownish; scutellum yellowish green, sparsely pubescent. Hemelytra: dark green; large areole coriaceous on inner 2/3, remainder infumate as is membrane, areolar veins greenish. Femora pale yellow; vestiture, both erect and recumbent; tibiae brownish yellow; tarsi yellow, becoming darker distally. Abdomen: sclerotized and blackish above genital atrium, and toothed there; genitalia and pygofer as illustrated (Fig. 87).

FEMALE (n = 18). Body: 6.20 (6.20–6.40) long; 1.50 (1.50–1.70) wide. Head: 1.05 (1.04–1.10) wide. Antennal segment lengths: I, 0.92 (0.83–0.93); II, 2.65 (2.60–2.67); III, 1.77 (1.77–1.85); IV, 1.70 (1.43–1.70). Rostrum: 1.94 (1.45–1.94) long. Pronotum: 0.94 (0.90–1.04) long; 1.50 (1.50–1.53) wide. Paler than golden; areolar veins yellowish.

NYMPHS (based on 1 third and 3 fifth instars). Third instar identical to that of *N. lanaiensis*; 5th instar with dorsal setae yellow, those on lateral margins of anterior wing bud evenly distributed and not concentrated in patches; lateral margins of pronotum convex; not otherwise differing from 5th instar of *N. lanaiensis*.

Distribution. W Maui [and E Maui][3].

Host. *Ilex anomala*.

Etymology. The name is a patronym for Maui I.

Type data. Holotype ♂ (BPBM 14,867), allotype ♀, paratypes 7♂, 3♀, **W MAUI**: Kaulalewelewe-Pu'u Kukui Trail, 1537 m, 28.X.1972, on *Ilex anomala*, W.C. Gagné. Additional paratypes with data differing only as follows: 6♂, 2♀, 915–1037 m, 25.VIII.1965, R.L. Usinger; 13♂, 8♀, 4N, Pu'u Kukui, 5000 ft. [1525 m], 18.II.1970; **E MAUI**: 9♂, 2♀, 6N, 'Ukulele, 5200 ft. [1585 m], 22.XII.1968 (BPBM)[3].

Discussion. This is the only 1 of 3 species on Maui in the *lanaiensis* subgroup that is apparently restricted to W Maui[3]. Both *N. byroniae* and *N. kawau* are found on Moloka'i as well as in E Maui.

44. *Nesiomiris kawau* Gagné, new species Figs. 26c, 88

The comparatively reduced arms of the claspers of the male most resemble those of *N. mauiensis* of W Maui from which it differs as noted above. It is placed here because reduction in the claspers would indicate that it is one of the most apomorphic species in the *lanaiensis* subgroup.

Description. **MALE** (n = 36). Body: 6.18 (6.18–6.65) long; 1.93 (1.53–1.93) wide. Head: 1.05 (1.05–1.13) wide; brownish; bifid; posterior carina indistinct; vestiture brown, very sparse. Antennal segment lengths: I, 1.13 (1.13–1.19), vestiture much shorter than diameter of segment, brown, recumbent, and erect hairs in about equal numbers, ratio of length to head width 9: 8.5; II, 3.19 (3.19–3.94), becoming brown from middle distally, dark brown at apex, vestiture brown; III, 2.05 (1.95–2.10); IV, 1.42 (1.42–1.91). Rostrum: 1.26 (1.26–1.50) long, reaching apex of mesosternum. Pronotum: 0.92 (0.83–0.92) long; 1.93 (1.53–1.93) wide; vestiture short, pale and sparse, a single brown bristle on anterior margins of callus. Hemelytra with vestiture short, sparse, and pale; large areole coriaceous on inner 1/3; smaller areole partially coriaceous. Abdomen: pygofer sclerotized dorsally but tooth lacking; genitalia and pygofer as illustrated (Fig. 88). Femora with vestiture short and pale; vestiture of tibiae shorter, pale; last tarsal segment brown, others yellowish.

FEMALE (n = 36). Body: 6.60 (6.20–6.65) long; 1.51 (1.45–1.75) wide. Head: 1.05 (1.04–1.10) wide. Antennal segment lengths: I, 1.05 (0.90–1.09); II, 2.94 (2.80–2.94); III, 2.04 (1.63–2.04); IV, 1.50 (1.50–1.70). Rostrum: 1.94 (1.35–1.94) long. Pronotum: 0.95 (0.90–1.05) long; 1.51 (1.45–1.75) wide. Differing from male as noted under *lanaiensis* subgroup and as follows: antennal segment II yellowish brown with a black apex; small areole wholly coriaceous; ratio of length of antennal segment I to head width as 10: 8.75.

NYMPHS. *1st to 3rd instars*. Unknown. *4th instar* (n = 1). Differing from 4th instar of *N. lanaiensis* as follows: dorsal thoracic setae pale. *5th instar* (n = 4). Differing from 5th instar of *N. la-*

---

3. Specimens of *mauiensis* from E Maui were sorted and labeled as paratypes by the author after submission of the original manuscript, but had not been entered into the final text and discussion before the author's death. The E Maui specimens have been added by the editors to the type data. — NLE.

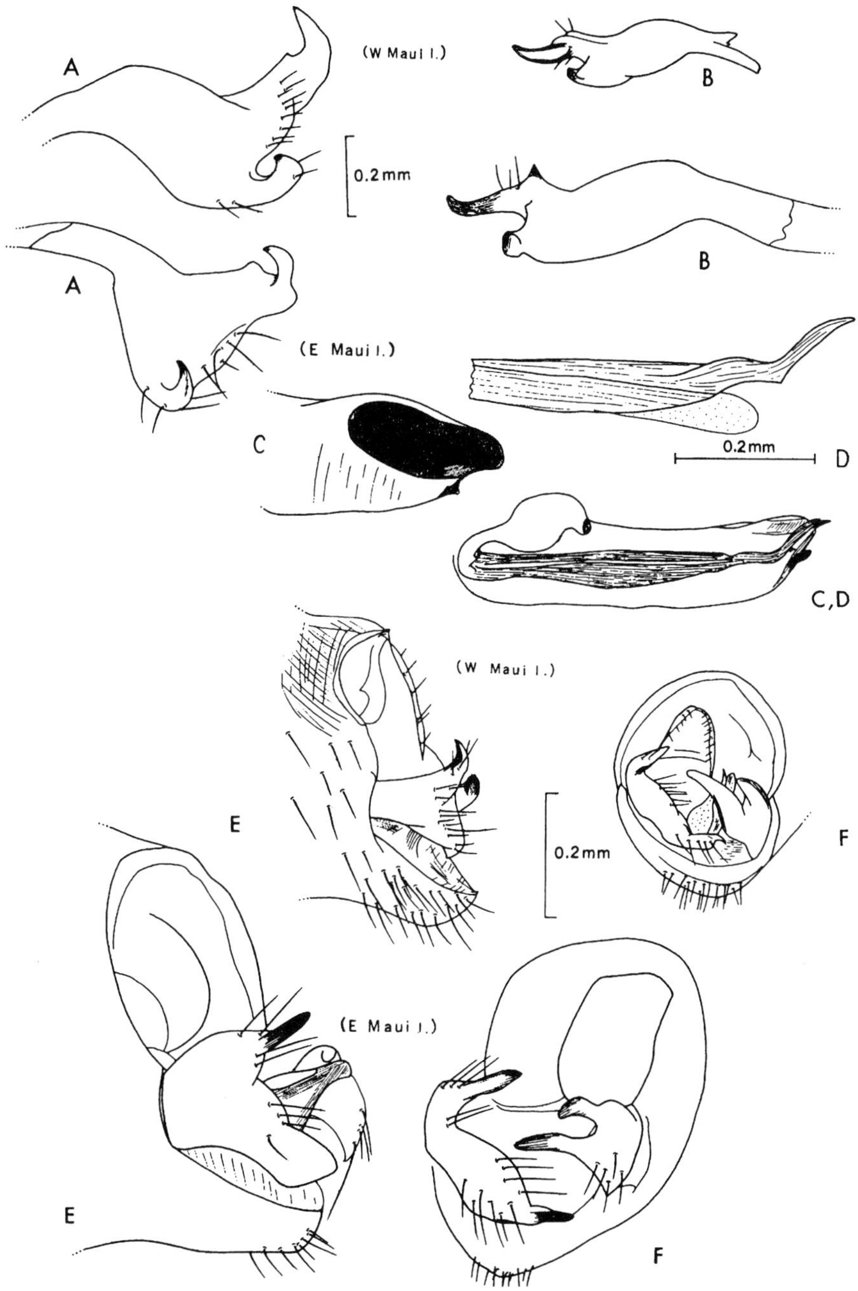

Fig. 88. *Nesiomiris kawau,* n. sp., male genitalia. **a**, left clasper, internal; **b,** right clasper, internal; **c, d**, aedeagus and spiculum, lateral; **c**, aedeagus, spiculum removed; **d**, spiculum, lateral; **e**, pygofer, left lateral; **f**, pygofer, terminal.

naiensis as follows: apex of head weakly indented; apices if mesonotal wing bud acuminate; rostrum reaching apices of hind coxae; hind tibiae with erect setae on all aspects.

Distribution. E and W Maui and Moloka'i Is.

Host. *Ilex anomala*.

Etymology. A Hawaiian name for *Ilex anomala* is *kāwa'u*.

Type data. Holotype ♂ (BPBM 14,868), allotype ♀, paratype 15♂, 6♀, **E MAUI**: 'Ukulele, 1585 m, 22.XII.1968, on *Ilex anomala*, W.C. Gagné. Additional paratypes with data differing only as follows: 4♂, 9♀, 4N, Kaupō Gap, Kahikinui Forest Reserve, [no elevation], 20.VII.1973; 3N, Kahana Stream, 2400 ft. [731 m], 16.IV.1971; 8♂, 7♀, 4N, Gulch E of Kaupō Gap, 1320 m, 21.V.1971; 5♂, 3♀, near Waikamoi Stream, 2035 m, 21.VII.1975; 1♂, 7♀, above Makawao Forest Reserve, 1737 m, 24.VIII.1975, H. E. Smith. **W MAUI**: 3♂, 12♀, 3N, Kaulalewelewe-Pu'u Kukui Trail, 915–1037 m, 8.VIII.1968; 2N, Kaulalewelewe-Pu'u Kukui Trail, 700-1300 m, 17–19.II.1970; 1♂, 1♀, Kaulalewelewe, 915 m, 27.V.1972; 2♂, 4♀, 3N, Kahoma Stream, 730 m, 16.VI.1971. **MOLOKA'I**: 1♂, 1♀, 914 m, [no date, host or collector; latter probably R.C.L. Perkins]; 3♂, Keōpukaloa, 610 m, 16.I.1973 (BPBM).

45. *Nesiomiris oahuensis* Gagné, **new species** Figs. 26l, 82e, 89

This is the most plesiomorphic species of the *lanaiensis* group on O'ahu. The strong sexual dimorphism for color, the shining dorsum, the infuscate membrane, the dorsal tooth on the pygofer (Fig. 89e), and other details of the male genitalia, as well as the unreduced membrane and the comparatively narrow body, place this species in the *lanaiensis* subgroup, although it is the only member on O'ahu. The remainder of the O'ahu species in the *lanaiensis* group are considered to belong to the *ilicis* subgroup that is restricted to O'ahu.

Description. **MALE** (n = 7). Body: 6.38 (6.00–6.38) long; 1.96 (1.45–1.96) wide. Head: 1.05 (1.02–1.07) wide. Antennal segment lengths: I, 0.92 (0.89–0.95); II, 3.18 (2.74–3.18); III, 1.95 (1.65–1.95); IV, 1.95 (1.43–1.95). Rostrum: 1.35 (1.31–1.35) long. Pronotum: 0.92 (0.86–1.00) long; 1.96 (1.46–1.96) wide. Differing from *N. byroniae* as follows: distinctly smaller; apex of head weakly bifid, brownish yellow; antennal segment I, greenish yellow, length shorter than head width (ratio 8: 8.5); II, yellowish brown with a blackish apex; III, brownish yellow; antennal segment I subequal to head width; collar greenish only on dorsum; propleuron greenish across apex; hemelytra with golden brown vestiture; mesoscutellum and basal 1/2 of scutellum yellowish brown, apical 1/2 of scutellum greenish; large areole coriaceous on inner 1/3, infuscate; large areolar veins mostly faded to brownish with a few greenish areas remaining; membrane clear adjacent to anal ridge and narrowly so along inner lateral border with a large, apically acute, strongly sclerotized tooth above the left clasper (Fig. 89e); upper arm of left clasper not reduced, its apex reaching at least roof of pygofer (Fig. 89f); upper arm of right clasper longer than lower arm and inwardly pointing (Fig. 89f); spiculum apically twisted and tooth on pygofer not reaching apex of pygofer (Fig. 89c).

FEMALE (n = 6). Body: 6.50 (6.30–6.50) long; 1.51 (1.51–1.59) wide. Head: 1.07 (1.03–1.09) wide. Antennal segment lengths: I, 0.96 (0.85–0.95); II 2.91 (2.74–2.91); III, 1.75 (1.75–1.93); IV, 1.79 (1.43–1.93). Rostrum: 1.44 (1.35–1.45) long. Pronotum: 0.94 (0.94–1.00) long; 1.51 (1.51–1.59) wide. Not differing from male other than as noted generally under description of *lanaiensis* subgroup. *K* and *L* structure as illustrated (Fig. 26l).

NYMPHS. *1st and 2nd instars*. Unknown. *3rd to 5th instars*. Similar to *N. lanaiensis* in all 3

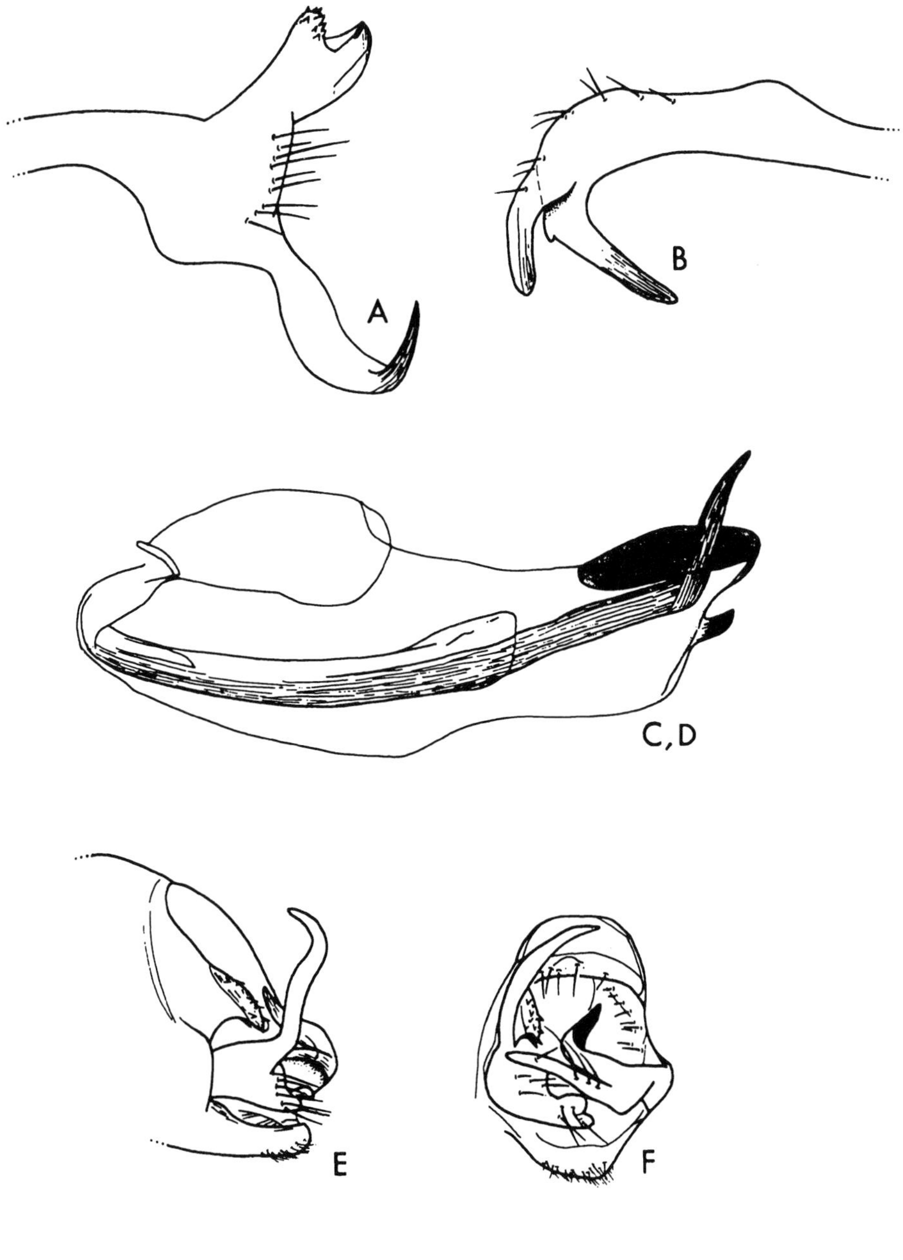

Fig. 89. *Nesiomiris oahuensis,* n. sp., male genitalia. **a**, left clasper, internal; **b**, right clasper, internal; **c**, **d**, aedeagus and spiculum, lateral; **e**, pygofer, left lateral; **f**, pygofer, terminal.

stadia except as follows: 3rd instar (n = 1) less robust, longest setae on antennal segment I shorter than greatest thickness of this segment; anterior margin of pronotum concave at middle; rostrum reaching bases of mesocoxae; 4th instar (n = 1) with pale dorsal thoracic setae; rostrum reaching bases of metacoxae; lateral margins of pronotum convex; setae on legs pale; 5th instar with apices of mesonotal buds acuminate, apex of head indentate.

Distribution. O'ahu I (Wai'anae Mts, summit of Ka'ala) (Fig. 19).

Host. *Ilex anomala*.

Etymology. A patronym for O'ahu.

Type data. Holotype ♂ (BPBM 14,869), allotype ♀, paratypes 19♂, 28♀, 10N, **O'AHU**: Mt Ka'ala, 1200 m, 18.VII.1968, *Ilex anomala*, W.C. Gagné. Additional paratypes with data differing only as follows: 1♂, 29.I.1970; 1♂, 2N, 4000 ft. [1219 m], 21.V.1970; 1♀, 4000 ft. [1219 m], 16.XI.1965, J.W. Beardsley; 1♂, 14.Iv.1966, [no elevation], J.W. Beardsley; 1♂, 20.Vi.1967, [no elevation], C.M. Yoshimoto; 11♂, 3♀, 16.V.1967, P.D. Ashlock, C.M. Yoshimoto; 2♂, 4♀, 1N, 23.X.1968; 2♂, 2N, 29.XII.1968; 9♂, 3♀, Dupont Trail, 1200 m, 21.V.1970; 1♂, 1♀, 6.VIII.1972, J. Obata (BPBM).

Discussion. This species is the most restricted geographically of the *lanaiensis* group on O'ahu, occupying only the summit area of Ka'ala at about 1200 m elevation, O'ahu's highest mountain, where it is the only species on *Ilex anomala* (Fig. 19). This suggests that it is now unable to compete with members of the *ilicis* subgroup on *Ilex* at lower elevations, and/or that it is adapted to a geographically restricted and biologically relict montane environment, and/or that there are physiological barriers within the host plant at lower elevations. The Moloka'i population of *N. byroniae* also has the tooth above the left clasper denticulated (Fig. 86e, f) as well as the ventral, apically located tooth on the aedeagus (Fig. 86c).

The *ilicis* Subgroup

Description. **ADULTS**. Convex-linear species; body usually widest across wings; apex of head strongly bifid (Fig. 45f); antennal segment I, base of antennal segment II, femora and tibiae with pale, suberect pilosity that is as long as greatest diameter of I, remainder of vestiture short, brown, recumbent, those on hemelytra still sparser and shorter; small areolar cell wholly, and large areolar cell sometimes wholly, coriaceous; membrane of forewing reduced, exceeding apex of cuneus by a length less than length of small areolar cell (Fig. 82f), weakly infuscate; hind femora lacking ventral brownish markings; tooth on pygofer situated above base of left clasper; upper arm of left clasper slender and elongate, its apex facing dorsum of pygofer; dorsum of male weakly shining; lower arm of right clasper curving dorsally over upper arm when viewed terminally; dorsum of pygofer yellowish brown; aedeagus with a subapical tooth on left side.

NYMPHS. Fifth instars pilose as in adults, setae on appendages of 1st to 4th instars not pilose, setae shorter than greatest diameter of antennal segment I.

Distribution. O'ahu I (Fig. 19).

Host. *Ilex anomala*.

Included species. Nos. 46–50.

Discussion. The following transformations are apparent in this subgroup; the most striking feature is the increasing complexity of the lower arm of the right clasper as it becomes ever more sinuate and elongate, while the upper arm is relatively constant. The tooth on the pygofer is initially apically rounded and denticulate (*N. ilicis*, Fig. 90e), then becomes acute and loses the denticles in the remainder of the subgroup; the left clasper is relatively constant; the tooth on the left apex of the pygofer is initially rounded (*N. ilicis*, Fig. 90e), but is more strongly expressed and apically acute in the remainder of the subgroup; no clear pattern emerges in the shape of the spicula nor in the reduction of the membrane of the forewing.

46. *Nesiomiris ilicis* Gagné, **new species** Fig. 90

This species is distinct in this subgroup of broad-winged species in that the left atrial tooth is flat, denticulate, scalelike, and apically rounded. The general conformity of the ♂ genitalia shows closest affinities to the remainder of the species in this subgroup in which the left atrial tooth is acute. It is placed here because the flat, broad, denticulate atrial tooth, and the general conformation of the right clasper most closely approximates that seen in *N. oahuensis* than any other member of this subgroup.

Description. **MALE** (n = 3). Body: 7.42 (7.42–7.50) long; 1.60 (1.60–1.90) wide. Head: 1.19 (1.19–1.21) wide; vestiture as for *N. planatus*. Antennal segment lengths: I, 1.26 (1.19–1.37), yellowish brown on proximal 1/2, greenish yellow on distal 1/2, vestiture brownish; II 3.60 (3.60–3.85), greenish at base, becoming browner distally, apex dark brown; III, 2.03 (2.03–2.17), light brown on basal 1/2, darkening distally; IV, 1.92 (1.87–2.07), dark brown. Rostrum: 1.55 (1.53–1.63) long, reaching slightly beyond middle of mesosternum. Pronotum: 1.00 (0.91–1.00) long; 1.60 (1.60–1.74) wide; yellow; vestiture as for *N. planatus*. Hemelytron: embolium uniformly greenish, clavus yellow on basal 1/2, corium yellow to base of anal ridge, remainder dark greenish; large areole completely coriaceous. Legs: femora yellowish, tibiae becoming browner on distal 1/2, tarsi greenish, last tarsal segment darker. Abdomen: tergites VIII and IX pale and membranous above genital atrium; left atrial tooth flat, scalelike nonacute; pygofer and genitalia as illustrated (Fig. 90).

FEMALE (n = 2). Body: 7.40 (7.15–7.40) long; 1.70 (all 1.70) wide. Head: 1.17 (all 1.17) wide. Antennal segment lengths: I, 1.28 (1.09–1.28); II, 3.40 (3.40–3.47); III, 1.95 (1.93–1.95); IV, 1.93 (1.78–1.93). Rostrum: 1.50 (1.50–1.55) long. Pronotum: 1.19 (1.00–1.19) long; 1.70 (all 1.70) wide. Differing from male as noted under *ilicis* subgroup and as follows: body shorter but considerably broader; antennal segments I and II yellowish, the latter dark brown at apex; tarsi yellowish; hemelytra yellowish green from apex of anal ridge, slightly more yellowish between base of wing and anal ridge; tip of ovipositor yellowish.

NYMPHS. *1st and 2nd instars*. Unknown. *3rd instar* (n = 1). Differing from 3rd instar of *N. planatus* as follows: apex of head rounded. *4th instar* (n = 2). Same as 4th instar of *N. planatus*. *5th instar* (n = 2). Differing from 5th instar of *N. planatus* as follows: apex of head strongly bifid; anterior humeral angle of pronotum not infuscate; apex of mesonotal wing bud weakly acuminate, its setae more concentrated along lateral margins.

Distribution. O'ahu I (Wai'anae Mts, S end) (Fig. 19).

Host. *Ilex anomala*.

Etymology. Named for its host association with *Ilex anomala*.

Type data. Holotype ♂ (BPBM 14,870), allotype ♀, paratypes 5♂, 2♀, **O'AHU**: Wai'anae

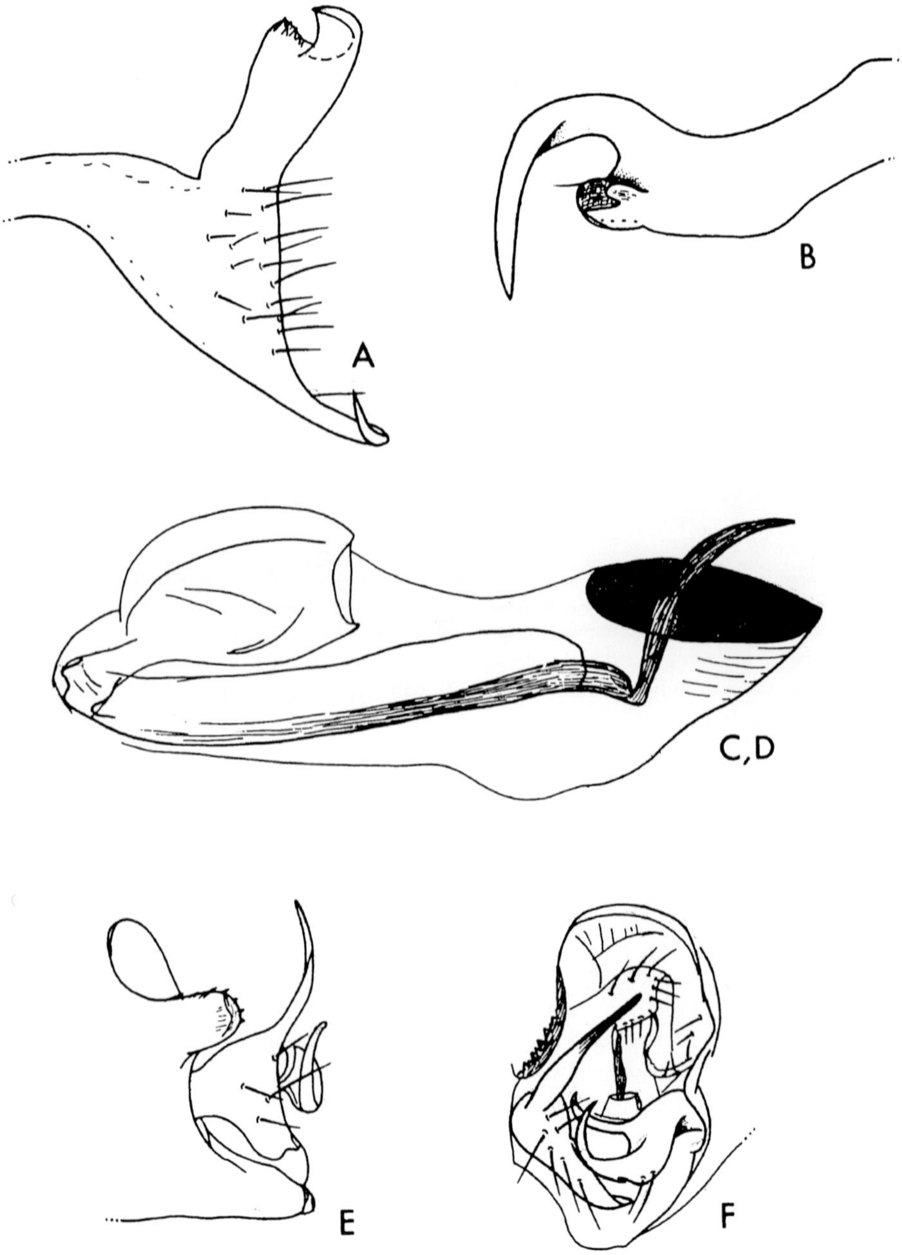

Fig. 90. *Nesiomiris ilicis,* n. sp., male genitalia. **a**, left clasper, internal; **b**, right clasper, internal; **c**, **d**, aedeagus and spiculum, lateral; **e**, pygofer, left lateral; **f**, pygofer, terminal.

Mts, Palikea Trail, 2600 ft. [788 m], 30.I.1972, on *Ilex anomala*, W.C. Gagné. Additional paratypes with data differing only as follows: 3♂, 1♀, Wai'anae Mts, XI.1954, on *Ilex*, E.J. Ford, Jr.; 1♂, Wai'anae Mts, VII.1955, [no elevation or host], E.J. Ford, Jr.; 1♂, 1♀, Wai'anae Mts, VI.1957, [no host or elevation], E.J. Ford, Jr.; 1♂, 1♀, Wai'anae Mts, XI.1954, [no elevation or host], E.J. Ford, Jr.; 2♂, 1♀, 3N, 840–930 m, 13.IX.1968; 1♂, 1N, 840–930 m, 18.IV.1970 (BPBM).

Discussion. This, one of the most restricted of any species in the complex, is limited to the extreme southeastern end of the Wai'anae Range. Here it may be narrowly sympatric to judge from Zimmerman's possibly mislabeled series of *N. planatus*.

47. ***Nesiomiris ilicivorus*** Gagné, **new species** Figs. 19, 26m, 82f, 91

This species differs from *N. planatus* in having the large areole coriaceous (Fig. 82f) and in having the upper arm of the right clasper evenly hooked toward the pygofer, over the lower arm (Fig. 91f); and from *N. restrictus* in having the aedeagal tooth distally located near the aedeagal orifice (Fig. 91c).

Description. **MALE** (n = 7). Body: 7.47 (7.35–8.06) long; 1.69 (1.59–1.70) wide. Head: 1.20 (1.17–1.24) wide, apex yellow; carinate across base; vestiture as for *N. planatus*. Antennal segment lengths: I, 1.26 (1.21–1.35), yellow; II, 4.00 (3.55–4.11), yellow on basal 1/4, becoming dark brown distally; III, 2.15 (2.14–2.30), pale near the base, remainder dark brown. Rostrum: 1.59 (1.57–1.61) long, slightly exceeding middle of mesosternum. Pronotum: 1.17 (1.09–1.17) long; 1.69 (1.59–1.74) wide, yellowish dorsally, faintly greenish on disc; subglabrous. Scutellum yellow with a yellowish orange apex. Hemelytra: greenish yellow to base of cuneus, darker green along the veins; cuneus and areole darker green; large areole completely coriaceous, vestiture as for *N. planatus*; membrane slightly infuscate posterior to large areole. Legs: femora and tibiae yellowish, latter slightly sinuate; last tarsal segment brownish; vestiture as for *N. planatus*. Abdomen: tergites 8 and 9 pale and membranous; genitalia and pygofer as illustrated (Fig. 91).

FEMALE (n = 7). Body: 7.85 (7.43–7.85) long; 1.70 (1.62–1.70) wide. Head: 1.19 (1.21–1.19) wide. Antennal segment lengths: I, 1.28 (1.09–1.31); II, 3.71 (3.37–3.71); III, 2.39 (1.78–2.39); IV, 2.23 (1.60–2.23). Rostrum: 1.62 (1.55–1.62) long. Pronotum: 1.18 (1.09–1.18) long; 1.70 (1.62–1.70) wide. Differing from male as follows: antennal segments I-III yellowish, brown at apex of II and III. *K* and *L* structures as illustrated (Fig. 26m).

NYMPHS. *1st and 2nd instars*. Unknown. *3rd instar* (n = 1). Differing from 3rd instar of *N. planatus* as follows: spiracular openings infuscate; lateral margins of the pronotum convex. *4th instar* (n = 2). Same as 4th instar of *N. planatus*. *5th instar* (n = 4). Differing from 5th instar of *N. planatus* as follows: lateral border of mesonotal wing bud somewhat linear; dorsal abdominal setae pallid; rostrum reaching between apex of mesosternum and apices of mesocoxae; femur, tibia, antennal segment I, and basal 1/2 of II, comparatively densely pilose; anterior humeral angle of pronotum pallid.

Distribution. O'ahu I (northern half of Wai'anae Mts) (Fig. 19).

Host. *Ilex anomala*.

Etymology. The name means "to eat *Ilex*."

Type data. Holotype ♂ (BPBM 14,871), allotype ♀, paratypes 3♂, 2♀, **O'AHU**: Mākaha Val, E side, 800 m, 19.VI.1972, on *Ilex anomala*, W.C. Gagné. Additional paratypes with data differing only as follows: 6♂, 2♀, Kukuiala [*sic* = Kukui'ula], 11.IV.1936, [no elevation or host], R.L.

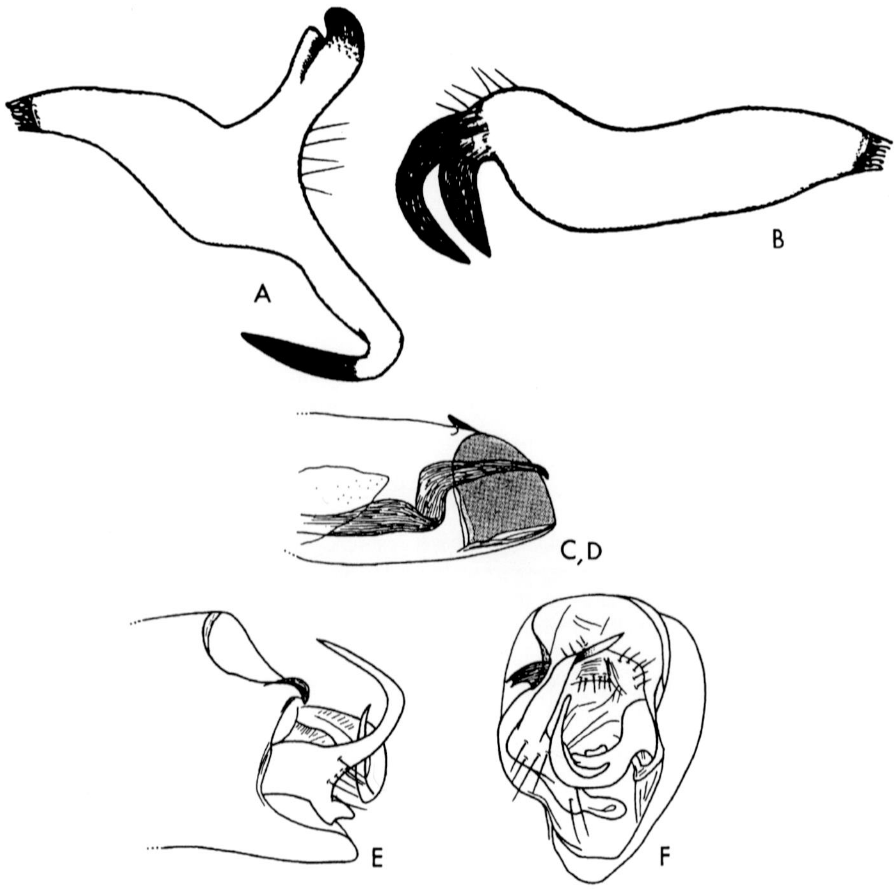

Fig. 91. *Nesiomiris ilicivorus*, n. sp., male genitalia. **a**, left clasper, internal; **b**, right clasper, internal; **c**, **d**, aedeagus and spiculum, lateral; **e**, pygofer, left lateral; **f**, pygofer, terminal.

Usinger; 1 ♀, Mt Ka'ala, 12.XII.1956, [no elevation], J.W. Beardsley; 1♂, Mt Ka'ala, Wai'anae side, VII.1957, [no elevation], on *Byronia*, J.W. Beardsley; 4♂, 2 ♀, Wai'anae Mts, W Makaleha-Makua Ridge, 2700 ft. [818 m], 20.III.1970; 1 ♀, Wai'anae Mts, Peacock Flats, 24.II.1970; 10N, Wai'anae Mts, Makaleha-Makua Ridge, 2700 ft. [818 m], 22.III.1970; 10N, Wai'anae Mts, ridge near end of Mokulē'ia Trail, 757 m, 23.II.1970; 5♂, 3 ♀, 10N, Wai'anae Mts, gulch near Pu'u Pane, 2200 ft. [666 m], 17.V.1970; 2♂, 4N, Wai'anae Mts, ridge above Aleiheihe Church, near Pu'u Pane, 560 m, 24.V.1970; 6♂, 4 ♀, 14N, Mākaha Valley, 800 m, 28.V.1970; 8_, Wai'anae Mts, gulch near Pu'u Pane, 2200 ft. [666 m], 29.V.1970; 3 ♀, Wai'anae Mts, Mokulē'ia Forest Reserve, Palikea Gulch, 300 m, 18.VI.1970, S.L. Montgomery; 4♂, 1 ♀, Wai'anae Mts, Mokulē'ia Forest Reserve, Palikea Gulch, 425 m, 1.VII.1970; 4♂, 1 ♀, 5N, Wai'anae Mts, Mākaha Valley, SE end, 800 m, 5.VII.1970; 11♂, 4 ♀, Wai'anae Mts, Mākaha Valley, N end, 800 m, 5.VII.1970 (BPBM).

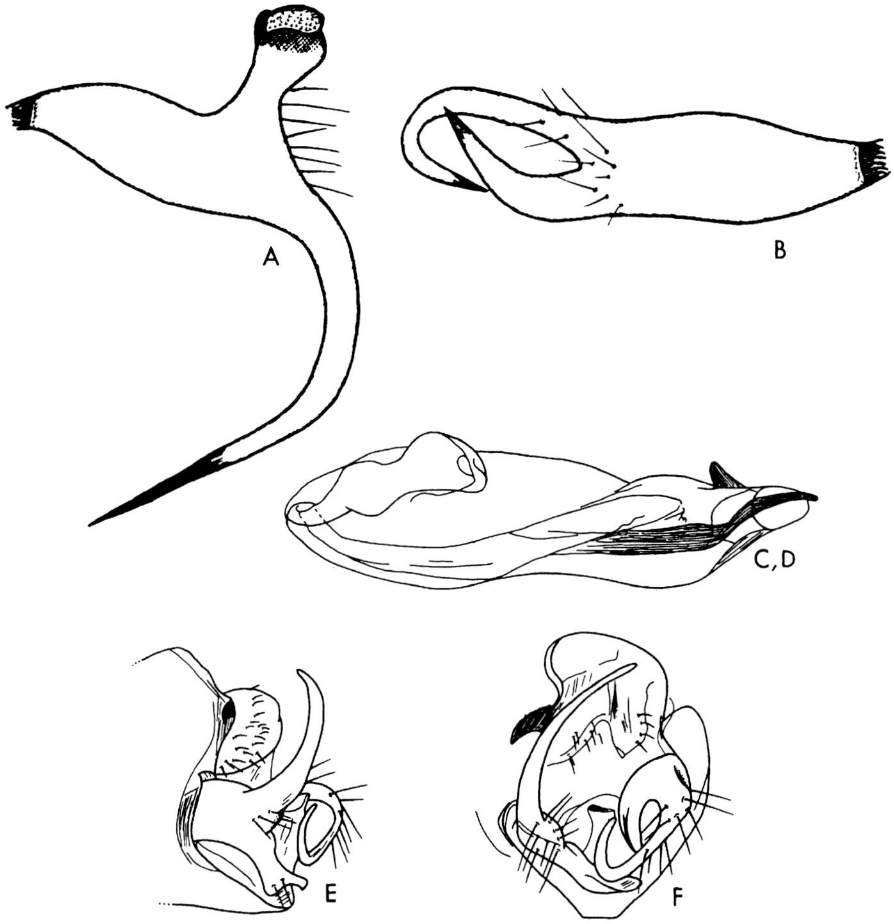

Fig. 92. *Nesiomiris planatus,* n. sp., male genitalia. **a**, left clasper, internal; **b,** right clasper, internal; **c, d**, aedeagus and spiculum, lateral; **e**, pygofer, left lateral; **f**, pygofer, terminal.

48. ***Nesiomiris planatus*** Gagné, **new species** Figs. 19, 26n, 45f, 92

This species differs from *N. ilicis* in having an acute left atrial tooth (Fig. 92f). It differs from *N. ilicivorus*, *N. contortus*, and *N. restrictus* in having the left atrial tooth farther from the base of the left clasper than the thickness of the base of the clasper when it is at rest against the base of the atrium, as well as in details of the right clasper (Fig. 92b).

Description. **MALE** (n = 7). Body: 7.85 (7.63–8.00) long; 1.75 (1.67–1.80) wide. Head: (Fig. 45f) 1.20 (1.19–1.24) wide; a few pale hairs on upper margins of eyes; apex not striolate. Antennal segment lengths: I, 1.35 (1.28–1.39), pale green; II, 3.80 (3.80–4.11), yellowish on proximal 1/2 with a dark brown tip; III, 2.23 (2.03–2.39), brownish with a dark brown tip; IV, 2.50 (1.78–2.65), brownish. Rostrum: 1.62 (1.62–1.74) long, reaching beyond middle of mesosternum. Pronotum:

1.21 (1.14–1.28) long; 1.75 (1.67–1.80) wide; basal 1/2 of hind lobe and its lateral margins green, remainder yellowish; glabrous dorsally. Mesonotum and scutellum, yellow; latter yellowish orange at apex and weakly wrinkled on disc. Hemelytra: yellowish green to apex of scutellum, remainder evenly greenish; vestiture composed of pale, sparse, adpressed hairs along the commissure, on the cuneus and the lateral margins of the embolium; large areole almost entirely coriaceous, excepting a narrow strip along the edge of the vein; membrane slightly infuscate. Legs: yellowish, fore femora slightly greenish above the junction of the femur and tibia; tarsi pale; vestiture on femora and tibiae pale, erect, shorter than the thickness of the segment on the femora; posterior tibia slightly sinuate proximally. Abdomen: pale and membranous above the pygofer; tergum VIII slightly sclerotized and darker on posterior portion; atrial tooth situated above the left clasper at a distance equal to the thickness of the base of the clasper (Fig. 92e); lower arm of the right clasper apically hooked (Fig. 92f); tooth on pygofer minutely serrate dorsally; aedeagus and spiculum as illustrated (Fig. 92).

FEMALE (n = 7). Body: 8.05 (7.21–8.05) long; 1.75 (1.61–2.17) wide. Head: 1.21 (1.15–1.21) wide. Antennal segment lengths: I, 1.35 (1.22–1.37); II, 3.70 (3.20–3.73); III, 2.20 (1.93–2.70); IV, 2.30 (1.78–2.35). Rostrum: 1.70 (1.53–1.71) long. Pronotum: 1.26 (1.00–1.26) long; 1.75 (1.61–1.77) wide. Differing from male as noted under *ilicis* subgroup and as follows: head subglabrous; membrane clear; rostrum reaching apex of mesosternum; *K* and *L* structure as illustrated (Fig. 26n).

NYMPHS. *1st and 2nd instars*. Unknown. *3rd instar* (n = 2). Spiracular openings not infuscate; lateral margins of pronotum subparallel; rostrum reaching base of mesocoxae. *4th instar* (n = 2). Rostrum reaching apex of mesocoxa; comparatively sparsely setose on all aspects. *5th instar* (n = 2). Lateral border of mesonotal wing bud evenly convex; dorsal abdominal setae light brown; rostrum reaching apex of mesosternum; femur, tibia, antennal segment I, and basal 1/2 of II, comparatively sparsely pilose; anterior humeral angle of pronotum weakly infuscate; setae on mesonotal wing bud not arranged along incipient wing veins.

Distribution. O'ahu I (central Wai'anae Mts and southern half of Ko'olau Mts) (Fig. 19).

Host. *Ilex anomala*.

Etymology. The broad wings give this species a flattened habitus, hence the name.

Type data. Holotype ♂ (BPBM 14,872), allotype ♀, paratypes 4♂, 1♀, **O'AHU**: Ko'olau Mts, Wiliwilinui Ridge, 1500 ft. [457 m], 9.V.1970, on *Ilex anomala*, W.C. Gagné. Additional paratypes with data differing only as follows: 1♂, Wailupe, 23.I.1915, [no collector or elevation]; 2♂, 1♀, Wai'anae Mts, Palikea, [no elevation], 11.XI.1936, on *Ilex*, E.C. Zimmerman; 1♂, Tantalus, [no elevation], 15.X.1951, J.W. Beardsley; 4♂, 3♀, 2N, 'Aiea Ridge Trail, 1600 ft. [499 m], 3.XI.1968; 2♀, 1N, 'Aiea Ridge Trail, 1800 ft. [549 m], 16.V.1970; 4♂, 500 m, 28.VI.1970; 17♂, 11♀, 7N, Wai'anae Mts, Kaua Trail, Pu'u Kaua, 666 m, 2.VIII.1970; 1♂, Wai'alae Nui Ridge, 425 m, 11.VII.1971; 6♂, 7♀, Wai'anae Mts, 'Ēkahanui Gulch, near Pu'u Kaua, 2200 ft. [666 m], 10.X.1971 (BPBM).

Variation. Sometimes the rostrum reached the mesocoxae; the areoles are sometimes completely coriaceous.

Fig. 93. *Nesiomiris contortus*, n. sp., male genitalia, showing variation of right clasper. **a**, left clasper, internal; **b**, **b-1**, right clasper, internal; **c**, aedeagus, spiculum removed; **d**, **d-1**, spiculum; **e**, pygofer, left lateral; **f**, **f-1**, pygofer, terminal.

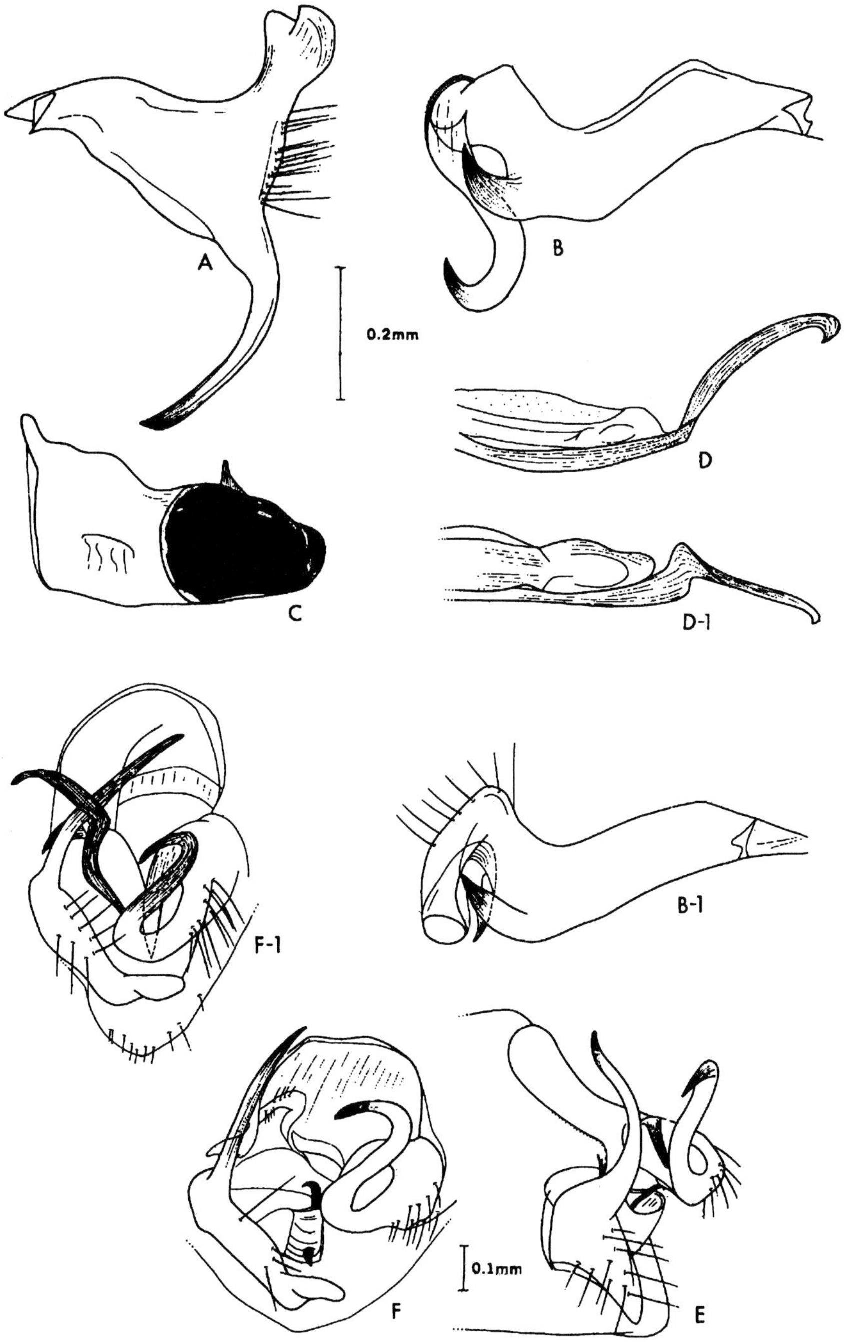
A
B
0.2mm
D
C
D-1
F-1
B-1
0.1mm
F
E

49. *Nesiomiris contortus* Gagné, **new species** Figs. 19, 22a, 93

This species is closest to *N. planatus* and *N. restrictus* from which it differs as noted in the key and as outlined below. It is also readily distinguished by the shape of the lower arm of the right clasper when viewed terminally (Fig. 93f). The shape of this clasper is intermediate between *N. planatus* (Fig. 92f) and *N. restrictus* (Fig. 94f); but its shape is constant.

Description. **MALE** (n = 7). Body: 7.25 (7.21–7.75) long; 1.70 (1.61–1.71) wide. Head: 1.18 (1.14–1.21) wide. Antennal segment lengths: I, 1.28 (1.28–1.35); II, 3.60 (3.55–3.95); III, 2.10 (1.89–2.30); IV, 1.95 (1.68–2.14). Rostrum: 1.53 (1.53–1.61) long. Pronotum: 1.18 (1.09–1.18) long; 1.70 (1.61–1.71) wide. Differing from *N. planatus* as follows: areoles wholly coriaceous; rostrum reaching middle of mesosternum; tooth on pygofer situated at a distance from base of clasper less than basal thickness of left clasper, not minutely serrate dorsally; lower arm of right clasper S-shaped (Fig. 93b, f); phallotheca and spiculum as illustrated (Fig. 93c, d, d–1).

FEMALE (n = 7). Body: 7.21 (7.10–7.75) long; 1.60 (1.60–1.74) wide. Head: 1.17 (1.14–1.19) wide. Antennal segment lengths: I, 1.35 (1.28–1.35); II, 3.60 (3.60–3.87); III, 2.15 (2.06–2.20); IV, 1.78 (1.43–1.85). Rostrum: 1.55 (1.45–1.71) long. Pronotum: 1.13 (1.09–1.13) long; 1.60 (1.60–1.74) wide. Differing from male as noted under *ilicis* subgroup and as follows: apex of antennal segment I yellowish green; apex of ovipositor pale.

NYMPHS. *1st to 3rd instars*. Unknown. *4th instar*. Differing from 4th instar of *N. planatus* as follows: more densely erect brown setose on dorsum, femur, coxa, antennal segment I, and basal 3/4 of antennal segment II, those on dorsum more dense along lateral borders. *5th instar*. Differing from 5th instar of *N. planatus* as follows: pubescence as described for 4th instar; anterior humeral angle of pronotum not infuscate; interior setae of mesonotal wing bud sometimes arranged along incipient wing veins.

Distribution. O‘ahu I (central Wai‘anae Mts and NW Ko‘olau Mts) (Fig. 19).

Host. *Ilex anomala*.

Etymology. Name refers to the contorted right clasper.

Type data. Holotype ♂ (BPBM 14,873), allotype ♀, paratypes 7♂, 4♀, 5N, **O‘AHU**: Wai‘anae Mts, Kānehoa-Hāpapa Trail, 2500 ft. [757 m], 25.I.1970, on *Ilex anomala*, W.C. Gagné. Additional paratypes with data differing only as follows: 1♀, Wai‘anae Mts, near Pu‘u Kānehoa, 27.III.1959, [no host or elevation], J.W. Beardsley; 10♂, 9♀, 5N, Lā‘ie Trail, 1800 ft. [549 m], 11.XI.1968; 1♂, 5♀, 5N, 2500 ft. [762 m], 26.XI.1968; 1♀, 1N, Pūpūkea-Kahuku Trail, 480 m, 10.II.1970; 3N, Wai‘anae Mts, ‘Ewa District, Honouliuli Forest Reserve, 1st ridge W of S fork of Kalua‘a Gulch, 800 m, 14.II.1970; 2♂, Wai‘anae Mts, Honouliuli Forest Reserve, S fork of Kalua‘a Gulch, 800 m, 15.II.1970; 3♂, 1♀, 3N, Wai‘anae Mts, Honouliuli Forest Reserve, W fork of Kalua‘a Gulch, 23.II.1970; 4♂, 1♀, Wai‘anae Mts, Honouliuli Forest Reserve, Kalua‘a Waieli Ridge, 760 m, 26.IV.1970; 1♀, 2N, Wai‘anae Mts, Honouliuli Forest Reserve, Kalua‘a Waieli Ridge, 850 m, 26.IV.1970; 2♂, 1N, Lā‘ie Trail, 459 m, 23.V.1970; 3♂, Kawailoa Trail, 9.VIII.1970; 1♂, Ko‘olau Mts, Pūpūkea Forest Reserve, 900 ft. [274 m], 31.X.1971; 1♂, 3♀, Kawailoa Forest Reserve, Pe‘ahinā‘a Trail, 700 m, 25.VIII.1973; 11♂, 9♀, 1N, Kawailoa Forest Reserve, Pe‘ahinā‘a Trail, 700 m, 8.IX.1974, on *Ilex anomala*; 10♂, 3♀, 3N, Ko‘olau Mts, ‘Ewa Forest Reserve, Mānana Trail, 530 m, 15.XII.1974 (BPBM).

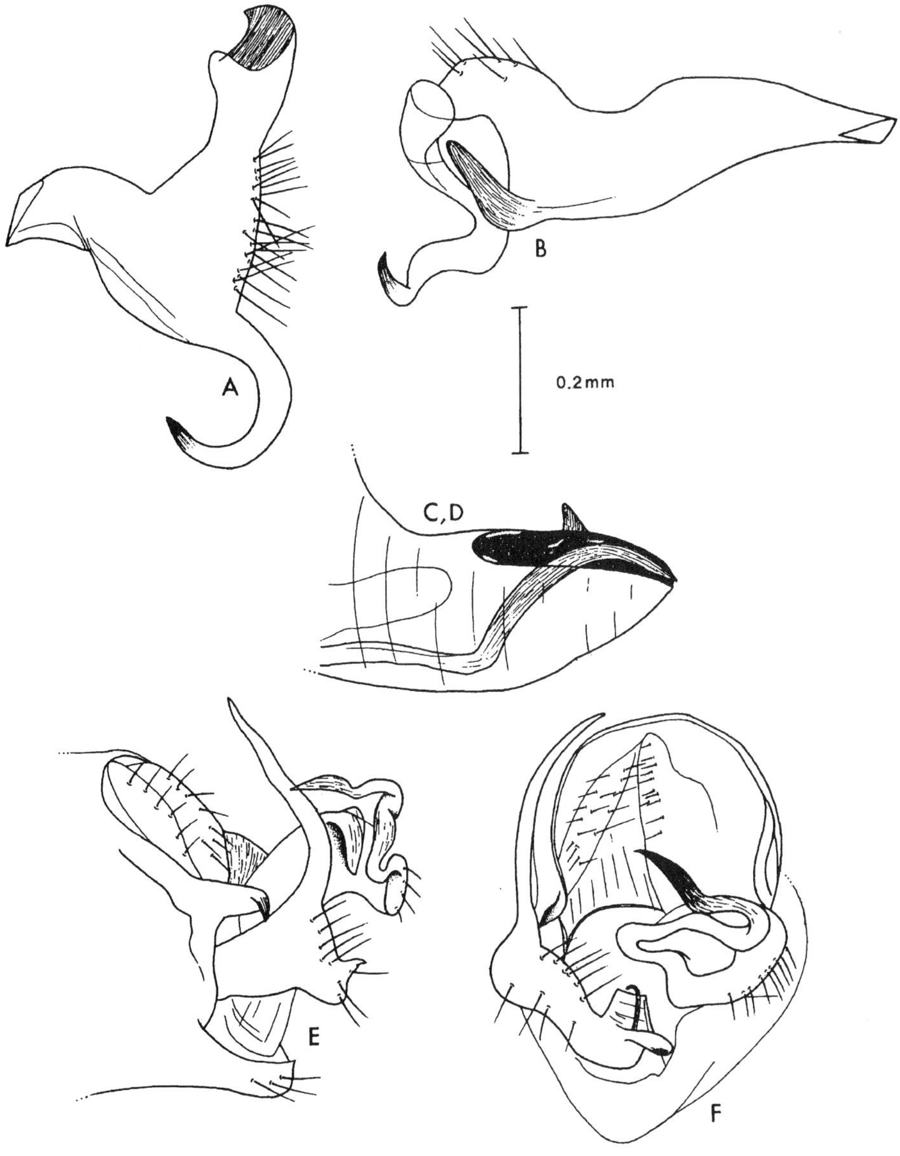

Fig. 94. *Nesiomiris restrictus,* n. sp., male genitalia. **a**, left clasper, internal; **b,** right clasper, internal; **c, d**, aedeagus and spiculum, lateral; **e**, pygofer, left lateral; **f**, pygofer, terminal.

50. ***Nesiomiris restrictus*** Gagné, **new species** Figs. 19, 94

This species is closest to *N. planatus* and *N. contortus*, but differs as noted in the key and as outlined below. The lower arm of the right clasper is more contorted than any other in the genus (Fig. 94f); only that of *N. contortus* approaches it in complexity (Fig. 93f).

Description. **MALE** (n = 4). Body: 7.53 (7.53–7.75) long; 1.65 (1.62–1.72) wide. Head: 1.19 (1.18–1.19) wide. Antennal segment lengths: I, 1.35 (1.33–1.38); II, 3.60 (3.60–3.70); III, 2.16 (2.16–2.35); IV, 1.95 (1.30–2.20). Rostrum: 1.51 (1.51–1.54) long. Pronotum: 1.13 (1.13–1.26) long; 1.65 (1.62–1.72) wide. Differing from *N. planatus* as follows: apex of head weakly striolate; areoles wholly coriaceous; rostrum reaching middle of mesosternum; tooth on pygofer situated above left clasper at a distance less than thickness of base of left clasper (Fig. 94e), not minutely serrate dorsally; lower arm of right clasper sinuate, almost forming double "S" (Fig. 94f); aedeagus and spiculum as illustrated (Fig. 94c, d).

FEMALE (n = 4). Body: 7.75 (7.85–8.05) long; 1.71 (1.71–1.79) wide. Head: 1.19 (1.19–1.21) wide. Antennal segment lengths: I, 1.35 (1.33–1.38); II, 3.60 (3.50–3.80); III, 2.10 (2.10–2.23); IV, 1.95 (1.93–2.46). Rostrum: 1.53 (1.53–1.55) long. Pronotum: 1.21 (1.19–1.26) long; 1.71 (1.71–1.79) wide. Differing from male as described under *ilicis* subgroup and as follows: calli once, rather than twice, impressed interiorly; apex of ovipositor dark brown.

NYMPHS. *1st and 2nd instars*. Unknown. *3rd instar* (n = 1). Differing from 3rd instar of *N. planatus* as follows: generally more setose on dorsum and appendages; rostrum reaching apex of mesocoxa. *4th instar* (n = 1). Differing from 4th instar of *N. planatus* as follows: setose as described for 3rd instar, rostrum reaching middle of mesosternum. *5th instar*. Differing from 5th instar of *N. planatus* as follows: rostrum reaching middle of mesosternum; lateral border of mesonotal wing bud sublinear; anterior humeral angle of pronotum not infuscate, femora comparatively slender.

Distribution. O'ahu I (extreme NW end of Ko'olau Mts) (Fig. 19).

Host. *Ilex anomala*.

Etymology. Named for its extremely restricted distribution.

Type data. Holotype ♂ (BPBM 14,874) allotype ♀, paratypes 3♂, 4♀, 3N, **O'AHU**: Waiāhole Forest Reserve, Turnover Peak, 2000 ft. [666 m], 51.V.1970, on *Ilex anomala*, W.C. Gagné (BPBM).

Variation. The setae on antenna 1 of the allotype are comparatively sparse, but this appears to be largely a result of abrasion in transit to the laboratory.

CLADISTIC ANALYSIS

Characters Examined

The taxonomic descriptions of the species were based on characters selected by prior experience in Miridae. These are listed with a summary of the cladistic states they exhibit in the taxa under study in Appendix I. For many it was impossible to reach any conclusions concerning apomorphy. This latter group of characters, which were omitted from the cladistic analysis (not marked with an asterisk [*] in Appendix I), were generally "simple characters," for which probable transformation series could not be established. On the other hand, a ground plan could be inferred for some characters, especially the more complex ones, such as those of the male genitalia. Deviation from this ground plan provided a basis for inferring synapomorphy. In addition, some of the "simple characters" obviously formed monophyletic groups. The latter 2 kinds of characters (marked with an asterisk [*] in Appendix I) were included in the cladistic analysis.

Schuh (1974: 259, table 1) compared 8 characters of Miridae with those of Nabidae for plesiomorphic and apomorphic states, because he considered them the most plesiomorphic family in the Cimicomorpha to which Miridae also belong. Similarly, he compared 10 characters in 8 subfamilies of Miridae (p. 260), including the Orthotylinae, and then 9 characters within the Orthotylinae (p. 270, table 3) to establish a phylogeny for Miridae as a whole and for the tribes of Orthotylinae in particular.

Basic Plan of Male Genitalia

The greatest diversity of structure found among the species was in the male genitalia. Since it is easier to make inferences concerning apomorphy where differences between attributes are large, it was expected that the genitalia would provide many such inferences.

First an attempt was made to establish the basic plan of male genitalia in Miridae. To accomplish this, the literature was surveyed for genitalia of 144 species in 99 genera of Miridae, and of representatives of all major subfamilies of Miridae, as illustrated by Kelton (1959). The male genitalia of these species were examined for structures that recur in the major groups and that could therefore be inferred to be primitive.

The Miridae

The elements of the structural scheme outlined below occur in a wide variety of subfamilies, although no group shows all elements at once. The basic plan is as follows:

Left clasper. The upper arm is longer than the lower arm and is apically hooked and acute. The lower arm is lobate.

Right clasper. This clasper is usually smaller, shorter, and less robust than the left clasper. Although previous authors have not identified which genera are primitive in the Orthotylinae based on male characters, there are a number of genera (e.g., *Ceratocapsus*, *Reuteria*, *Heterotoma*, *Heterocordylus*, *Lopidea*, *Ilnacora*, and *Ilnacorella*) illustrated by Kelton (1959) that have the right clasper at least trifurcate, the most parsimonious plesiomorphic character state for *Nesiomiris*. Emphasizing the development of the *K* structures of females, however, Schuh (1974) placed the Nichomachini and Halticini as ple-

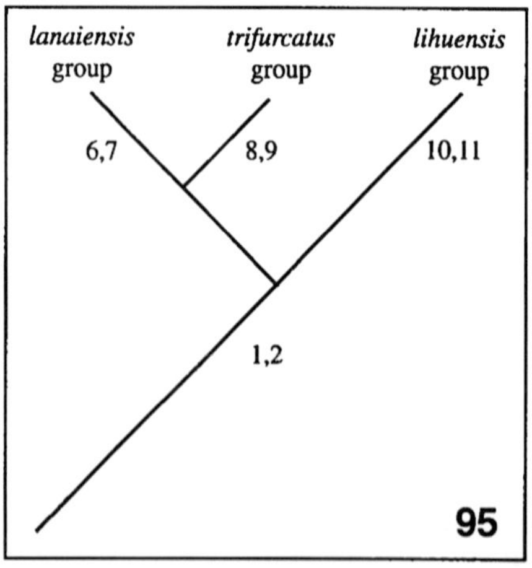

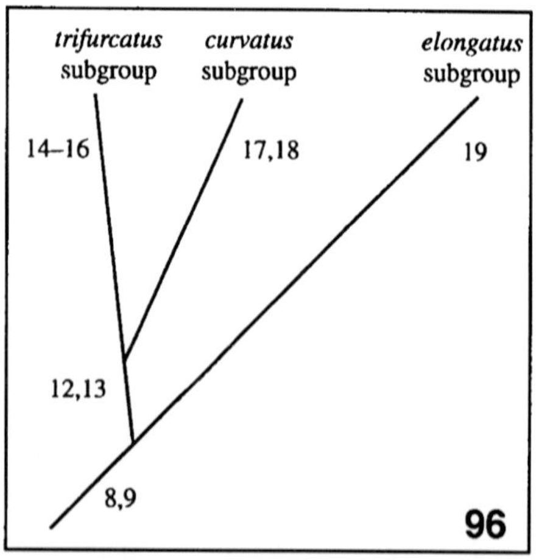

Figs. 95–96. Phylogenetic relationships of *Nesiomiris*. **95,** species groups; **96,** species subgroups of the *trifurcatus* group. Numbers refer to synapomorphic characters discussed in the text.

siomorphic to the Orthotylini. In the male genitalia that he and Kelton (1959) illustrated for the 2 former tribes, the right clasper is usually simply lobate and uniramous.

Aedeagus. The transformation series in the family is from membranous aedeagus and vesica to a sclerotized aedeagus and a vesica with 1 or more sclerotized spiculi.

Armature of pygofer. No uniform attention appears to have been paid to this character from a phylogenetic aspect, so no transformation series can be inferred.

The Orthotylinae and Orthotylini

Genera with type 1 parempodia (*sensu* Schuh 1974) that is parempodia "distinctly fleshy, convergent apically, recurved (lyre-shaped)" have 2 types of male genitalia—most genera have a membranous vesica that may or may not possess sclerotized spiculi, while others have a rigid sclerotized vesica. The "basic plan" of the claspers in the Orthotylini is opposite to that outlined above; that is, the right clasper is generally larger than the left.

The Orthotylus *Group*

Genera placed in the *Orthotylus* group (*sensu* Schuh 1974) have well-developed, long, heavily sclerotized spiculi on the vesica, and the claspers are highly variable. Of the 40 genera considered by Schuh to belong to this group, the male genitalia of 13 genera containing 20 species were illustrated by Kelton (1959). I postulate a transformation series upon examination of Kelton's illustrations as follows: (1) single simple spiculum (*Labopidea*, *Parthenicus*; (2) 1 spiculum with a subapical flabellum (*Globiceps*, *Mecomma*), plus *Nesiomiris*; (3) 2 spiculi, 1 of which is bifid (*Hadronema*, *Lopidea*, *Ilnacora*, *Ilnacorella*, *Melanotrichus*); (4) 3 spiculi, some serrate (*Heterocordylus*, *Pseudopsallus*, *Heterotoma*, *Melanotrichus*); and (5) 4 straplike, serrate spiculi (*Orthotylus*). If this is the correct phylogenetic sequence, *Nesiomiris* stands near the base of the transformation series, with *Orthotylus* at the apex. Unfortunately, the aedeagus (= phallotheca of Kelton) was not illustrated for the majority of these genera, nor for any genera possessing the single, subapically flabellate spiculi.

Cladistic Procedure

I used the following procedure, adapted from that outlined by Munroe (1974), to reconstruct the cladistic relationships of the species of *Nesiomiris*. Probable transformation series and inferred apomorphic states are from the distribution of characters discussed in the next section and are marked with an asterisk (*) in Appendix I. Monophyletic groups described by the inferred synapomorphic states were successively added to a cladogram. Each time 2 monophyletic groups were inferred to form a 3rd monophyletic group containing both, 1 mode of the cladogram was established. In this way, the sister-group relationships were determined as the cladogram was constructed. The operation, then, consisted of a search for monophyletic groups. When conflicts arose, the most parsimonious alternative was chosen; that is, the choice that would require the fewest steps in a transformation series to arrive at a particular character state. In some cases, this involved a simple reversal of the direction or transformation of 1 character minimizing the number of conflicts.

Transformation Series and Monophyletic Groups

Generic Level

1. The generic diagnosis for *Nesiomiris* is believed to define a monophyletic group since there is no reason to believe that occurrence of its assemblage of characters is due to convergence. The species are considered to have arisen from 1 successful colonizer of the archipelago.

2. *On Araliaceae*. All species on Araliaceae are considered as belonging to a monophyletic group. There are no species groups on other families of host plants that could be considered plesiomorphic to this assemblage.

Supra-species Group Level (Fig. 95)

3. *Coriaceous areolar cells*. The plesiomorphic state of the partially to wholly membranous areolar cells is inferred because of its distribution in a large number of species. The transformation series ranges almost continuously from partially coriaceous on the inner margin of the large areole to completely coriaceous in both cells. In the latter cases its presence is apparently correlated with low vagility. All the species associated with *Ilex anomala* and many of those associated with Araliaceae have variously coriaceous areoles. This inferred synapomorphy unites the *lanaiensis* group on *Ilex* and the *trifurcatus* group on Araliaceae into a monophyletic group.

4. *Color of areolar veins*. Two character states were evident: species with green veins and species with reddish veins. Since species with green veins also have partially to wholly coriaceous areolar cells, the 2 species groups with these character states are considered to belong to a monophyletic series. The character state of a green areolar vein is the more

widely distributed and so is considered plesiomorphic to character states 10 and 11 (described below).

5. *Apical area of aedeagus*. Two groups of species have the left apical area of the venter of the apex with a thickened wall or tooth. Within the species of these groups are transformation series from a simple thickening on the left apex to more elaborate chelate and spatulate projections and another series culminating in ventral, subapical teeth. The former elaborations appear to correlate with reduction in size and complexity of the right clasper. This inferred apomorphy also serves to unite the group on *Ilex* with the *trifurcatus* group on Araliaceae.

Species Group Level (Fig. 95)

The group on Araliaceae shows 2 evolutionary pathways, both with plesiomorphic characters compared to the group on *Ilex anomala*. The members of the *lihuensis* group on Araliaceae have in common rufescent areolar veins and a sinuate spiculum sometimes hooked at the apex. The membrane is not coriaceous. The members of the *trifurcatus* group on Araliaceae have a sinuate spiculum, usually with a tooth or flabellum near the apex. The propleural stripe is present in the bulk of the species. The base of the head is emarginate.

The *Ilex*-Associated Group (Fig. 95).

6. *Sexual dichromism*. All of the species that occur on Aquifoliaceae are sexually dimorphic for color: the males are dark green, while the females are light green. This feature is less widely distributed and is considered a synapomorphy.

7. *Apex of venter of male*. Males of species that are sexually dimorphic for color and occur on Aquifoliaceae have a tuft of denser setae on the apex of the venter that is another inferred synapomorphy.

The *trifurcatus* Group

8. *Spiculum of male*. Species with green areolar veins that are found on Araliaceae almost always have a subapically flabellate spiculum.

9. *Upper arm of left clasper*. Members of this group also have a long upper arm of the left clasper that is variously sinuate and reaches the dorsum of the pygofer, rather than facing apically into the atrium of the pygofer.

The *lihuensis* Group (Figs. 95, 99)

10. *Membranous areolar cells*. This character state is less widely distributed than coriaceous areolar cells. One species, *N. bifurcatus*, with otherwise apomorphic characters has a propleural stripe, green areolar veins, but membranous areolar cells indicating a transition from the *trifurcatus* group.

11. *Red areolar veins*. All species with clear areolar cells, except *N. bifurcatus* have red areolar veins. This character is considered to be apomorphic because it is less widely distributed.

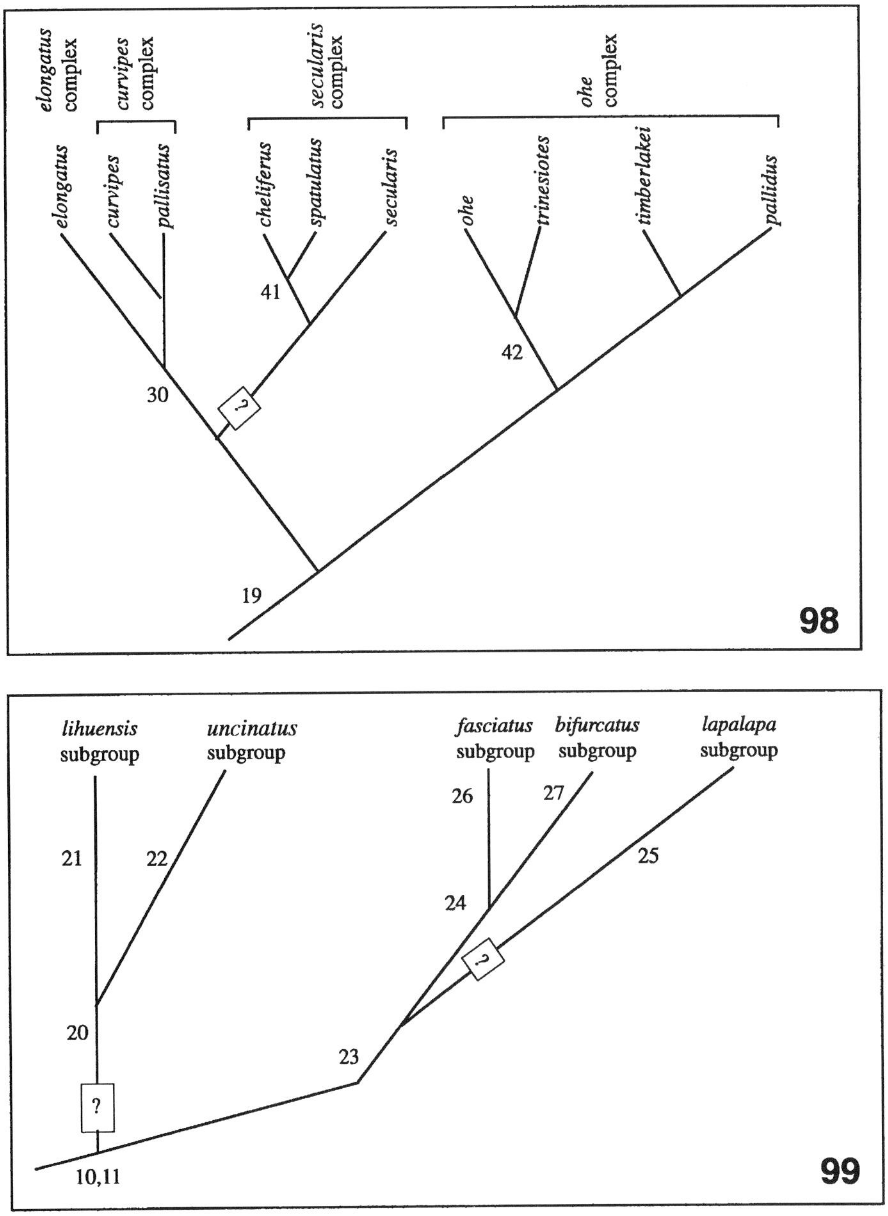

Figs. 98–99. Phylogenetic relationships of *Nesiomiris*. **98**, *elongatus* subgroup and its complexes. **99**, *lihuensis* group and its subgroups. Numbers refer to synapomorphic characters discussed in the text.

Species Subgroup Level

12. *Upper arm of right clasper.* Species in the *trifurcatus* and *curvatus* subgroups have the upper arm of the right clasper elongated and unreduced. The lower arm ranges from elongate to a short, blunt knob.

13. *Propleural strripe of nymphs.* Nymphs of the *trifurcatus* and *curvatus* subgroups lack propleural stripes.

The *trifurcatus* Subgroup (Fig. 96)

14. *Propleural stripe of adults.* Occurrence of a propleural stripe that extends from the collar onto the hind lobe of the pronotum, as contrasted to a character state in which it extends posteriorly only to the mesal constriction, was inferred to be apomorphic.

15. *Head and portions of appendages.* The head, antennal segment I, and portions of the legs are rufescent.

16. *Armature of right claspers.* The right clasper is trifurcate; a condition considered to be plesiomorphic in the genus as a whole, but expressed unique here as an autapomorphy in the *trifurcatus* group.

The *curvatus* Subgroup (Fig. 97)

17. *Third prong of right clasper.* In this species group the 3rd prong of the right clasper appears to be represented by a small knob between the bases of the outer arms. This is thought to represent a transformation series from the *trifurcatus* group, and its reduction is a synapomorphy.

18. *Upper arm of left clasper without teeth.* The upper arm is uniform, without denticles or subapical arms. This represents the apomorphic condition.

The *elongatus* Subgroup (Fig. 98)

18. *Right clasper.* The arms are much reduced with at most 2 blunt arms present. This appears to represent another abrupt transformation series from the *trifurcatus* and *curvatus* subgroups and is inferred to be synapomorphic.

The *lihuensis* and *uncinatus* Subgroups (Figs. 99, 100)

Several of the characters defining these subgroups are of restricted occurrence in the *lihuensis* group but are widespread in the other species groups. Because of their possession of a number of widespread or plesiomorphic character states, these subgroups are placed at the base of the *lihuensis* group. Both subgroups are placed in a monophyletic group based on the possession of character state 20.

20. *Pygofer wall and spiculum.* Usual possession of a subapically flabellate spiculum and a tooth or teeth on the pygofer infers a monophyletic group.

The *lihuensis* Subgroup (Fig. 99)

21. *Dorsal vestiture.* Because of the unique, dense, erect, golden brown pubescence, *lihuensis* is placed in its own subgroup.

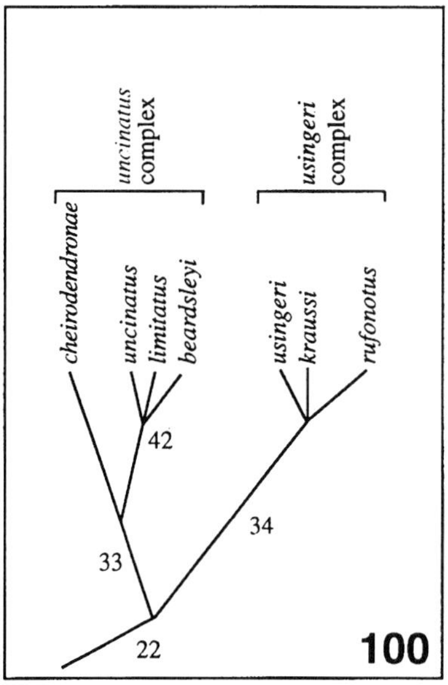
cheirodendronae
uncinatus
limitatus
beardsleyi
usingeri
kraussi
rufonotus
uncinatus complex
usingeri complex
42
33
34
22
100

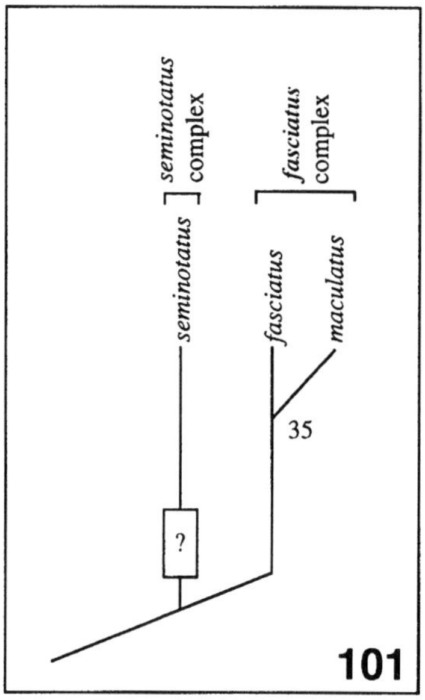
seminotatus
fasciatus
maculatus
seminotatus complex
fasciatus complex
?
35
101

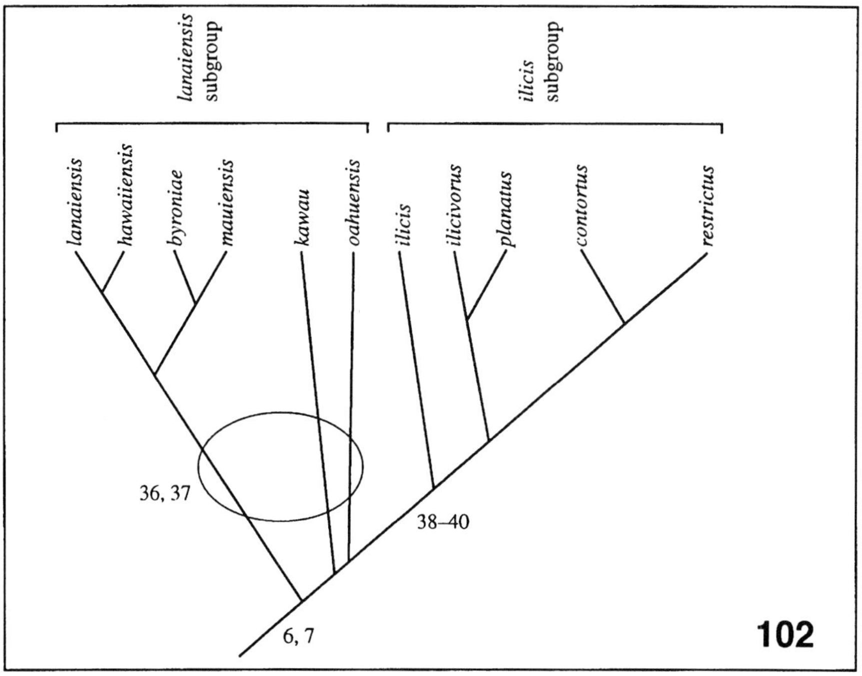
lanaiensis
hawaiiensis
byroniae
mauiensis
kawau
oahuensis
ilicis
ilicivorus
planatus
contortus
restrictus
lanaiensis subgroup
ilicis subgroup
36, 37
38–40
6, 7
102

The *uncinatus* Subgroup (Fig. 99)

22. *Spiculum apically uncinate*. The "basic plan" of those genera with a single spiculum is for a single sinuate, nonuncinate, spiculum. Thus, this derived condition indicating a monophyletic subgroup.

The *fasciatus*, *bifurcatus*, and *lapalapa* Subgroups (Fig. 99)

23. *Upper arm of left clasper*. This arm is usually biramous or apically toothed.

The *fasciatus* and *bifurcatus* Subgroups (Fig. 99)

24. *Apex of head*. In these subgroups, the species have distinctly bifid apices of the head. This is a weak character on which to base a monophyletic group since it is not a unique apomorphy at the major clad level, reoccurring in species of the *lanaiensis* group.

The *lapalapa* Subgroup (Fig. 99)

25. *Apex of right clasper*. Species of this subgroup have the apices of the right clasper hooked or uncinate. This also is a weak character in which to base a holophyletic group for the same reasons given under character 23, but there is otherwise close correspondence in male genitalia, which constitute an obviously related transformation series.

The *fasciatus* Subgroup (Fig. 99)

26. *Upper arm of left clasper*. Species in this subgroup have the upper arm of the left clasper uniramous. This is a weak character on which to base a synapomorphy for the reasons given under character 24, but these species show close structural conformity in male genitalia.

The *bifurcatus* Subgroup (Fig. 99)

27. *Upper arm of left clasper*. This is a unique character that is thought to unite these species into a monophyletic species subgroup. One species in the *lapalapa* subgroup has an apically bifid upper arm of the left clasper, which may be synplesiomorphic with this subgroup.

The *bifurcatus* and *lapalapa* Subgroups

The characters used to define the species complexes within the *bifurcatus* subgroup all appear to represent non-unique apomorphies, as discussed below. No further cladistic analysis was attempted in this subgroup. This was also found to be the situation in the *lapalapa* subgroup. There is thought to be retention of a striking propleural stripe in 1 species in each of these 2 subgroups, *N. lineatus* in the former, *N. legnotus* in the latter. To unite *lineatus* and *legnotus* in a supposed monophyletic group would be infeasible.

Figs. 100–102. Phylogenetic relationships of *Nesiomiris*. **100**, *uncinatus* subgroup and its complexes. **101**, *fasciatus* subgroup and its complexes. **102**, *lanaiensis* group and its species and species complexes. Numbers refer to synapomorphic characters discussed in the text.

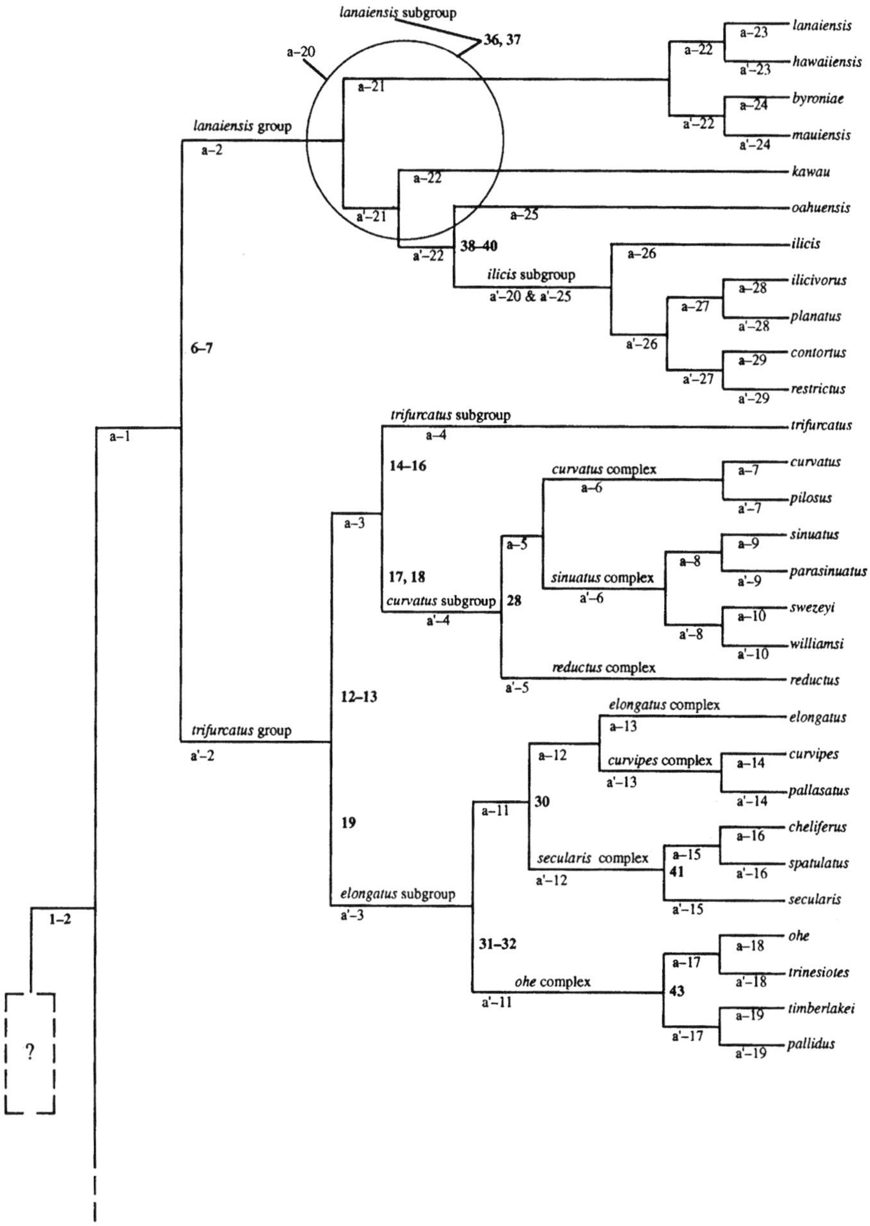

Fig. 103a. Summary of phylogenetic relationships of the species of *Nesiomiris.* Boldface numbers beside vertical lines refer to synapomorphic characters discussed in the text. Letters a and a´ below horizontal lines refer to inferred sister groups and associated numbers with these letters indicate an approximation of the evolutionary sequence.

These 2 species fit comfortably into transformation series, which culminate with *N. legnotus* in the *lapalapa* subgroup and with *N. lineatus* in the *bifurcatus* subgroup (see Fig. 99). The propleural stripe is thought to be retained in these 2 unrelated species because of some adaptive significance or through convergence.

The *curvatus* and *sinuatus* complexes

28. *Upper arm of right clasper.* This external arm is unreduced and hooked in species of both complexes. this is thought to represent a synapomorphy. I have placed the curvatus complex lower on the cladogram because of the closer structural resemblance of the right clasper to that of the trifurcatus subgroup.

29. *Reduced right clasper.* All arms of the right clasper are reduced but still discernable in these complexes. This is interpreted as representing a synapomorphy, and is a sister group to the above 2 complexes.

The *elongatus* and *curvipes* complexes

30. *Incrassate hind legs.* Members of these complexes possess incrassate hind femora, with the base of the hind tibia also bent somewhat like a shepherd's crook. This striking deviation from the basic plan is inferred as synapomorphic. This feature is incipient in *N. elongatus* of the *elongatus* complex, and the complexes form a sister group from which a transformation series can be inferred.

The *ohe* Complex)

31. *Right clasper* single lobed. The *ohe* complex is unique in that the right clasper consists of a single, simple lobe without any vestiges of other arms. This contrasts with the biramous condition in the remainder of the species in the *elongatus* subgroup. On the basis of overall similarity, the *elongatus* and *curvipes* complexes form a sister group with the *secularis* complex.

32. *Twisted apex of aedeagus.* The atrium of the aedeagus of the *ohe* is twisted apically to the left so that the atrium faces the left clasper. This condition deviates from the basic plan of the aedeagus of other orthotylines in which the apex is not twisted. It is presumed therefore that the *ohe* complex forms a monophyletic group.

The *uncinatus* Complex (Fig. 100)

33. *Spiculum apically uncinate and subapically flabellate.* This combination of characters presents a unique situation inferred to be synapomorphous.

The *usingeri* Complex (Fig. 100)

34. *Spiculum apically uncinate.* Only this character appears to define a group of species derived from the *uncinatus* complex. It unites a group of small species restricted to *Cheirodendron trigynum* on Moloka'i and Maui. This assumed synapomorphy is based on the fact that the spiculum, while it is apically uncinate, is never subapically flabellate. The inferred transformation series is presented in Figure 100.

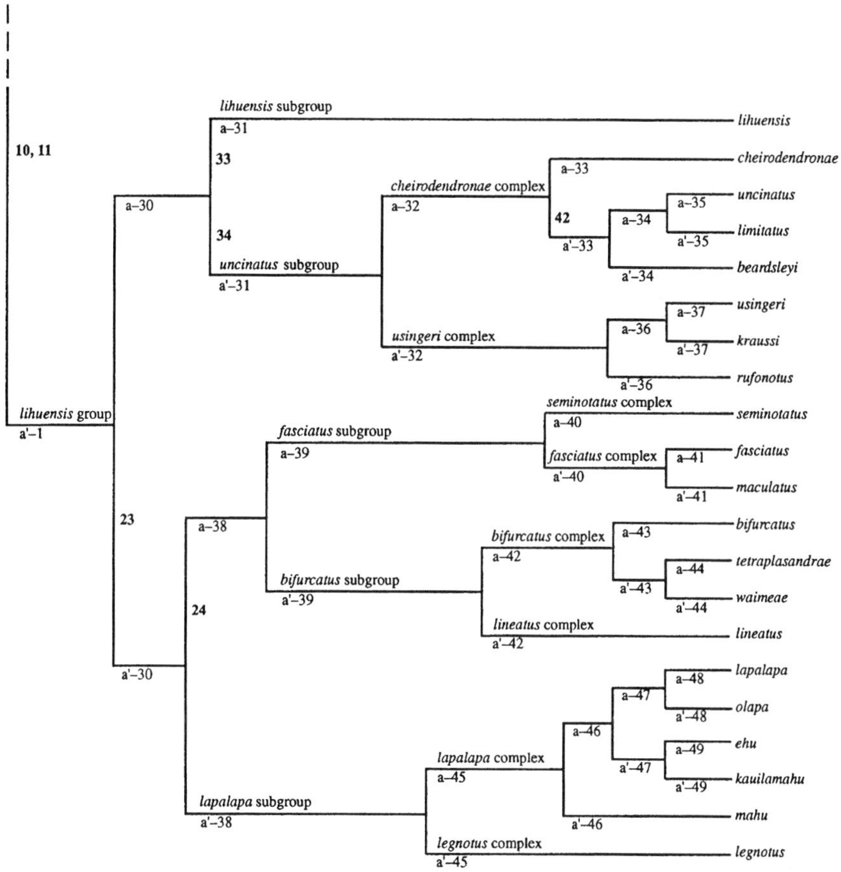

Fig. 103b. Detail of clade A. Boldface numbers beside vertical lines refer to synapomorphic characters discussed in the text. Letters a and a´ below horizontal lines refer to inferred sister groups and associated numbers with these letters indicate an approximation of the evolutionary sequence.

The *fasciatus* Complex (Fig. 101)

35. *Hemelytra green spotted.* This character unites the 2 species, *fasciatus* and *maculatus*, which are associated with *Tetraplasandra kavaiensis* on Kaua'i, in a unique inferred synapomorphic group.

The *lanaiensis* Subgroup (Fig. 102)

36. *Dorsum of male.* The males of this complex have polished dorsa, a unique sexually dimorphic, synapomorphic attribute.

37. *Dorsum of pygofer.* The males of this subgroup have the dorsa of the pygofers blackish; a unique feature considered to be synapomorphic.

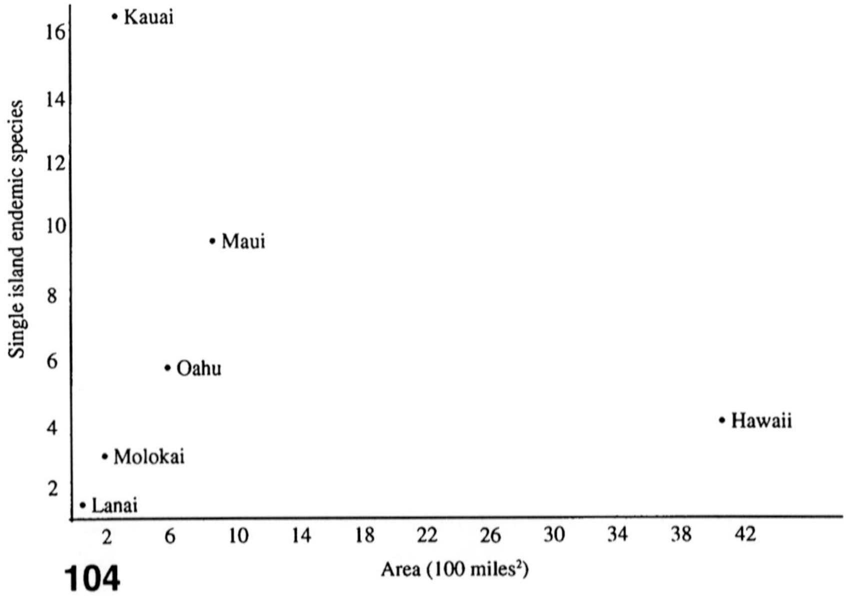
Kauai
Maui
Oahu
Hawaii
Molokai
Lanai
Single island endemic species
2
4
6
8
10
12
14
16
2 6 10 14 18 22 26 30 34 38 42
Area (100 miles²)

104

Kauai
Waianae Mts Koolau Mts
Oahu
East Maui
Hawaii
West Maui
Molokai
Single island endemic species
2
4
6
8
10
12
14
16
0.5 1.0 1.5 2.0 2.5 3.0 3.5 4.0 4.5 5.0 5.5
Age (m yrs)

105

The *ilicis* Subgroup (Fig. 102)

38. *Body shape.* Adults have the lateral margins of the hemelytra convex linear, rather than linear for the remaining species in the genus. This defines a unique apomorphy.

39. *Lower arm of right clasper.* In the males of this subgroup, the lower arm of the right clasper curves dorsally over the upper arm when viewed terminally. This unique attribute unites the subgroup. A transformation series is apparent in the increasing sinuosity and lengthening of this lower arm.

40. *First segment of antennal segment I and femora.* Antennal segment I and the femora of the adults and the 5th-instar nymphs have long hirsute, pale setae. This is considered a unique synapomorphy.

The *seminotatus* Complex

While the general conformation of male genitalia would indicate that the *seminotatus* complex is a sister group of the *fasciatus* complex, no characters were discovered that were uniquely synapomorphic to it. Only *seminotatus* in the *lihuensis* group possesses green areolar veins; I believe that this represents retention of a plesiomorphic feature. This would appear to make this group polyphyletic, but *seminotatus* otherwise fits readily into a transformation series that contains the *fasciatus* complex.

Intra-species Complex Level

Monophyletic groups within species complexes are more difficult to infer. This is because the complexes are usually so homogenous phenetically that transformation series and plesiomorphies are often difficult to recognize, especially in the small complexes. It was not possible to establish the relationship of every species within its particular complex, but a number of monophyletic groups were inferred.

The species are arranged within the complexes in Table 1 so that the transformation series can be read from top to bottom. Generally, these series are constructed of associated sequences of species, also utilizing distribution (e.g., older to newer island), host-plant relationships, and the detailed correspondence of male genitalia. Some of the attributes of the latter are shape and placement of the tooth (if any) on the pygofer; shape, length, and armature of the claspers and the process (if any) on the aedeagus; and shape of spiculum.

Species Within *secularis* Complex

41. *Apical appendage on aedeagus.* Within the *secularis* complex, *N. spatulatus* and *N. cheliferus* form a species pair possessing a spatulate to chelate appendage on the apex of the aedeagus. This condition appears to have arisen only once and is considered to be a synapomorphy.

Figs. 104–105. 104, Relationship between the number of monoinsular endemic species of *Nesiomiris* in the Araliaceae "guild" and island area. **105,** Number of monoinsular endemic species of *Nesiomiris* in the Araliaceae "guild" as a factor of island age. Bar indicates range in age as determined by potassium-argon method (after Macdonald & Abbot 1970).

Species Within *uncinatus* Complex

42. *Fleshy lobe beneath gonopore*. In the *uncinatus* complex, *uncinatus*, *beardsleyi*, and *limitatus* appear to form a holophyletic group based on the presence of a fleshy, denticulate lobe beneath the gonopore. This is a unique synapomorphy making these 3 species a sister group of *N. cheirodendronae*.

Species Within *ohe* Complex

43. *Denticulate and lumpy left clasper*. Within the *ohe* complex, *ohe* and *trinesiotes* appear to form a holophyletic group in which the inner face of the upper arm of the left clasper has a series of lumps and/or denticles, particularly evident in the latter species. This unique synapomorphy would make these 2 species a sister group of *timberlakei* + *pallidus*.

Non-unique Apomorphies and Major Clades

The phylogenetic scheme described in the foregoing pages and summarized in Figures 103a and 103b incorporates a number of apomorphies that are inferred to have arisen more than once. Some of these defined major clades, that is, above the species-complex level. Since these major clades could be important in placing additional species that may yet be found, they seem worthy of discussion. These attributes and the various groups in which they occur are as follows:

1. *Head and antennal segment I rufescent*. *Trifurcatus*, *uncinatus*, and *lihuensis* subgroups, as well as scattered species in the *lihuensis* group have these characters.

2. *Apex of head*. The *ilicis* subgroup and *N. tetraplasandrae* have distinctly bifid apices of the head. All intermediate degrees to rounded head apices are found. In the *curvatus* and *lapalapa* subgroups, the *curvipes* and *usingeri* complexes have distinctly rounded head apices.

3. *Apex of hind femur*. *Nesiomiris trifurcatus* and *N. elongatus* usually have the apex of the hind femur rufescent in adults and nymphs.

4. *Propleural stripe*. A propleural stripe is lacking in the *lanaiensis* (except some males of *lanaiensis*), *lihuensis*, and *uncinatus* (except obsolescent in *N. uncinatus*) subgroups, as well as the *bifurcatus* and *lapalapa* complexes. In the latter complexes it is retained conspicuously in *N. lineatus* and *N. legnotus*.

5. *Loss of pygofer tooth*. The presence of a sclerotized tooth or lobe on the edge of the pygofer has roughly the same enigmatic pattern as character 4 above. It disappears and reappears throughout the genus, rendering an otherwise conspicuous feature virtually useless for cladistic purposes. It is present in all species groups. Its shape and placement are valuable, however, in species recognition. It appears to have some function in conjunction with the claspers during copulation. Within species complexes, it is useful in ascertaining the directionality of transformation series. It has been lost once each in the *lapalapa* subgroup, the *lanaiensis* subgroup, the *curvatus* subgroup, and the *secularis* complex.

6. *Loss of flabellum on spiculum*. The subapical flabellum appears to have been lost once in each of the species groups. All members of the *lanaiensis* group lack it, as does the *usingeri* complex and the *fasciatus*, *lapalapa*, and *bifurcatus* subgroups, in addition to *N. pallasatus* and some populations of *N. ohe*.

Table 8. Heirarchical arrangement for hosts of increasing distribution of the Araliaceae guild (vertical column) showing numbers of species of *Nesiomiris* per host genus.

Hosts	Parameters			Islands					
	Total acres (thousands)	Ecological range	Total associated *Nesiomiris* spp.	Hawai'i	Maui	Moloka'i	Lāna'i	O'ahu	Kaua'i
Munroidendron	120	2	1	—*	—	—	—	—	1
Reynoldsia	1110	3	3	1	1	1	0	0	—
Tetraplasandra									
Eutetraplasandra	690	3	6	2	3	2	1	—	—
Pterotropia	1000	4	7	1	0	0	0	3	3
Nothotetraplasandra and *Neoplasandra*	1980	5	12	2	3	1	1	2	5
Cheirodendron	1980	5	23	3	6	2	1	4	8

* host genus absent

The criterion for deciding which characters show homoplasy, and which do not, is essentially one of parsimony. The characters listed above represent the "fewest independent origins" that were required by the remainder of the system of inferences.

Cladism and Island Biogeography

> "One of the most controversial of Hennig's (1966) rules, which forms the basis for the cladistic approach to biogeography, states that species possessing the most primitive characters are found within the earliest occupied part of the area, i.e., the center of origin for the group" (Briggs 1974).

In contrast to the above quote, Briggs subsequently argued that, "in general, the more advanced species will be found in the center of origin." It would then appear that my cladistic arrangement of *Nesiomiris* with the more plesiomorphic species at the Kaua'i end of the archipelago would be in conflict with Briggs. However, an analysis of his theory of the function of zoogeographic barriers, be they "major" as between continents, or on a less grand scale as between isolated islands in an archipelago, as well as the realization that he and I are working with vastly different time scales, indicates to me that cladistics and zoogeography, at least on the systematics of insular biota, are not incompatible.

The following observations seem pertinent to a resolution of this case: (1) Kaua'i is the oldest of the islands occupied by *Nesiomiris*; (2) there is reason to believe from the resistance to island colonization, as well as the diversity, distribution, and availability of hosts there, that this island has long offered great ecosystem continuity; (3) the most parsimonious arrangement of the genus would indicate that Kaua'i could have been founded at most only once from a younger island (O'ahu); (4) all other pylogenetic lineages can be derived from Kaua'i that has proportionately the most plesiomorphic species; and (5) a relatively young island might be expected to retain more archaic species and hence show less extinction, since it could be expected to resist founding from other islands.

Two other biogeographic concepts, which seem best discussed here, are (1) the theories of island biogeography advanced by MacArthur and Wilson (1967), and (2) the "host plants as evolutionary islands in time" concept advanced by Swezey (1925) and Southwood (1960a, b, c) and as refined notably by Opler (1974a).

MacArthur and Wilson (1967) present extensive data to support their argument that island size is a strong force in determining the diversity of an island's biota and in turn affects the turnover of species by affecting the rate of extinction. They claimed that smaller islands will have smaller biotas and greater turnover through colonization and extinction. The distribution of *Nesiomiris* species, however, would indicate that geological age might be of paramount importance in archipelagos, which have a number of suitable host plants relatively evenly distributed throughout the archipelago. I would postulate that island size will be a factor that differs according to the taxonomic group, for example, vertebrates versus invertebrates and their differing demands upon the habitat in terms of the dimensions of their niche requirements. Judging from the distribution of the monoinsular endemic species of *Nesiomiris* (Figs. 14, 104), island size would not appear to be a strong factor in determining species number. Hawai'i Island, which is equal in size to the remainder of the archipelago altogether, has only 4 monoinsular endemic species. Moloka'i and Lāna'i have 2 monoinsular endemic species each, and although these are the smallest of the islands with *Nesiomiris*, as pointed out earlier their geological history is complicated by the fact that these have been joined with Maui at various times (Stearns 1966). In fact, several species are shared with Maui (Fig. 14). If there is a lower size limit operative that affects the number of species on an island, as would appear to be the case with vertebrates, for example, this asymptotic limit would not appear to have been reached with Hawai'i, Kaua'i, O'ahu, or Maui for *Nesiomiris* species.

The colonization and equilibrium effects of this "guild" of plant-feeding insects, which forms a component of the phytophagous insect community in Hawai'i, can be correlated and contrasted with findings applicable to the biota of islands as proposed by MacArthur and Wilson (1967) in other ways. Opler (1974a: 67) theorized that:

> "As a starting point, we might hypothesize that a host plant of increasingly large distribution will expand into a larger and larger range of ecological habitats. This allows for (1) an increasing probability that insect populations requiring specialized habitats will survive, while reducing the probability of extinction due to interspecific competition or extremes of the physical environment; and (2) an increasing probability of successful colonization by insects new to that host. Because the niche characteristics of each species differ, the potential available resource level for each species will also differ. Those species with the broadest niche will tend to occupy more habitats, while those with narrower niches will occupy a smaller number of habitats."

From the above postulates, we might expect the probability of successful colonization to be a direct function of the insect's niche breadth and the number of phytophagous insect species to be a function of the abundance or distributional area of their host plant. This phenomenon of an increase in the kinds of insects or in insect diversity (here referring to the balance of population numbers in communities of more than 1 species) correlated with increasing distributional area of host plants is analogous to the phenomenon explained by the theories of island biogeography (MacArthur & Wilson 1967).

The data presented below result from a 4-year survey of the distribution and host specificity of 50 species of *Nesiomiris*, which feed on native members of Araliaceae and Aquifoliaceae. The relationships and the distributional areas of 34 taxa of the Araliaceae guild are examined.

Within the main islands (Kaua'i, O'ahu, Moloka'i, Lāna'i, Maui, and Hawai'i) there are 34 species of the *Araliaceae*: 6 *Cheirodendron*, 8 *Reynoldsia*, 1 *Munroidendron*, and 19 *Tetraplasandra*, as well as 1 species of Aquifoliaceae (*Ilex anomala*). The following genera are found throughout the main islands of the archipelago: *Cheirodendron*, *Tetraplasandra*, and *Ilex*, while *Reynoldsia* is found on all but Kaua'i, and *Munroidendron* is restricted to Kaua'i. At the specific level, only *Ilex anomala* is found on all the main islands. *Nesiomiris* species do not appear to "distinguish" the so-called species of *Cheirodendron*, excepting *C. platyphyllum* of O'ahu that is apparently ecologically unsuitable, nor species of *Reynoldsia*, nor between *Munroidendron racemosum* and *Tetraplasandra kavaiensis*, and only at the sectional level of *Tetraplasandra*.

Unfortunately, there is no fossil record that allows me to ascertain the minimal age for most extant host species. Thus, I cannot assume approximately equal ages for the host islands. To judge from the disparity in the distribution of the monoinsular species of *Nesiomiris* (Figs. 14, 105), the ages of the islands of the archipelago might indicate established host plant associations on these islands well before some of the younger islands such as Hawai'i and Maui even reached the surface of the ocean.

The rain forest and mesic communities, which extend from 300 m to 2000 m in elevation, have *Cheirodendron*, *Ilex*, and *Tetraplasandra* species as major components; while *Reynoldsia* is a denizen of the now largely decimated lowland dry forests (Rock 1913: 351). *Munroidendron* is a geographically restricted ecological equivalent of *Reynoldsia* on Kaua'i. The area effectively occupied by populations of each host varies widely and is due not only to the ecological optima of each host, as enforced by interspecific competition, but to historical opportunity and/or accident. The area occupied by more or less continuous populations of each host genus or section was calculated by extrapolating from the figures presented by Nelson (1973: 150) for the "major forest types" to account for additional area as it may have been occupied before human alteration or obliteration (Table 8). The "ecological range" of each taxon was calculated by scoring occurrences in each of 300 m elevation intervals from sea level to tree line (2400 m). Other than the summer-deciduous *Reynoldsia* and *Munroidendron*, other taxa appeared to be suitable for feeding throughout the year and are not considered to differ significantly in phenological characteristics.

The data presented above suggest that in island archipelagos, which have their component islands sufficiently closely adjacent to allow considerable island founding between them, the number of monoinsular endemics is proportional to island age. Other biogeographic factors such as ecological range and distributional area of the host plant seem of secondary importance in influencing species numbers. Island size seemed to have a lower asymptotic limit, Moloka'i and Lāna'i being cases in point. There did not appear to be any effect of increasing island size. For example, Hawai'i Island had only 4 monoinsular endemics. In light of this, it would appear that island biogeographic theory should initially compare similar taxonomic groups, or better, ecologically equivalent "guilds" at the same trophic level as Opler (1974a) has pointed out, before proceeding to make broader implications, for example, as for island avifaunas (cf. Mayr 1965).

LITERATURE CITED

Akinbohungbe, A.E. 1974. Nymphal characters and higher classification analysis in the Miridae (Hemiptera: Heteroptera) with a subfamily key based on the nymphs. *Can. Entomol.* **106**: 687–94.

Ashlock, P.D. 1966. New Hawaiian Orsillinae (Hemiptera-Heteroptera: Lygaeidae). *Pac. Insects* **8**: 805–25.

Bianchi, F. [1967]. Notes and exhibitions: *Nesiomiris* sp. *Proc. Hawaii. Entomol. Soc.* (1966) **19**: 132.

Bock, W.J. 1970. Microevolutionary sequences as a fundamental concept in macroevolutionary models. *Evolution* **24**: 704–22.

Briggs, J.C. 1974. Operation of zoogeographic barriers. *Syst. Zool.* **23**: 248–56.

Carlquist, S. 1965. *Island life: A natural history of the islands of the world.* Nat. History Press, New York. 451 p.

Carson, H.L. 1970. Chromosome tracers of the origin of species. *Science* **168**: 1414–8.

———. **& H.D. Stalker.** 1968. Polytene chromosome relationships in Hawaiian species of *Drosophila*: I. The *D. grimshawi* subgroup. Studies in Genetics IV. *Univ. Texas Publ.* **6818**: 337–54.

Carvalho, J.C.M. 1952. On the major classification of Miridae (Hemiptera) (with keys to subfamilies and tribes and a catalogue of the world genera). *An. Acad. Bras. Cienc.* **24**(1): 31–110.

———. 1955. Keys to the genera of Miridae of the world (Hemiptera). *Bol. Mus. Goeldi* **11**(2): 151 p.

———. 1959. Catalogo dos mirídos do mondo. *Arq. Mus. Nac.* (Rio J.) **48**, 384 p.

Cobben, R.H. 1968. *Evolutionary trends in Heteroptera.* Part I. Eggs, architecture of the shell, gross embryology and eclosion. Centre for Agricultural Publishing and Documentation, Wageningen. 475 p.

Davis, C.J. 1974. Recent introductions for biological control XVIII. *Proc. Hawaii. Entomol. Soc.* **21**: 355–58.

Dethier, V.C. 1954. Evolution of feeding preferences in phytophagous insects. *Evolution* **8**: 33–54.

Eyles, A.C. & J.C.M. Carvalho. 1975. Revision of the genus *Dolichomiris*, with a revised key to the genera of the Stenodemini (Heteroptera: Miridae). *J. Nat. Hist.* **9**: 257–69.

Fosberg, F.R. 1948. Derivation of the flora of the Hawaiian islands, p. 107–19. *In*: E.C. Zimmerman, *Insects of Hawaii*. Vol. 1. Introduction. Univ. Hawaii Press, Honolulu.

Gagné, W.C. [1969]. New species and a revised key to the Hawaiian *Cyrtopeltis* Fieb. with notes on *Cyrtopeltis* (*Engytatus*) *hawaiiensis* Kirkaldy (Heteroptera: Miridae). *Proc. Hawaii Entomol. Soc.* (1968) **20**: 35–44.

Giffard, W.M. 1908. Presidential address. *Proc. Hawaii. Entomol. Soc.* **1**: 176–84.

Gillett, G.W. 1972. Genetic affinities between Hawaiian and Marquesan *Bidens* (Asteraceae). *Taxon* **21**(4): 479–83.

Grant, P.R. 1972. Convergent and divergent character displacement. *Biol. Linn. Soc.* **4**(1): 39–68.

Gressitt, J.L. & C.J. Davis. [1972]. New plagithmysines from Kauai, Molokai and Hawaii (Coleopt.: Cerambycidae). *Proc. Hawaii. Entomol. Soc.* (1971) **21**: 67–77.

Hennig, W. 1966. *Phylogenetic systematics.* Univ. Illinois Press, Urbana. 263 p.

Henry, T.J. 1985. Newly recognized synonyms, homonyms, and combinations in the North American Miridae (Heteroptera). *J. N.Y. Entomol. Soc.* **93**: 1128–36.

Hering, C. 1951. *Biology of the leaf miners.* W. Junk, The Hague. 420 p.

Kelton, L.A. 1959. Male genitalia as taxonomic characters in the Miridae (Hemiptera). *Can. Entomol. Suppl.* **11**.

Kirkaldy, G.W. 1902. *Fauna Hawaiiensis.* Vol. III, Part II. Hemiptera. Cambridge Univ. Press, Cambridge.

———. 1906a. Notes on the classification of the hemipterous Superfamily Miroidea. *Can. Entomol.* **38**: 369–76.

———. 1906b. List of the genera of the pagiopodous Hemiptera-Heteroptera, with their type species from 1758 to 1904 (and also of the aquatic and semi-aquatic Trochalopoda). *Trans. Am. Entomol. Soc.* **32**: 117–56.

———. 1908. A list of the described Hemiptera (excluding Aleyrodidae and Coccidae) of the Hawaiian Islands. *Proc. Hawaii. Entomol. Soc.* **1**: 186–208.

———. 1910. Supplement to Hemiptera. *In*: D. Sharp, ed., *Fauna Hawaiiensis* **2**(6): 531–99.

Kullenberg, B. 1947. Über Morphologie and Funktion der Kopulationsapparats der Capsiden und Nabiden. *Zool. Bidr. Fran. Uppsala* **24**: 218–418.

MacArthur, R. H. & E.O. Wilson. 1967. *Theory of island biogeography.* Princeton Univ. Press, Princeton. 203 p.

Macdonald, G.A. & A.T. Abbott. 1970. *Volcanoes in the sea: The geology of Hawaii.* Univ. Hawai'i Press, Honolulu. 442 p.

Mayr, E. 1963. *Animal species and evolution.* Belknap Press of Harvard Univ. Press, Cambridge. 797 p.

———. 1965. Avifauna: Turnover on islands. *Science* **150**: 1587–88.

Munroe, D.D. 1974. The systematics, phylogeny, and zoogeography of *Symmerus* Walker and *Australosymmerus* Freeman (Diptera: Mycetophilidae: Ditomyiinae). *Mem. Entomol. Soc. Can.* **92**: 1–183.

Nelson, R.E. 1973. Forestry, p. 150-51. *In*: R. W. Armstrong, ed., *Atlas of Hawaii.* Univ. Hawai'i Press, Honolulu. 222 p.

Opler, P.A. 1974a. Biology, ecology and host specificity of microlepidoptera associated with *Quercus agrifolia* (Fagaceae). *Univ. Calif. Publ. Entomol.* **75**.

———. 1974b. Oaks as evolutionary islands for leaf-mining insects. *Am. Sci.* **62**: 67–73.

Reuter, O.M. 1910. Neue Beiträge zur Phylogenie und Systematik der Miriden nebst einleitenden Bemerkungen über die Phylogenie der Heteropteren-Familien Mit einer Stammbaumstafel. *Acta Soc. Sci. Fenn.* **37**(3), iv + 167 p.

Rock, J.F. [1913] 1974. *The indigenous trees of the Hawaiian Islands.* Privately published under patronage. 517 p. Reprint, Charles E. Tuttle Co., Rutland, Vermont, and Tokyo, Japan.

Schuh, R.T. 1974. The Orthotylinae and Phylinae (Hemiptera: Mirida) of South Africa with a phylogenetic analysis of the ant-mimetic tribes of the two subfamilies for the world. *Entomol. Am.* **47**: 1–332.

Sherff, **E.E.** 1952. Further studies of Hawaiian Araliaceae: Additions to *Cheirodendron Helleri* Sherff and a preliminary treatment of the endemic species of *Reynoldsia* A. Gray. *Bot. Leaflets* **6**: 6–19.

———. 1954. Revision of the genus *Cheirodendron* Nutt. ex Seem. for the Hawaiian Islands. *Fieldiana Bot.* **29**(1): 3–45.

———. 1955. Revision of the Hawaiian members of the genus *Tetraplasandra* A. Gray. *Fieldiana, Bot.* **29**: 49–142.

Slater, **J.A.** 1950. An investigation of the female genitalia as taxonomic characters in the Miridae (Hemiptera). *Iowa St. Coll. J. Sci.* **25**(1): 1–81.

Southwood, **T.R.E.** 1960a. The abundance of the Hawaiian tree and the number of their associated insect species. *Proc. Hawaii. Entomol. Soc.* **17**: 299–303.

———. 1960b. The evolution of the insect-host tree relationships—-a new approach. *Proc. XI Int. Congr. Entomol.* (Vienna) **1**: 651–55.

———. 1960c. The number of species of insects associated with various trees. *J. Anim. Ecol.* **30**: 1–8.

Spieth, **H.T.** 1966. Courtship behavior of endemic Hawaiian *Drosophila*. Studies in Genetics III. *Univ. Texas Publ.* **6615**: 245–313.

———. 1968. Evolutionary implications of the mating behavior of the species of *Antopocerus* (Drosophilidae) in Hawaii. Studies in Genetics IV. *Univ. Texas Publ.* **6818**: 319–23.

Stearns, **H.T.** 1966. *Geology of the State of Hawaii*. Pacific Books, Palo Alto. 266 p.

Swezey, **O.H.** 1925. The insect fauna of trees and plants as an index of their endemicity and relative antiquity in the Hawaiian Islands. *Proc. Hawaii. Entomol. Soc.* **6**: 195–209.

———. 1954. Forest entomology in Hawaii: An annotated check-list of the insect faunas of the various components of the Hawaiian forests. *B.P. Bishop Mus. Publ.* **44**, ix + 226 p.

———. & **E.H. Bryan, Jr.** [1929]. Further notes on the forest insects of Molokai. *Proc. Hawaii. Entomol. Soc.* (1928) **7**: 293–314.

———. & **F.X. Williams**. [1932]. Some observations on the insect faunas of native forest trees in the Olinda forest on Maui. *Proc. Hawaii. Entomol. Soc.* (1931) **7**: 493–504.

Thorsteinson, **A.J.** 1960. Host selection in phytophagous insects. *Ann. Rev. Entomol.* **5**: 193–218.

Timberlake, **P.H.** 1921. Notes and exhibitions: *Nesiomiris*. *Proc. Hawaii. Entomol. Soc.* **4**: 472.

Usinger, **R.L.** 1942. The genus *Nysius* and its allies in the Hawaiian Islands (Hemiptera, Lygaeidae, Orsillini). *B.P. Bishop Mus. Bull.* **173**.

———. & **P.D. Ashlock**. 1966. Evolution of orsilline insect faunas on oceanic islands (Hemiptera, Lygaeidae), p. 233–35. *In*: R.I. Bowman, ed., *The Galapagos*. Proceedings of the Galapagos International Scientific Project. Contrib. No. 30. Univ. Calif.

Press, Berkeley & Los Angeles.

———. 1972. Robert Leslie Usinger: Autobiography of an entomologist. *Mem. Pac. Coast Entomol. Soc.* **4**, xiii + 330 p.

Zimmerman, E.C. 1948a. *Insects of Hawaii.* Vol. 1. Introduction. Univ. Hawai'i Press, Honolulu. 206 p.

———. 1948b. *Insects of Hawaii.* Vol. 3. Heteroptera. Univ. Hawai'i Press, Honolulu. 255 p.

APPENDIX

Characters Examined in Study

Characters are of adults and nymphs, unless indicated otherwise, or are genital characters. Characters marked with an asterisk (*) are those for which apomorphies were inferred in the cladistic analysis.

*1. Color of antenna I of adult: greenish fading to yellows; or rufescent.

*2. Color of apices of antennal segments II and III of nymphs: reddish annulate; or brownish fuscate; or pallid.

*3. Vestiture of antennal segment I: setae mostly reclined, brownish, length shorter than greatest diameter of antennal segment I; or erect, pale pilose, as long as, or longer than the greatest diameter of antennal segment I.

*4. Color of dorsum and tylus: green when alive, fading to yellows and yellowish browns after death; or rufescent, not fading.

5. Apex of head: bifid; or weakly bifid; or rounded.

6. Length of rostrum: reaching from middle of mesosternum to apices of hind coxae.

7. Color in adults: basal lobe darker green, collar sometimes green; or totally pallid.

8. Pubescence: usually reclined, sparse, inconspicuous, brownish; rarely erect, dense, golden brown.

*9. Proplueral stripe of adults: when present, extending across collar to base of pronotum or from collar to base, sometimes obsolescently so, or from collar to mesal constriction, sometimes interrupted at middle or represented by a spot on the mesal constriction; or absent.

*10. Propleural stripe of nymphs: when present, extending from apex to base of propleuron, sometimes interrupted at middle; or absent.

*11. Color in life: dark to light green; or pallid with a pattern of green spots; or totally pallid, sometimes with darker spots at bases of some setae; sometimes males darker green than females; apex of corium sometime rufescent.

12. Vestiture: usually recumbent, brown, inconspicuous; or rarely erect, long, golden brown.

*13. Body shape: lateral margins linear, less than width across base of pronotum; or rarely, convex-linear, wider than width across base of pronotum.

*14. Color of areolar veins: pale to dark greenish, fading to yellowish brown after death; or rufescent, not fading after death.

*15. Structure of areolar cells: partially to wholly coriaceous; or totally membranous when areolar veins are rufescent.

16. Color of membrane: infuscate; or infuscate rufescent; or clear.

*17. Size of membrane: normal, exceeding apex of cuneus by a length greater than length of small areolar cell; or reduced, exceeding apex of small areolar cell by a length less than length of small areolar cell.

18. Hind wings: 1A and SC+R infuscate, rest of vein clear; or all veins clear.
*19. Color: usually greenish, fading to yellowish browns after death; rarely, apices of femora reddish annulate.
20. Vestiture: inclined brownish, setae usually shorter than greatest diameter of appendage on which setae occurred; or setae pilose, paler brown, equal to or longer than greatest diameter of appendage on which the setae occurred.
21. Shape: normal, diameter of femora and tibiae gradually decreasing apically; or hind femora and bases of tibiae only weakly to strongly incrassate.
22. Venter of hind femur: yellowish; or with semiannulate brown markings.
23. Color of tarsi: all tarsi blackish; or only last tarsus blackish or all tarsi pallid.
24. Color of pygofer: yellowish; or yellowish with blackish dorsum.
*25. Armature of pygofer: usually with a tooth or sclerotized projection dorsally or laterally above the base of the left claspers, rarely above base of right clasper; or lacking tooth or sclerotized projection.
26. Apical venter of pygofer: with or without a tuft of short, denser setae.
*27. Upper arm of left clasper: usually longer than any arm of right clasper; armed with teeth; or apically toothed; or biramous; or simple, and variously hooked apically.
*28. Armature of right clasper: rarely bifid; or usually bifid; sometimes with 3rd arm indicated by a reduced lump; or rarely uniramous.
*29. Aedeagus or phallotheca: with various apical or subapical process from simple thickening of left apical wall, flanges, spatulate or chelate processes; ventral or lateral teeth; or simple.
30. Atrium of aedeagus: opening dorsally; or twisted such that atrium faces left clasper.
*31. Shape of spiculum: simply sinuate with or without subapical flabellum; or uncinate, with or without subapical flabellum.
32. Fleshy area about gonopore: unarmed, no denticles; or with denticles dorsally; or with fleshy denticulate lobe ventrally.
*33. Lobes of *K* structure: outer lobe much shorter than inner lobe; or outer lobe equal to or longer than inner lobe.
*34. Host plant family: on Araliaceae; or on Aquifoliaceae.
*35. Host plant genus in Araliaceae: on *Tetraplasandra* and *Munroidendron* species only, or on *Tetraplasandra* and *Cheirodendron* species only, or on *Cheirodendron* species only, or on all genera of Araliaceae.
*36. Host plant section association in *Tetraplasandra*: on species of section *Pterotropia* only; or on species in section *Eutetraplasandra* only; or on species in section *Eutraplasandra* only; or on species in section *Pterotropia, Neotraplasandra*,and *Nothotetraplasandra*; or on species in sections *Neotraplasandra* and *Nothotetraplasandra*.

INDEX TO SPECIES DESCRIPTIONS

Wayne Charles Gagné (1942–1988)